Refractories for the Chemical Industries

Prasunjit Sengupta

Refractories for the Chemical Industries

Springer

Prasunjit Sengupta
Technical Director
SKG Refractories, Ltd.
Nagpur, Maharashtra, India

ISBN 978-3-030-61242-9 ISBN 978-3-030-61240-5 (eBook)
https://doi.org/10.1007/978-3-030-61240-5

This Springer imprint is published by the registered company Springer Nature Switzerland AG
The registered company address is: Gewerbestrasse 11, 6330 Cham, Switzerland

Dedicated to my teachers who always inspired, guided, and blessed me

Preface

I cherished the desire to write a book on this subject for long because no such book is really available in English. The COVID-19 pandemic made it easier for me because of the nationwide lockdown when we had to confine ourselves at home and ample time was available to write this book.

It is needless to explain the importance of chemical industries in the modern world. Can we think of a world without petrochemicals, fertilizers, pharmaceuticals, different organic chemicals, cement, glass, detergent, and hundreds of other chemicals? Almost all the industries mentioned involve high-temperature operation at some stage or another. In fact, metallurgical process, for example, smelting of iron in a blast furnace, is a chemical process and basically the blast furnace is nothing but a high-temperature reactor where iron oxide is reduced to metallic iron in the presence of carbon. Most of the metallurgical processes are, in reality, high-temperature chemical processes. Those processes are not considered here because they come under the domain of metallurgists. For controlling the high temperature, refractories are an absolute necessity and one of the most important materials for the construction of high-temperature reactors. Here it needs to be clarified that in chemical industries, furnaces or kilns are used to carry out certain chemical reactions only, for example, cement rotary kiln or glass tank furnace, and can be called reactors as well.

Chemical industries, including cement and glass industries, consume more than 15% of the refractories produced. They consume all types of refractories and some very high-end refractory products. Although refractory consumption per ton of finished products is low and the cost involved may be insignificant, the fact remains that the high-temperature chemical processes cannot run without it and the refractory breakdown in such a plant is very expensive because of production loss, and it requires quite a few days to replace the refractory lining and to restart the unit.

It is a pity that the refractory as a subject has never been taught in chemical engineering curriculum. This book is written, firstly, with an aim to make the chemical engineering students aware of the importance of refractory material as a material of construction for designing and operating high-temperature reactors. This book will provide them an idea about refractory raw materials and products, their properties and characteristics, manufacturing process outline, process of selection, and the method of installation and maintenance.

Very scanty information is available in some journals and magazines about the design, selection, application, and performance of refractories in chemical industries. For petrochemical industries, manuals and guidelines for refractory installation are available from API (American Petroleum Institute); however, no such documents are available for other chemical industries. A second thought behind writing this book was to compile information on the existing refractory practices in different chemical industries.

This book is designed for chemical engineering students, who are going to run chemical industries in the future. Chemical industries are of numerous types, and they demand refractories with different thermo-chemical and thermo-mechanical properties, depending on the chemical and physical process to be carried out in the reactor or furnace. Wide variety of high-temperature chemical industries have been selected including both organic and inorganic chemical, process industries. In operating under oxidizing and reducing environment, industries with solid feed, and gaseous feed with solid and gaseous products and processes working under high pressure, hydrogen and acidic gas atmosphere etc. for discussion, in this book.

Three inorganic chemical industries are chosen, namely Limestone calcination, Portland cement manufacturing, and Glass manufacturing. The two organic chemical industries chosen are synthetic gas manufacturing and coal gasification, both are very important for clean energy and for synthesizing heavy chemicals such as methanol, Fischer-Tropsch liquid, and ammonia. Petroleum refinery is a complex integration of many different units, and out of that, process heaters, fluidized catalytic cracking unit (FCCU), and sulfur recovery unit are chosen, which operate at high temperature. Carbon black industry is a downstream industry of petroleum refinery and is chosen as it is a heavy industrial chemical, and the reactors used operate at very high temperatures. Steam is an important ingredient for many chemical processes. Boiler is used to generate steam, and waste heat boilers are also used widely today to utilize the waste heat energy from the industry, which saves the environment and improves the economic viability of the industry. A chapter of Refractories is dedicated to discuss the operating condition, refractory selection, and requirement for boilers. There is a myth that refractory performance depends on refractory quality only, but this is not true. The last chapter has discussed in detail the factors that are responsible for refractory performance. It is shown that it depends more on other factors besides quality. The influence of installation quality on the ultimate performance of the refractory lining is addressed in detail. The steel components used along with refractory undergo sigma transformation and are attacked by hydrogen gas at high temperature with the effect of reduction in its mechanical strength and ultimately failure of refractories, which is discussed. The chapters dedicated to address refractories for different industries mentioned are placed at random with no specific sequence.

This book will be very helpful to process engineers working in chemical industries who use refractory in a routine way without knowing much about the reason why a particular refractory is used for certain application. The important process parameters of individual industries, mentioned earlier, are described. This will help them to analyze and understand the process of selection of the refractory for a particular application. Monitoring refractory lining is a very important part of the refractory management to avoid any untoward incident in the plant, and this book has addressed that issue. The different techniques of monitoring the lining and its maintenance, both in running furnace and after taking shutdown, are discussed. The use of suitable insulation is very important to the economic viability of the process and here the properties and characters of different insulation material are explained.

This book will help refractory product development engineers and refractory marketing engineers to know about the operational parameters of respective chemical process industries which are absolutely necessary to develop or recommend a product for certain application.

I do hope that this endeavor to make chemical engineers understand the importance of refractories in designing and maintaining the lining of reactors or furnaces will be successful. The reader of this book will definitely benefit from the contents of the book and will be able to form a concept about the different aspects and perspectives of the refractory lining requirements of reactors or furnaces.

I am indebted to many individuals who supported me to write this book, and it is impossible to thank every individual who helped me make this attempt successful. Nevertheless, some of them deserve special attention for their valuable help rendered in its preparation. The illustrations and figures in this book are prepared by Mr. S. Banerjee, to whom I am grateful. I could not have written this book without the support of my life companion Mrs. Shyamali Sengupta. Mr. Debarpan Sengupta and Mrs. Madhurima Mukherjee provided valuable editorial suggestions. Special gratitude must go to Mrs. Bhadra Sengupta for her continuous inspiration. I am also thankful to the management of SKG Refractories Ltd for providing me many inputs.

Nagpur, Maharashtra, India Prasunjit Sengupta

Contents

1 Refractory: Characterization 1
1.1 Introduction 1
1.1.1 Refractory Materials 2
1.1.2 Refractory Properties 4
References 28

2 Classification of Refractories 31
2.1 Classification of Refractories 31
2.2 Basis of Classification 31
2.2.1 By Basicity 31
2.2.2 By Form 32
2.2.3 Insulating Property 40
2.2.4 Special Refractories 41

3 Manufacturing and Properties of Refractories 43
3.1 Introduction 43
3.2 Shaped Refractories 43
3.2.1 Silica Refractories 44
3.2.2 Aluminosilicate Refractories 46
3.2.3 Basic Refractories 49
3.2.4 Fused Cast Refractories 56
3.3 Unshaped Refractories 58
3.3.1 Castables 58
3.3.2 Phosphate-Bonded Castable 63
3.3.3 Ramming Mass 68
3.3.4 Plastic Mass 68
3.3.5 Mortars 69
3.4 Insulation Refractories 69
3.4.1 Insulation Shapes 70
3.4.2 Calcium Silicate 70
3.4.3 Insulating Castables 71
3.4.4 Ceramic Fiber Insulation 73

3.4.5 Microporous Insulation Product ... 78
References ... 82

4 Design, Installation, and Maintenance of Refractories ... 85
4.1 Introduction ... 85
4.2 Refractory Selection Criteria ... 85
4.3 Shaped Refractories ... 86
4.3.1 Features of Lining with Shaped Refractories ... 88
4.4 Installation of Shaped Refractories ... 88
4.4.1 Basic Rules of Refractory Lining ... 89
4.4.2 Vertical Wall ... 90
4.4.3 Expansion Joints ... 91
4.4.4 Roof design ... 93
4.5 Installation of Monolithic Refractories ... 97
4.5.1 Conventional Castables ... 97
4.5.2 Low Cement/Ultralow Cement/No Cement Castables ... 98
4.5.3 Shotcreting ... 104
4.5.4 Gunning ... 106
4.5.5 Pumping ... 107
4.5.6 Monitoring and Safety ... 107
4.5.7 Joints ... 107
4.5.8 Choice of Installation Method ... 108
4.5.9 Anchor ... 110
4.5.10 Anchor Spacing ... 114
4.6 Drying and Preheating ... 117
4.6.1 Preheating of Refractories ... 119
4.6.2 Preheating of Castable ... 121
4.7 Pressurization and depressurization ... 123
4.8 Carbonation ... 124
4.9 Insulation Refractory ... 125
4.9.1 Installation of Ceramic Fiber Lining ... 126
4.10 FEA (Finite Element Analysis) ... 129
4.11 Refractory Lining Monitoring ... 130
References ... 133

5 Refractories for Lime Calcination ... 135
5.1 Introduction ... 135
5.2 Process ... 135
5.3 Vertical Shaft Kilns ... 136
5.3.1 Single Shaft Kiln ... 137
5.3.2 Twin Shaft Kiln ... 137
5.3.3 Refractory ... 140
5.4 Rotary Kiln ... 141
5.4.1 Refractory ... 142
5.4.2 Insulation ... 144
References ... 146

6 **Refractories for Syngas Manufacturing** ... 149
6.1 Introduction ... 149
6.2 Chemical Process ... 150
6.2.1 Steam Methane Reforming (SMR) ... 151
6.2.2 Partial Oxidation (POX) and Catalytic Partial Oxidation (CPO) ... 155
6.2.3 Autothermal Reforming (ATR) ... 156
6.2.4 Combined Reforming ... 157
6.3 Refractories ... 158
6.3.1 Primary Reformers ... 159
6.3.2 Refractory for Secondary Reformer ... 163
6.3.3 Precautions in Refractory Lining ... 166
6.3.4 Case Study ... 169
References ... 170

7 **Refractories for Coal Gasification** ... 173
7.1 Introduction ... 173
7.2 Process ... 175
7.3 Gasifier Design ... 179
7.3.1 Characterization of Slag ... 180
7.4 Refractories Used for Lining ... 185
7.4.1 Refractories for Membrane Wall ... 185
7.4.2 Concept of Freeze Lining ... 186
7.4.3 Refractories for Air-Cooled Wall ... 191
7.4.4 Corrosion of Refractory ... 191
References ... 195

8 **Refractories for Carbon Black Manufacturing** ... 197
8.1 Carbon Black ... 197
8.2 Process Principle ... 197
8.3 Manufacturing Process ... 198
8.4 Reactor Design ... 198
8.5 Temperature at Different Stages of Carbon Black Manufacturing ... 200
8.6 Selection of Refractories ... 201
References ... 203

9 **Refractories for Petroleum Refinery** ... 205
9.1 Introduction ... 205
9.2 Process Heaters ... 206
9.2.1 Refractory for Process Heaters ... 207
9.2.2 Refractory Selection ... 207
9.3 Fluid Catalytic Cracking Unit ... 212
9.3.1 Process Description ... 212
9.3.2 Refractories for Riser Pipe ... 215
9.3.3 Refractories for Reactor and Regenerator wall ... 217
9.3.4 Refractories for Transfer Lines ... 219

9.3.5 Refractories for Cyclones 219
9.3.6 Refractories for Flue Gas Line.......................... 222
9.4 Sulfur Recovery Unit.......................... 223
9.4.1 Introduction 223
9.4.2 Process.......................... 223
9.4.3 Refractory Lining Pattern 226
9.4.4 Checker Wall 227
9.4.5 Refractory or Ceramic Ferrule.......................... 230
9.4.6 Tail Gas Incinerator and Waste Heat Recovery Boiler..... 234
References.......................... 235

10 Refractories for Glass Manufacturing 237
10.1 Introduction 237
10.2 Container Glass Manufacturing Process 238
10.2.1 Glass Melting Furnace.......................... 238
10.3 Refractories 241
10.3.1 Bottom 242
10.3.2 Side Wall Lining 244
10.3.3 Throat.......................... 246
10.3.4 Superstructure 246
10.3.5 Crown.......................... 248
10.4 Float Glass Process 250
10.4.1 Refractory Selection.......................... 252
10.4.2 Tweels 254
10.5 Refractory Corrosion by Molten Glass 255
10.6 Defects in Glass due to Refractories 258
10.7 Regenerator 262
10.7.1 Ceiling 263
10.7.2 Side Wall 265
10.7.3 Upper side wall 265
10.7.4 Target Wall.......................... 265
10.7.5 Checker Refractory 265
10.8 Monitoring and Maintenance.......................... 272
10.8.1 Ceramic Welding 274
References.......................... 275

11 Refractories for Portland Cement Manufacturing 277
11.1 Introduction 277
11.2 Process of Manufacturing 277
11.3 Fuel 279
11.4 Process Design.......................... 279
11.5 Selection of Refractories 281
11.6 Refractories for Different Equipment 282
11.6.1 Cyclones and Ducts.......................... 282
11.6.2 Calciner 285
11.6.3 Riser Duct 286

11.6.4 Smoke Chamber 286
11.6.5 Meal Pipes 287
11.6.6 Insulation 287
11.6.7 Kiln Hood 287
11.6.8 Tertiary Air Duct and Cooler 288
11.6.9 Cooler 289
11.6.10 Kiln 289
11.7 Kiln Refractory Performance 299
References 301

12 Refractories for Boiler and Waste Heat Recovery 303
12.1 Introduction 303
12.2 Boiler Components 303
12.3 Refractories for Different Components 303
12.3.1 Combustor Wall 305
12.3.2 Combustor Upper Wall and Ceiling 306
12.3.3 Bottom Lining 308
12.3.4 Cyclone 309
12.3.5 Seal Pot 310
12.3.6 Ash Cooler 311
12.3.7 Flue Gas Duct Between Cyclone and Backpass 312
12.4 Waste Heat Utilization for the Power Generation 312
12.5 Waste Heat Recovery 313
References 315

13 Refractory Performance and Mechanism of Damage 317
13.1 Introduction 317
13.2 Thermo-chemical Effects 320
13.2.1 Interaction with Alkalis 322
13.2.2 Interaction with Sulfur and Chlorides 326
13.2.3 Interaction with Vanadium Oxides 329
13.2.4 Redox Reactions 331
13.2.5 Hydration 333
13.2.6 Ruby Formation 336
13.3 Thermal Load 337
13.3.1 Overheating 337
13.3.2 Flame 338
13.4 Thermo-mechanical Load 340
13.4.1 Ovality 341
13.4.2 Strain-Controlled Load 345
13.4.3 Thermal Shock 347
13.4.4 Abrasion 348
13.4.5 Effect of Kiln Diameter 348
13.5 Lining Tightness 351
13.6 Refractory Failure due to Anchor 352
13.6.1 Sigma-Phase Embrittlement 352

13.6.2 Remedy 353
13.6.3 High-Temperature Attack of Hydrogen (HTHA) 354
13.7 Refractory Performance 356
13.8 Refractory Lining Failure Statistics 356
13.9 Analysis of Refractory Lining Failure 357
References 360

Glossary 363

Index 367

Chapter 1
Refractory: Characterization

1.1 Introduction

Different types of chemicals are occupying a considerably important space in our daily life. Modern civilization cannot think of a world without gasoline, diesel, polymers, fertilizers, cements, and glass which all belong to the category of chemical industry. In fact many other industries can also be treated as chemical industries, e.g., production of alumina from bauxite by Bayer's process or smelting of iron ore in the blast furnace, direct reduction of iron, etc., but as those specifically fall in the domain of metallurgy, those are excluded from our consideration. The process flow sheet of the chemical industries, mentioned, involves high-temperature operation in some or other stages. Whenever a process is carried out at a temperature beyond 400 °C, it requires the use of refractories to protect the steel as the material of construction because like other metals steel also loses its strength at elevated temperature (Fig. 1.1).

Refractories basically serve two purposes. Firstly, refractories can withstand high temperature in presence of corrosive solid,liquid and gases. Furnaces or reactors operate at high temperature are constructed with steel and its alloys. Steel loses its mechanical strength drastically above 400 °C (Fig. 1.1) and cannot be used [1] unless protected by Refractory lining over it.

Secondly, refractories inhibit the flow of energy out from the reactor system, which ensures the saving of the fuel and the safety of the process. Use of other high-temperature withstanding metal alloys will be prohibitively expensive and hence economically unviable. Refractory material lining over the steel casing of the furnace, kiln, or reactor will protect the steel shell from direct contact of high temperature inside and will retain the structural integrity. Refractory lining used inside is not the same for all the furnaces, kilns or reactors. It depends upon the chemistry and physics of the gas, liquid, or solid being processed inside. Through thorough study and long-time experience, the engineers have standardized the refractory practices

P. Sengupta, *Refractories for the Chemical Industries*,
https://doi.org/10.1007/978-3-030-61240-5_1

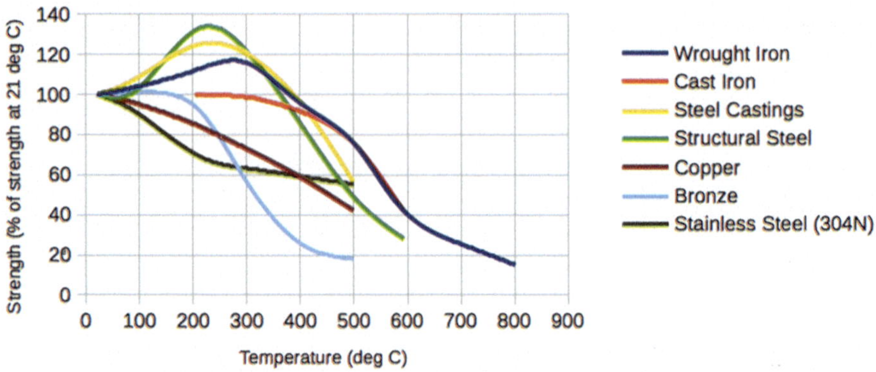

Fig. 1.1 Mechanical strength of steel with increase of temperature

for different process industries. Nevertheless, many a times, it is very difficult to analyze the reason of the abnormal behavior of the refractory, theoretically, with the so far accumulated knowledge. There are instances that the same refractory is not behaving identically in an exactly similar equipment in the same plant. Through a constant effort, technologists are trying to identify the influences of different factors on refractory performances, although it is not an easy job. The major hurdle towards this is the impossibility of laboratory simulation of the exact condition that the refractories are put under in actual use. The in-depth studies of refractory-related problem requires knowledge of high-temperature chemistry, physics, material science, mineralogy, mathematics, civil engineering, chemical engineering, etc., and in most of the cases, the combination of all these knowledge is not readily available to solve the problem. Here, it is necessary to mention that refractories are tailor-made products and should be designed and made for a particular application. The refractory which is very good for steel melting furnace may not work well in an application in a reactor producing carbon black. We shall discuss in this book the different aspects of refractory materials and their properties and behavior in context to various chemical industries.

1.1.1 Refractory Materials

The refractory materials are inert inorganic solid materials which are stable at high temperature in contact with corrosive solid, liquid, and gas and can retain its physical shapes and structural strength at high temperature. These are mainly oxides, carbides, nitrides, and borides of aluminum, silicon, alkaline earth metals, and transition metals. Table 1.1 furnishes a comprehensive chart of different refractory materials with very high melting point.

Out of all these materials, very few qualify for use in industrial scale, because of their instability under normal atmospheric condition or because of the rare avail-

Table 1.1 Melting point of different refractory materials

	Melting point (°C)			
Elements	Oxide	Nitride	Carbide	Boride
Aluminum	2050	2200	2200	–
Silicon	1702	1900[a]	2300	–
Beryllium		2200	2100[a]	–
Magnesium	2800	1500	–	–
Calcium	2600	–	2160	–
Barium	1923	–	–	–
Titanium	1843	2950	3067	3230
Vanadium	–	2177	2648	2673
Chromium	2275	–	1810	2400
Zirconium	2700	2980	3450	3246
Niobium	1512	2400	3400	3273
Molybdenum	–	–	2600	2823
Hafnium	2758	3387	3928	3250
Tantalum	1872	3093	3983	3310
Tungsten	1700[a]	–	2776	2365
Boron	–		2450	2973

[a]Decomposes

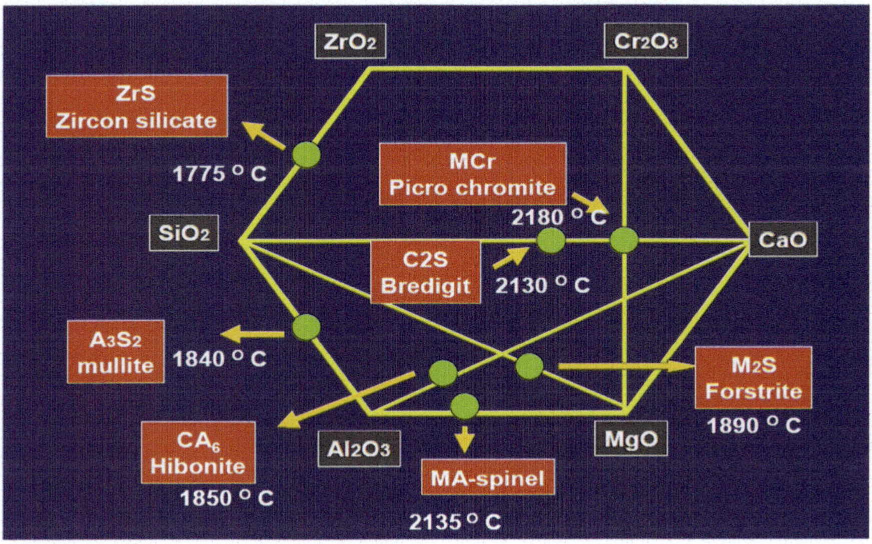

Fig. 1.2 Formation of different refractory mineral phases

ability and high cost. For example, barium oxide or calcium carbide and aluminum carbide react very fast with atmospheric moisture. Vanadium, niobium, molybdenum, hafnium, etc. are too expensive to be considered for refractory application. Finally, oxides like Al_2O_3, SiO_2, MgO, CaO, Cr_2O_3, ZrO_2, and carbon in different mineralogical forms, individually or in combination (Fig. 1.2), are used most widely to manufacture refractories for all metallurgical and chemical process industries and in other applications. The criteria of selection, of the abovementioned materials, are their abundance in nature, stability, and ease of processing to manufacture refractory products.

The source of the raw materials can be natural or synthetic. The raw materials used for refractory manufacturing are mainly naturally occurring minerals like bauxite, magnesite, clay, etc. which are mined and processed before being used for refractory manufacturing. Some synthetic materials like mullite (3 $Al_2O_3.2SiO_2$), fused alumina (Al_2O_3), silicon carbide (SiC), spinel ($MgO.Al_2O_3$), etc. are also being used widely in refractories for cement industry. Properties of the naturally occurring raw materials may vary considerably from one deposit to another, because of its geological formation and associated impurities.

1.1.2 Refractory Properties

Refractories are characterized by their chemical and physical properties and are used to correlate its behavior in actual high-temperature application.

1.1.2.1 Specific Gravity

All refractory minerals have different densities and it is a fundamental property of the material. Refractory materials can be identified by their specific gravities. Specific gravity can be determined by making a powder of the sample of a specific size and using a specific gravity bottle and a balance. Table 1.2 gives the specific gravity value of some refractory bricks and refractory minerals.

Table 1.2 Sp. gravity of refractory

Material	Sp. gravity
Fused quartz	2.20
Fireclay brick	2.56
Sillimanite brick	2.86
Bauxite brick	3.65
Corundum brick	3.97
SiC brick	3.1
Magnesia spinel brick	3.58
Magnesite-chrome brick	3.8
Zirconia	5.7

1.1.2.2 Bulk Densities

It is the mass of the material per unit volume including the pores. For the same kind of refractory, the bulk density can vary. The higher is the bulk density, the lesser is the porosity and normally higher is the mechanical strength. Bulk density is different from true density in a way that the total volume considered in the calculation is the sum of the volume of both material and pores. Therefore, the value of true density is always more than that of bulk density. Bulk density is an important property needed to calculate the weight of the total refractory required for lining a furnace, kiln, or reactor.

1.1.2.3 Apparent Porosity

Refractories contain pores, and some of the pores are open and connected and some are closed (Fig. 1.3). Total volume of a refractory body = volume of the matter + volume of the open pores + volume of the closed pores. The apparent porosity of refractory, expressed in %, is defined as

$$\text{The apparent porosity} = \left(\text{Volume of open pores} \times 100\right) / \text{Total volume} \quad (1.1)$$

It is a very important property and influences the mechanical strength, corrosion resistance, and thermal conductivity of a refractory. Porosity and bulk density of a refractory are inversely related. The lower is the apparent porosity, the higher is the bulk density, mechanical strength, thermal conductivity, and corrosion resistance of the body. Besides the total pore volume, the pore sizes are also very important to influence the corrosion resistance and thermal conductivity of the refractory. The smaller the pore sizes, the better is the corrosion resistance and the lower is the

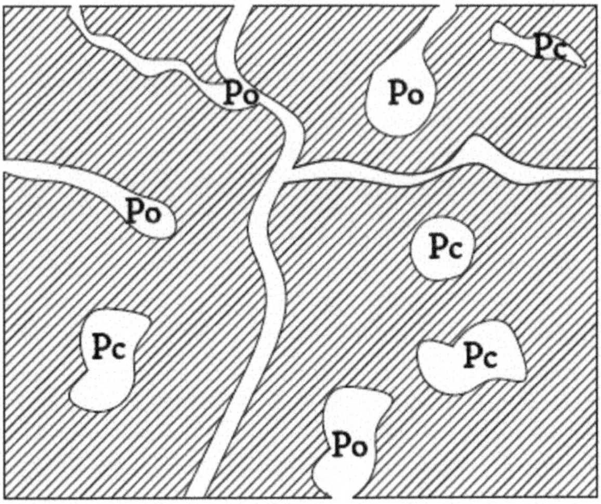

Fig. 1.3 Pores inside the refractory body. P_c – closed pores, P_o – open pores

thermal conductivity. True porosity is the total volume of open and closed pores. It is expressed in % and is defined as

$$\text{True Porosity} = 100 \times (\text{Sp.Gravity} - \text{Bulk density}) / \text{Sp.Gravity} \quad (1.2)$$

The test can be done as per the method, ISO (International Organization for Standardization) 5017.1988.

The volume, size, and structure of the pores have a close relationship with the penetration of slag and the permeation of gases inside the refractories. The pore diameter in dense and fired refractory materials is in the range of 0.1–100 millimicron (μm). Larger pores are found in insulating refractories. Refractory castables have high share of pores below 1 μm.

1.1.2.4 Permeability

It is the measure of flow of gases through the pores within the refractory body, and it indicates the extent of pore linkage. Permeability of refractories gives an indication of how well the refractories will stand up to a molten slag, a melt, or a gas penetration.

Specific gas permeability is defined by the equation given below, with laminar gas flow:

$$\mu = \frac{v}{t} \cdot \eta \cdot \frac{\delta}{A} \cdot \frac{1}{p_1 - p_2} \cdot \frac{2p}{p_1 + p_2} \quad (1.3)$$

where

μ = permeability of the refractory
V/t = volumetric flow of the gas, through the refractory, in m^3/s
η = dynamic viscosity of the gas at the test temperature in Pa.s
δ = thickness of the refractory
A = cross-sectional area that gas flows through
p = absolute pressure of the gas
p_1 = absolute pressure of the gas at the entry point
p_2 = absolute pressure of the gas at the exit point

The factor $2p/(p_1 + p_2) = 1$ for small pressure differences.

The unit of gas permeability is m^2. The value for the gas permeability of refractories is usually very low and is normally expressed as μm^2. The previously used unit was perm or nanoperm and 1 μm^2 = 10 nPm.

Gas permeability of the refractories is determined by the share of pores with diameter greater than 10 μm. Gas permeability decreases substantially with increasing viscosity of the gases at higher temperature. A decrease of 50% permeability can be expected at 500 °C temperature. At higher temperature, the closure of micro-cracks also brings down the permeability.

The permeability of a refractory to gas can be determined by ISO standard method 8841.

1.1.2.5 Mechanical Properties

Mechanical strength at ambient temperature required to transport and handle the shaped refractory products from manufacturing site to work sites.

Cold Crushing Strength (CCS)

In this test, the cube cut from the brick sample is subjected to increasing load until it gets crushed and reported the value as load per unit area. It indicates the adequacy of firing temperature, for shaped refractory products, required for proper sintering and for developing the required microstructure and the quality of hydraulic or chemical bond in case of unshaped refractories. In the unshaped products, the CCS does not remain the same after heat treatment, and it decreases or increases with temperature of heat treatment. The high cold crushing strength of shaped refractories protects them from damages during handling and also from mechanical abuses in service. CCS can be determined by following the ISO standard methods 10059-1 and 10059-2. There is a myth that refractory having high CCS is better. But it is not true, strength at ambient temperature has nothing to do with its strength at high temperature. On the contrary, sometimes higher CCS is obtained due to more glassy phase in the refractory body, and it may get deformed easily at high temperature under load and may lead to bad performance.

Modulus of Rupture

The test is conducted by putting the bar of a specified cut size from the refractory body on two point supports and applying load in the middle of the two supports till the bar breaks. It is calculated as

$$\mathrm{MOR} = \frac{2WL}{bd^2} \tag{1.4}$$

where W = load at which the sample breaks, L = length between the two supports, b = breadth of the sample, and d = depth of the sample. It actually shows the strength of the body under tensile stress.

It can be tested following the method given in ISO 5014.1986. The test actually indicates the tensile strength of the refractory. The higher is the MOR value, the more it can withstand the tensile force.

Modulus of Elasticity

It defines the stress-strain relationship and is a fundamental property of materials. Like any other material, the refractories also obey Hooke's law, and accordingly, it exhibits a linear relationship between applied stress (σ) and mechanical deformation (strain = ε). The proportionality constant between the two is modulus of elasticity, E, when the stress is compressive or tensile. If shear stress or torsional stress (τ) is applied, the strain is γ and the proportionality constant is called shear modulus, G. Thus, the mathematical relations are

$$\sigma = E.\varepsilon \tag{1.5}$$

$$\tau = G.\gamma \tag{1.6}$$

E and G of a refractory give an idea about its stiffness. Poisson's ratio, a dimensionless factor, connects E and G by the expression below [15]:

$$G = \frac{E}{2(1-\mu)} \tag{1.7}$$

where μ = Poisson's ratio.

Dynamic modulus of elasticity of the refractory body can be determined using ASTM standard method C1259-95 [2]. The principle of the test method is shown in Fig. 1.4 [3]. E (in GP_a) is calculated as $E = \rho.v^2$ where ρ = density (kg/m^3) and v = ultrasonic speed (m/s).

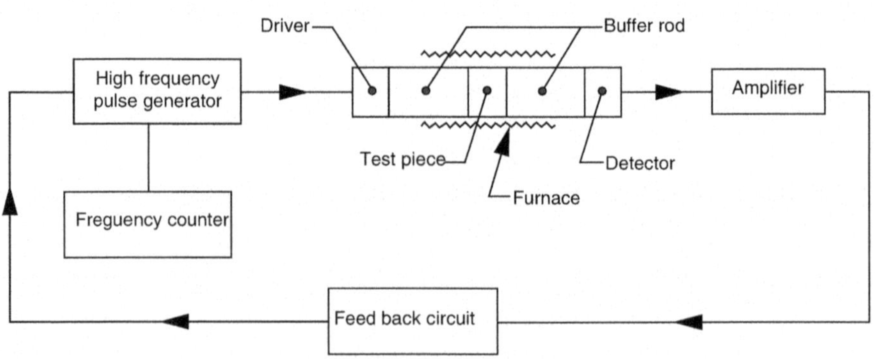

Fig. 1.4 Schematic diagram of the test method of dynamic modulus of elasticity

The lower is the modulus of elasticity, the better it can withstand mechanical deformation and thermal shock.

Fracture

Refractories are composite brittle materials at ambient temperature, and its fracture process is different from that of high-strength single-phase ceramic material with fine grain size. The refractory lining is to withstand different mechanical and thermo-mechanical stresses generated during its use, for example, mechanical stresses generated and applied on the refractory lining because of ovality in the kiln shell or thermo-mechanical stresses generated during heating and cooling of the refractory lining.

The brittleness of a material can be visualized from the load displacement curve of that material under three-point bending as is done in modulus of rupture test (Fig. 1.5a). If the load displacement curve shows a pattern like that in Fig. 1.5b, it will be considered as a brittle material [4].

But it is to be borne in mind that refractories are not very strong by nature and need not to be very strong also, because it is not used to carry any load except its own weight. The main purpose of refractory lining is to protect the equipment steel casing or shell from high temperature. The strength of refractory at higher temperature is more of concern to the refractory engineers than its strength at ambient temperature. When the HMOR of refractory is plotted against temperature, it exhibits a maximum value and then decreases rapidly with increase of temperature.

This maximum strength is obtained at a temperature between 600 and 1400 °C, depending upon the type of refractory. The displacement curve of a refractory if plotted against the load at high temperature shows a curve as in Fig. 1.5c, which indicates there is no more brittleness and plastic flow in the material. Thus, it is very important to understand that refractories are brittle materials that have high strength at low temperature and becomes weaker at elevated temperature, but it develops plasticity above 600 °C and becomes much less brittle [5, 6].

Microcracks always exist in the refractory body with coarse grains. When a stress is applied on the refractories, some of the cracks may propagate to cause fail-

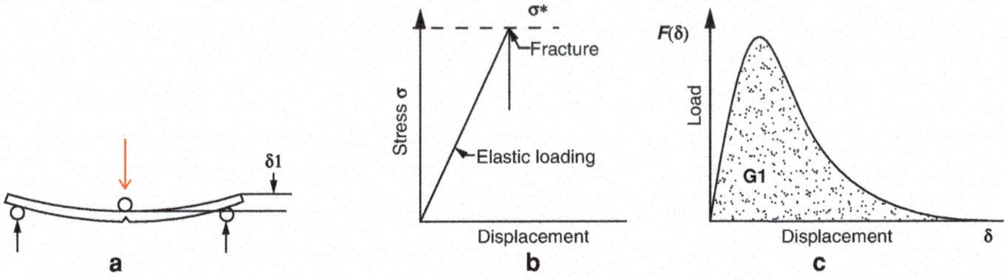

Fig. 1.5 (**a**) Three-point bending of a bar, (**b**) the load displacement curve of a brittle material, (**c**) the load displacement curve of a refractory at high temperature

ure of the refractory. The strength of the refractory body depends on the dimension of the crack. The fracture toughness K_{IC} is proportional to the square root of the critical crack length, Eq. 1.4 [5]:

$$K_{IC} = \sigma Y C^{0.5} \quad (1.8)$$

where σ is the critical stress, Y is the geometric factor, and C is the critical crack length. The higher is the fracture toughness, the more difficult it is to initiate and propagate a crack.

For refractories, the fracture toughness remains within the value 0.5–1.2 MPa.m$^{0.5}$.

Wedge-Splitting Test

Technologists have devised a test called wedge-splitting test which gives the idea about the toughness or its resistance to crack under the influence of mechanical stresses. The schematic diagram of the devise is shown in Fig. 1.6 [6]. A sample of 100 mm × 100 mm × 75 mm is cut out of the brick and is provided with a starter notch, and two side-guided notches are taken and are put into the testing machine, and the load is applied on the wedge vertically [7, 8].

The application of the vertical force F_v develops two horizontal forces F_H and causes the splitting of the sample. During the testing process, the displacement is recorded. From the load- displacement diagram, the specific fracture energy can be determined by integration:

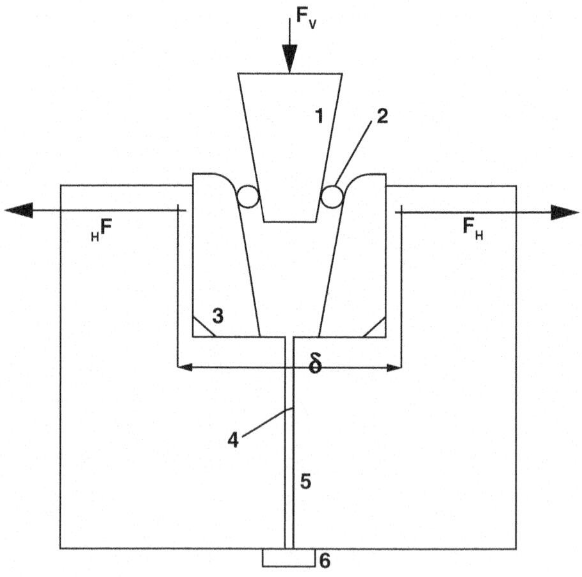

Fig. 1.6 Schematic representation of wedge-splitting test. Here F_v = the vertical force, F_H = the horizontal force, and δ = displacement. (*1*) Wedge, (*2*) rollers, (*3*) sample under testing, (*4*) notch, (*5*) side groove, (*6*) linear support

$$G_F = \frac{1}{A}\int_0^{\delta_{ult}} F.d\delta \tag{1.9}$$

δ_{ult} is the ultimate displacement before splitting of the test sample, A is the area of projection of the fracture surface, and G_f is the specific fracture energy in N/m^2.

$$\text{Here also, } F = F_v / 2\tan(\beta / 2) \tag{1.10}$$

where β = angle of the wedge in Fig. 1.5.

The higher the value of G_F, the tougher will be the refractory for an application. The same test can also be carried out in higher temperature. This test becomes very useful to predict the behavior of basic refractories in the application of burning zone in cement rotary kiln. The typical load displacement graph of wedge-splitting test is shown in Fig. 1.7 [9, 10].

Abrasion Resistance

This test becomes important for the application where the refractory lining is exposed to moving gas, liquid, or solid. The higher is the velocity of the moving particles, the higher is the abrasion. The abrasion results in the wearing of refractory. Two standard testing methods are followed to compare the abrasion resistance of the refractory body, namely, ASTM C704 or BS 1902.4 and EN 993-20 – the grinding method according to DIN 52108 or DIN EN 102. The tests are comparative methods to test the abrasion resistance of two or more products under identical condition.

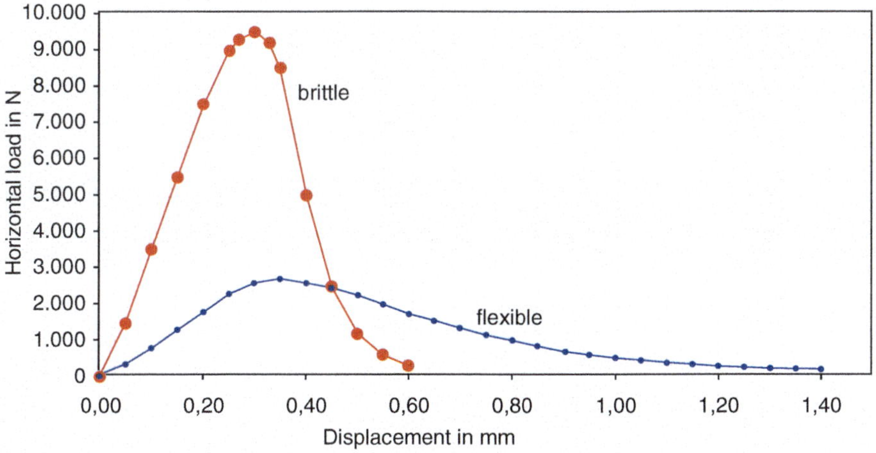

Fig. 1.7 Typical load-displacement graph indicating flexible and brittle material properties

The abrasion resistance of refractories depends on the intrinsic hardness of the grains in the bonded structure and the microstructural features, i.e., grain size, porosity, pore sizes, etc. Correlations exist between porosity, cold crushing strength, and cold modulus of rupture, which can be utilized for the rough evaluation of abrasion resistance. But the prediction of abrasion resistance, based on strength factor alone, is insufficient, because the bond phase of the refractory, the abrasive media grain size, the grain morphology, and the angle of impingement of the grains have tremendous influence on abrasion resistance [11–13, 15]. Abrasion generally decreases in a fired brick with increasing temperature [14, 16]. It can be zero when the brick surface attains a visco-plastic state.

In monolithic refractories, a general relation is known to exist between abrasion resistance and the modulus of rupture, i.e., the higher is the modulus of rupture, the better is its resistance to abrasion (Fig. 1.8) [11]. Abrasion loss is highest when the impingement of the particles on the refractory surface is at the right angle, but it does not hold good when the impinging particles are much smaller compared to the aggregate size of the monolithic refractory. When the size of the impinging particles is much smaller than the aggregates, then the matrix of the monolithic refractory undergoes abrasion first, although the abrasion resistance of the aggregate may be very good, and loosens the aggregates, which falls off. Therefore, in that case, to withstand the abrasion of dust-laden gases, the matrix must have to be abrasion resistant, and use of abrasion-resistant aggregate alone will not be effective. If the impinging grains are larger than the average aggregate size, then both the aggregate and the matrix are removed together and both need to be abrasion resistant.

The abrasion resistance of a refractory surface in service can change drastically as a result of corrosion or by a coating on the surface. Extensive abrasion and erosion can also occur when hot gases (even dust unladen) pass over the lining at high speed. Table 1.3 shows the variations of abrasion of some of the refractories with temperature.

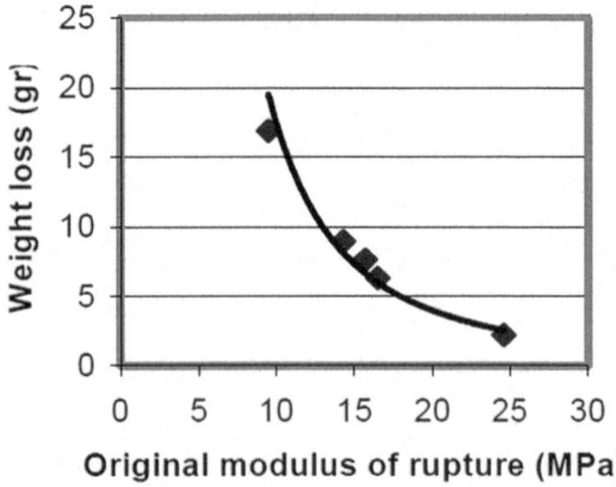

Fig. 1.8 Relation between modulus of rupture and abrasion resistance of refractory castable

Table 1.3 Relation between temperature and abrasion of some refractory material

Material	Index abrasion count		
	20 °C	1000 °C	1400 °C
Normal fireclay brick	300	70	0
Dense fireclay brick	45	25	0
Sillimanite brick	230	70	10
85% Al_2O_3 brick	60	25	0
Fused cast alumina	70	45	15
Silicate bonded SiC	90	20	NA
Fireclay castable	100	210	NA

1.1.2.6 Thermal Properties

Pyrometric Cone Equivalent (PCE)

Most of the refractories are made of naturally occurring raw materials, which contain some inherent impurities. Sometimes the presence of the impurities brings down the softening point of the refractory. Refractory products are normally a combination of different raw materials and do not have sharp melting point similar to pure crystalline material. Depending upon the quality and quantity of impurities, the liquid phases form at elevated temperature. The quantity of the liquid and its viscosity dictate the softening behavior of refractory at high temperature. The PCE test gives an idea about the softening temperature and behavior of the refractory material. From the idea of softening point, we can roughly estimate the MST (maximum service temperature) which can be considered as 200 °C below the PCE. The PCE value can also be used to compare the refractoriness of two refractory products from different sources or two similar raw materials from different sources. In this test, the refractory material is finely grounded and made in the form of a cone of a specific size. The cone is mounted on an alumina plate (Fig. 1.9) along with few standard cones having a definite softening temperature and put in a furnace, and the temperature is gradually raised till the test cone starts bending along with another standard cone. The softening point of the test cone is reported as the cone number of the standard cone along with which it bends. The test method in ISO528.1983 can be followed to determine the PCE.

Reversible Thermal Expansion (RTE)

Like all other materials, the refractory also expands when heated and comes back to its original dimension on cooling. The reversible thermal expansion values of different type of refractories are different. For construction of the furnace lining, this is an important parameter to be taken into consideration, because the provision has to be kept to accommodate the expansion of the refractories in service.

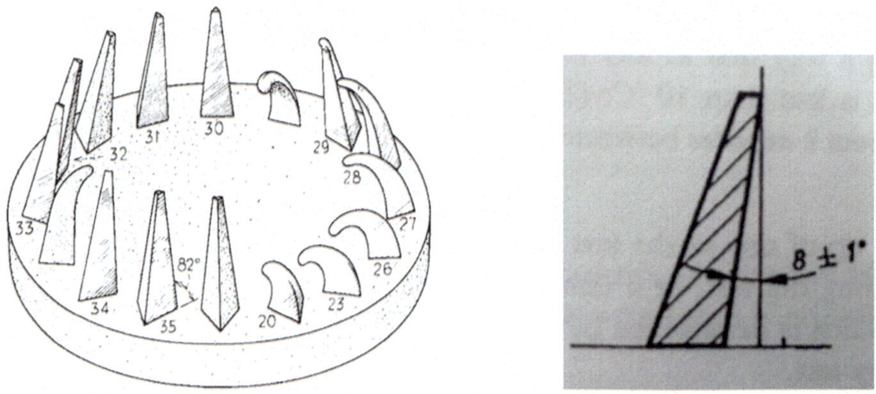

A B

Fig. 1.9 (**a**) Test cones and method of mounting on a plaque. (**b**) The test cone

It is expressed as % change in L_1 length from room temperature to a certain temperature. For example, if the length of a refractory bar on heating to a certain temperature, becomes L_2, then the thermal expansion of the body at that temperature is

$$100(L_2 - L_1)/L_1\% \tag{1.11}$$

If the provision for expansion is not kept in a lining, tremendous amount of stress can be generated which may lead to collapse of the lining.

The thermal expansion is measured by the instrument called dilatometer.

Permanent Linear Change (PLC)

Refractory bricks, during the course of manufacturing, when fired at high temperature, generally change in dimensions. It either shrinks or expands. It is because of the solid-state reaction forming liquid phases or because of some phase transformation. The kinetics of the solid-state reactions, generally, are very slow and never attains equilibrium in the course of the high-temperature firing process, adopted during manufacturing. In many instances, the refractories are fired at lower temperature than the service temperature of the refractory. Therefore, the refractories, when exposed to high temperature for a long duration in service, change dimension often, which is irreversible and is known as permanent linear change (PLC). It is also expressed in %, similar to RTE. This property is determined by heating a bar, cut out from a refractory body, at a specified temperature for a specified time. The initial length and the length after the firing are measured to determine the % PLC. Refractories, sometimes, are designed purposefully to have a positive or negative PLC. The PLC of a refractory product can be determined following ISO standard method ISO2478.1987.

Thermal Conductivity

It is the fundamental property of a material and can give the estimation of heat flow through a material. Heat energy flows from high temperature, inside the furnace, to low-temperature ambience, outside the furnace, through the refractory lining of the furnace. In a steady-state operation of a furnace, after a part of the thermal energy is absorbed by the furnace construction material, there is a steady outflow of energy from the system, and it is a very important factor to decide the economy of the process. The thermal conductivity of the refractory lining material is thus an important parameter for designing the furnace lining.

The thermal conductivity is an inherent property of materials. It is different for different materials, and it is a function of temperature which means it varies with temperature for the same material. That is why the thermal conductivity at mean temperature is used for design purposes. The major factors that affect the thermal conductivity of the refractory material are the mineral composition, the amount of amorphous material, and its pore volume, pore size distribution, microstructure, and temperature. Thermal conductivity also depends upon the composition of the gases surrounding the refractory. The thermal conductivity of an aluminosilicate refractory is much higher in hydrogen atmosphere than what it is in normal atmosphere. Thermal conductivity is a difficult property to measure in a steady-state method, because it takes a very long time to attain steady state and to maintain it. Therefore, unsteady-state methods are easier to perform, saves time, and are widely used and one of these methods is hot wire method.

Thermal conductivity is, specially, important for the designing of the insulation lining to maintain the outside shell temperature of the reactor or the furnace to a permissible temperature limit.

Thermal conductivity can be measured following the method given in ISO-8894.2.1990 or 8894.1.1987.

Heat Capacity (C_p)

It is the fundamental property of the material. This gives an estimation of the heat stored in refractory structure in a furnace. C_p value is important because it is required to calculate the heat required to raise the temperature of the refractory to a certain temperature.

Thermal Diffusivity (D)

Thermal diffusivity is a measure of how quickly a body can change its temperature. It is defined as

$$D = \frac{K}{C_p \rho} \tag{1.12}$$

where K = thermal conductivity, C_p = heat capacity, and ρ = Density. Values of both K and C_p depend on temperature.

1.1.2.7 Thermo-mechanical Properties

The refractory materials are used in high temperature and different types of compressive, tensile, and shear stresses that act on the refractory body at high temperature. The refractory must be able to withstand those stresses at high temperature. The thermo- mechanical properties of the refractory, gives an idea about the capability of the refractory to withstand those stresses at elevated temperatures. The thermo mechanical properties and some tests, devised to measure those properties will be discussed here.

Hot Modulus of Rupture (HMOR)

This is similar to three-point bending test performed in cold modulus of rupture, but it is performed at higher temperature inside a furnace, and the HMOR value is calculated using the same formula. HMOR value indicates the tensile stress the refractory can withstand at certain temperature. It gives an idea about the bonding strength between the matrix and the grains, in the refractory body. It is influenced by the amount of liquid phase generated at the test temperature by its viscosity by the microstructures, i.e., the nature of porosity, microcracks, etc. and its distribution in the refractory body. This test can be carried out following the method given in ISO 5013.1985.

Refractoriness Under Load (RUL)

Refractories deform when put under a compressive stress at elevated temperature. The deformation is mainly due to the formation of liquid phases in the matrix of the refractory body at elevated temperature. The extent of deformation depends upon the temperature, the compressive stress, and the quantity and viscosity of the liquid phases formed. The higher is the volume of the liquid phases formed and the lower

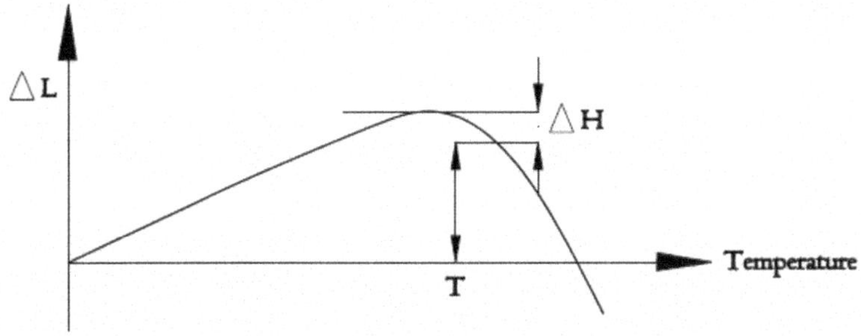

Fig. 1.10 RUL, temperature versus deformation curve

is their viscosity, the higher is the deformation at high temperature. Higher temperature and higher compressive stresses lead to more deformations.

RUL measures the temperature at which a specific deformation of refractory occurs and gives an idea about the load bearing capacity of the refractory at high temperature. RUL can be determined following the test method given in ISO1893.1989.

In this test, the cylindrical sample of diameter 50 mm and height 50 mm is drilled out of the refractory body and is put into a vertical tubular furnace, on one of its flat side. A load of 0.2 MPa is applied on the surface through a graphite rod, and the temperature is raised at a rate of 5 °C/min. The change of length of the sample is measured during heating and plotted against temperature as a curve shown in Fig. 1.10. The temperatures at which deformations of 0.2%, 2%, and 5% occur are denoted on the temperature vs deformation curve as $T_{0.2}$, T_2, and T_5. The deformation shown in figure is

$$\left[\left\{\Delta H\left(\text{in mm}\right)\right\}/50\right]\times 100\% = 2\Delta H\% \tag{1.13}$$

Creep in Compression

This property characterizes the time-dependent deformation of the refractory at high temperature. When a refractory is subjected to a compressive stress for a long time, then its deformation behavior over that period can be predicted by this property. In this test, the cylindrical sample of 50 mm diameter and 50 mm height is drilled out of the refractory under testing. The sample is put into a cylindrical furnace and the temperature is raised to test the temperature and a load of 0.2 MPa is applied. The temperature and the load are kept constant. The deformation of the sample is measured against time over a long period of time, to get its creep behavior. The creep in compression can be measured following ISO method 3187.1989.

Thermal Shock Resistance or Spalling

The refractory lining faces temperature cycling during its campaign for various reasons, for example, because of start-up and shutdown of the kiln, because of variation in fuel rate, etc. In most of the cases, because of exigency of production, the heating and cooling are made fast. The fast temperature variation in the kiln or furnace causes the development of thermo-mechanical stresses in the refractory lining. Differential temperature in the lining results in differential expansion in the same body and develops mechanical stresses, either compressive or tensile in nature, which ultimately damages the refractory.

The refractory lining develops compressive stress on fast heating and tensile stress on fast cooling of the furnace. Normally, the ceramic materials are weaker under tensile stress, and therefore fast cooling damages the refractory more than fast heating.

The thermal shock resistance of a refractory is related to its other properties like thermal conductivity, thermal expansion coefficient, modulus of elasticity, and rupture strength. The higher the thermal conductivity and rupture strength, the better is the thermal shock resistance. The lower value of modulus of elasticity and thermal expansion coefficient makes the thermal shock resistance property of the body better.

Rapid temperature change on the surface of a refractory body causes large temperature gradient that leads to considerable stress which can be expressed by the equation below [3]:

$$\sigma = \frac{E\alpha}{1-\mu}\Delta T \tag{1.14}$$

Here, σ = stress produced, E = Young's modulus, α = coefficient of linear expansion, μ = Poisson's ratio, and ΔT = temperature difference.

If a material has a breaking tensile strength S (MOR) and can resist maximum temperature difference $\Delta T_{\max}$ when the stress developed exceeds the breaking strength, then

$$\Delta T_{\max} = \frac{S(1-\mu)}{E\alpha} \tag{1.15}$$

A term R' was introduced to indicate the crack initiation resistance of a body under thermal temperature change, as

$$R' = \frac{S.\lambda}{E.\alpha} \tag{1.16}$$

where λ = thermal conductivity

This equation shows that a refractory with high strength and high thermal conductivity, low modulus of elasticity, and low thermal expansion has high resistance to crack initiation.

To test the thermal shock resistance property in normal laboratory, the refractory samples of definite shape are cut out of the refractory bricks and the samples are heated at a definite temperature for a definite time, and then it is quenched either in cold air or water for a certain time, and the total process is called one cycle and the same process is repeated and continued till the refractory suffers a specified damage, and it is reported as the number of cycle under the specified condition of heating and cooling. The thermal shock resistance testing is carried out to compare between two and more refractory products. The PRE/R5 method of the European Federation of Refractories can be used for this comparison.

Structural Spalling

In many occasions, an altered zone is formed in the refractory lining, near the working surface during service, and it tends to crack or fall off in the form of thin or thick layer called "peeling" or "slabbing." This happens due to the formation of an internal stress caused by, even, small temperature fluctuation. The stress builds up, due to the mismatch of properties between the original and altered zones of the refractory as a result of mineralogical changes and liquid phase formation. The major cause of the formation of altered zone is the chemical reaction of the refractory at elevated temperature with different gas, liquid, and solid which remain in contact with refractory forming different mineral phases. The penetration of the liquid slag also causes the alteration of the refractory texture and initiates the structural spalling. Overheating accelerates those reactions and formation of the liquid phases and causes densification of the refractory body in the altered zone. The altered zone formation can be reduced if the slag penetration can be reduced. It has been found that addition of small phosphate-based material in the chrome-magnesite refractory in a gasifier can reduce the slag penetration by 1/5 and can improve the refractory performance [17]. The thermal shock resistance testing is carried out to compare between two and more refractory products. The PRE/R5 method of the European Federation of Refractories can be used for this comparison.

1.1.2.8 Corrosion Resistance

It is the property of the refractory which indicates its inertness to chemical reaction with the specified solid, liquid, or gas at high temperature. Due to the chemical reactions mentioned, the refractory forms different products or phases which are not suitable as a refractory material, and the refractory stops functioning for the purpose it is used. The reactions are more prominent with liquids and gases, and the reaction rate increases fast with increasing temperature. The process of corrosion is very complex in nature and is a combination of chemical and physical processes. Besides the chemical reaction, the pores in the refractory allow the molten liquid or gas to permeate inside and react further [18]. The reaction products may develop stress because of its higher volume and causes cracks in the body, which increases the

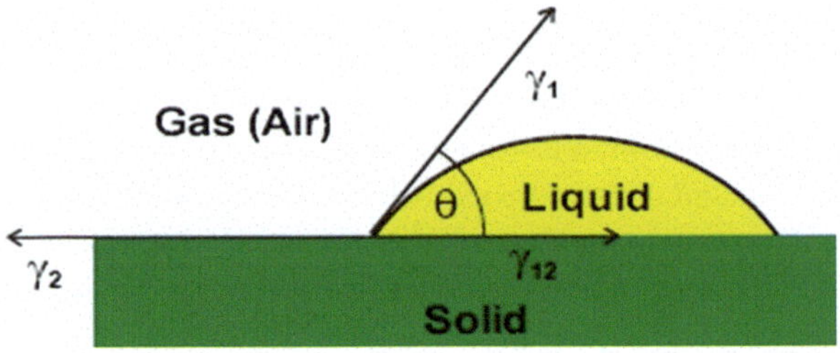

Fig. 1.11 Wetting or contact angle, θ

contact area with further enhancement of corrosion. For the physical penetration of the liquid, the rate of penetration, dl/dt, is governed by the following equation:

$$dl/dt = (r.\gamma.\cos\theta)/4\eta l \tag{1.17}$$

where r = radius of pore, γ = surface tension of liquid, η = viscosity of the melt, and θ = wetting angle or contact angle (Fig. 1.11). It shows that higher pore size of refractory, lower wetting angle of the liquid on the refractory, and lower viscosity increase the rate of penetration. The reaction rate increases fast with increasing temperature and the type of flow of the fluid (laminar or turbulent) in contact.

When liquid slag reacts with refractory, the reaction product(s) can directly dissolve in the liquid in contact, or an insoluble solid can form at the refractory slag interface which then can dissolve in the liquid slag (indirect dissolution). In direct dissolution when the diffusivity of the reaction products is faster than the rate of the chemical reaction at the refractory liquid interface, the initial dissolution rate can be described by the following equation [19]:

$$J = K(A_c/A_o).C_m \tag{1.18}$$

where J = dissolution rate, A_c = actual area of the refractory, A_o = apparent area of the refractory, and C_m = concentration of the reactant in the melt.

In case of indirect dissolution, the corrosion rate can be expressed by Nernst equation as

$$J = D(C_s - C_m)/\delta \tag{1.19}$$

where J = corrosion rate, D = diffusion coefficient, C_s = saturation concentration of reactant in the liquid, C_m = concentration of reactant in the liquid, and δ = boundary layer thickness at the interface.

A local corrosion is very often encountered at the refractory, liquid glass/slag, and the gas/air interface as shown in Fig. 1.12. A local movement of slag is found originated because of a surface tension gradient set up at the junction called Marangoni effect. This effect takes place when there is a gradient of surface tension

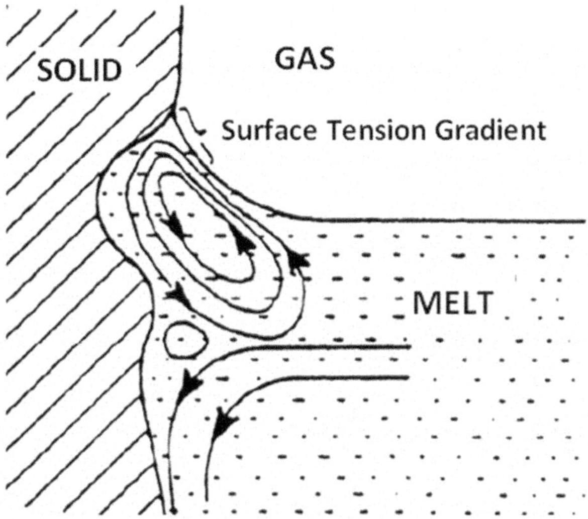

Fig. 1.12 Liquid convection due to Marangoni effect

at the liquid-gas interface, due to variation in solute concentration, and a strong convection motion can be generated depending upon the liquid character.

Extensive computerized thermodynamic databases have been prepared for solid and liquid metals, slags, and solid oxide solutions over the last three decades by critical evaluation of all available phase equilibrium diagram and thermodynamic data. Sophisticated thermo-chemical computing systems such as FactSage, CEQCSI, and MTDATA are available and used to understand the interaction between molten slag, refractory, inclusion, and gas phases [20].

It is very difficult to simulate the conditions of the refractory and melt reaction in the laboratory. The corrosion resistance test is a qualitative test, designed to compare the affinity of two or more refractory products to react with a liquid, solid, or gas at a certain temperature. However, to get the idea of the compatibility of a refractory to a particular liquid phase at a certain temperature, tests are carried out. There are two different types of test, static and dynamic.

In static test, a cube of 75 mm size is cut out of the refractory under testing. A drill is made of 25 or 50 mm diameter and depth of 40–50 mm in one face of the cube. The cup made is filled with the solid with which the reaction of the refractory is to be studied, when it is in molten condition. The cube is then put in the furnace and fired till the desired temperature is reached. The sample is allowed to reside inside the furnace at that temperature for a specific time, allowing the reaction to happen, and then the furnace is put off and cooled. The refractory cube is taken out and the cup is cut vertically to observe the extent of the reaction. It is a comparative test and difficult to quantify. The judgment of corrosion resistance is made by visual observation only after a refractory is allowed to undergo through a corrosion reaction at a specified temperature for a specified time.

In dynamic slag test, the melt remains in the furnace in a crucible, and an attachment holds a small bar of a specified dimension which rotates on its axis in the melt at a desired test temperature for a specified time. Then, the sample is taken out of the melt, cooled, and visually observed for the reaction to happen.

1.1.2.9 Alkali Resistance Test

It is a qualitative test and compares the alkali resistance of a refractory with others. In this test, a 40 mm hole is drilled in a refractory brick to form a cup. The cup is filled with potassium carbonate, and the test piece is put into a furnace and the temperature is raised to 1100 °C and kept for 2 h. The furnace is put off and cooled. The test piece is observed. The test piece will develop a crack if it is not resistant to alkali attack. The extent of crack determines its resistance towards alkali attack. The photograph in Fig. 1.13 shows the test samples after alkali resistance test.

Different test methods are available to compare the corrosion resistance of two or more refractory products under identical condition. It can be tested by C454-83 of ASTM standard.

Vapor attack can be restricted by the pore size design of the refractory. Decrease of mean pore size distribution of the refractory is very beneficial to reduce the alkali corrosion [21].

Addition of zircon ($ZrSiO_4$) in mullite brick improves the alkali corrosion resistance because of the densification of the brick and reduction in porosity [22].

Fig. 1.13 Alkali test of high alumina brick with K_2CO_3

1.1.2.10 Stability in the Gaseous Atmosphere

Common refractory materials used are mostly oxides. Many of these oxides are found not stable in reducing atmosphere, e.g., in the presence of CO or H_2 as exists in the reformers, catalytic crackers, FCCU, etc. [23, 24]. Hydrogen-rich process gases can react with aluminosilicate refractories at high temperature and can remove silica from the system by the reaction given below:

$$SiO_2(s) + H_2(g) = SiO(g) + H_2O(g) \tag{1.20}$$

At lower temperature, this SiO reacts with hydrogen to deposit metallic silicon as per the reaction given below:

$$SiO + H_2 = Si + H_2O \tag{1.21}$$

When exposed to CO atmosphere over a certain temperature and time, carbon is formed as per Boudouard reaction (1.22), and carbon starts getting deposited inside the refractory where any form of free iron or iron oxide acts as nucleation site. The deposition of carbon inside the refractory needs extra space to accommodate it and develops the stress which causes refractory damage by cracking called CO disintegration.

$$2CO = CO_2 + C \tag{1.22}$$

It has been found that Boudouard reaction is highly favored by the presence of the catalyst Fe_xO_y and the presence of H_2 in the gas mixture [25]. The grain size of iron oxide also influences the reaction.

It has been reported [26] that the presence of S or its compound like $BaSO_4$ in the aluminosilicate refractory can prevent this reaction to occur or may decelerate the kinetics of the reaction to a great extent.

Oxides are major group of refractory products, and some of which are not thermally stable in strong reducing atmosphere and tend to get reduced. Gibbs free energy (ΔG) of a reaction is the measure of the driving force that makes a reaction occur. A negative value of ΔG indicates that the reaction can proceed spontaneously without external inputs. The equation for Gibbs free energy is

$$\Delta G = \Delta H - T\Delta S \tag{1.23}$$

where ΔH = change in enthalpy of the reaction, T = absolute temperature, and ΔS = change in entropy.

Since ΔH and ΔS are essentially constant with respect to temperature, the plot of ΔG with temperature, unless a phase change occurs, gives a series of straight lines which is called Ellingham diagram (Fig. 1.14). In this straight-line graph, ΔS is the slope and ΔH is the intercept on the Y-axis.

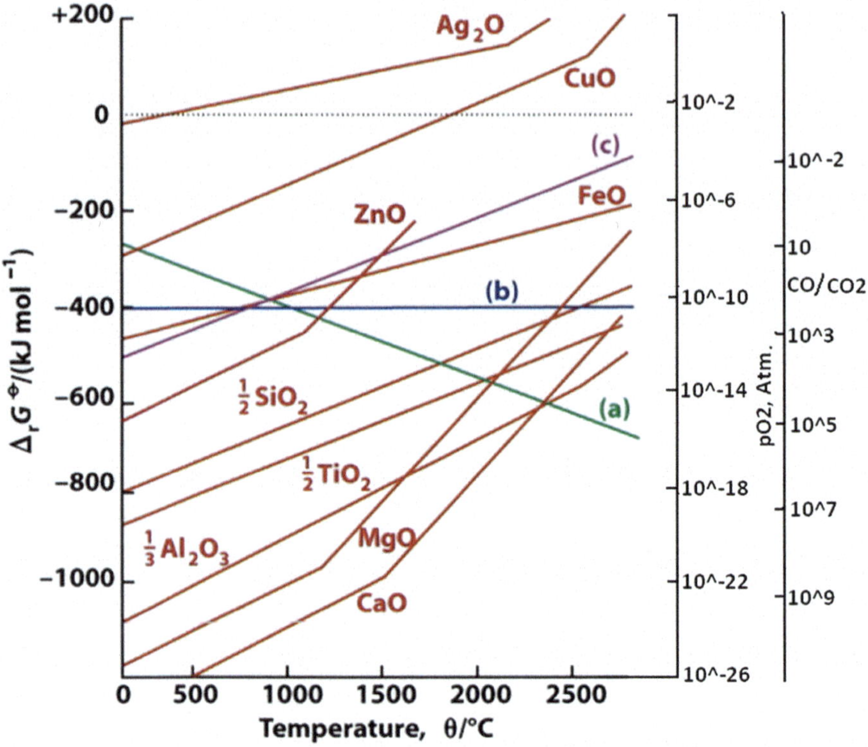

Fig. 1.14 Ellingham diagram. (*a*) Graph for $2C + O_2 = 2CO$, (*b*) graph for the reaction $C + O_2 = CO_2$ (*c*) Graph for $2CO + O_2 = 2CO_2$

This diagram can be used for three main purposes:

1. To get the idea of the relative ease of reducing a metal oxide to metal
2. To determine the partial pressure of oxygen in equilibrium with a metal oxide at any temperature
3. To determine the ratio of CO to CO_2 that will be able to reduce the metal oxide

The diagram shows all metal oxide curves has an upward slope. If the material melts or vaporizes, the slope changes. There can be three situations:

1. When Line 'a' is below M → MO line, then carbon reduces the metal oxide to produce CO.
2. When Line 'b' is below M → MO line then C reduces metal Oxide to produce CO_2.
3. When CO → CO_2 line is below M → MO line CO reduces the MO to produce CO_2.

1.1.2.11 Microstructure

This tells about the physical structure of the refractory, e.g., the grain sizes and morphology, pore sizes and morphology, orientation and distribution of the grains, matrix and pores, presence of different mineral phases, presence of microcracks, etc. All the microstructural features mentioned earlier influence the physical properties of the refractory. The microstructure of the refractory is revealed through different instrumental methods as mentioned below. Two refractory products having the same chemical analysis may have very different physical properties due to different microstructures. For example, a refractory brick in aluminosilicate category has much better RUL, creep, and thermal shock resistance if it contains higher percentage of mullite, and even the properties differ based on the crystal growth of mullite ($3Al_2O_3.2SiO_2$). But the presence of mullite phase cannot be known from normal chemical analysis or other physical tests mentioned so far.

Optical Microscopy

The polished section or the thin section of the refractory can be seen under high-resolution microscope either by reflecting light or by transmitting light. The photo can be taken to record the microstructural features. It shows the grain sizes, pore sizes, and their distribution along with that of the matrix and morphology of grains and pores (Fig. 1.15).

Fig. 1.15 Optical micrograph of an aluminosilicate refractory showing pores (*P*), aggregate (*A*), and matrix (*M*)

Electron Microscopy

In this instrument, electron beam is used in place of ordinary light. Magnetic field does the function of lens in electron microscope. The resolution in this microscope is very high and 1000 times more than the optical microscope. It enables us to see the microstructure with much larger magnification (Fig. 1.16). The different methods of testing, using electron beam, are scanning electron microscopy (SEM), electron probe microanalyzer (EPMA), and transmission electron microscopy (TEM). By EPMA, any point on the section of the refractory can be chemically analyzed, and the chemical map of a selected area on the sample under microscope can be obtained which is very useful to have an idea of mineral phases present and their distribution (Fig. 1.17).

By exposing the powder of a refractory sample to X-ray, a pattern is obtained. When that X-ray pattern is analyzed, then different mineral phases present in the refractory body can be known by the presence of their characteristic peaks (Fig. 1.17). The presence of the peaks indicated the presence of the particular mineral, and the intensity of the peak gives an idea about its quantity present.

Fig. 1.16 Scanning electron micrograph of aluminosilicate refractory shows mullite crystals

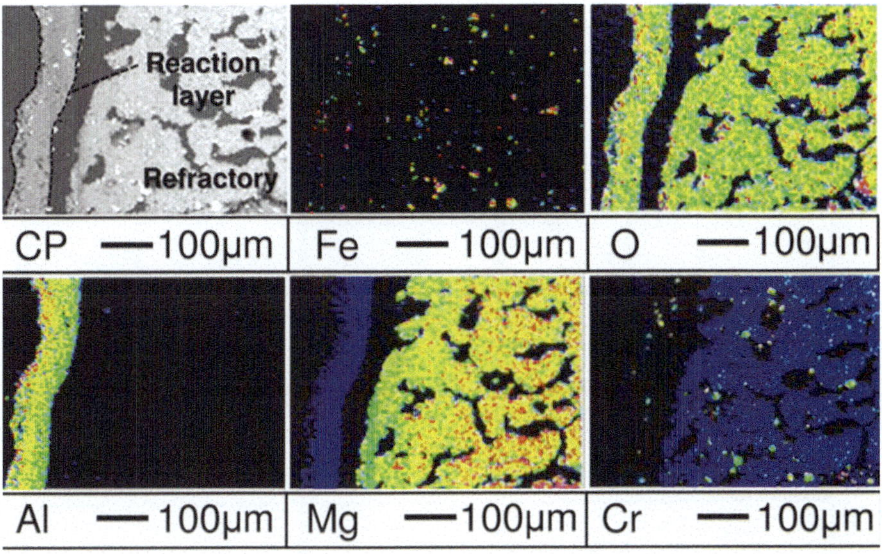

Fig. 1.17 EPMA image of a refractory reacted with slag. The chemical mapping clearly shows the distribution of the elements and helps in identifying the reaction mechanism

X-Ray Fluorescence Method (XRF)

X-ray fluorescence method (XRF) is used for the chemical analysis of any refractory material. In this method, the powder of the test sample is used to form a pallet which is fed to the XRF machine, and on running the machine, it produces a diffraction pattern (Fig. 1.18). The standard X-ray diffraction pattern is already there for different minerals. If the produced X-ray diffraction pattern is matched with the diffraction pattern of the known minerals, then the mineral phases present in the sample under investigation can be ascertained. All these comparisons are done by the software incorporated in the instrument, and as an output, the final chemical analysis of the material is obtained.

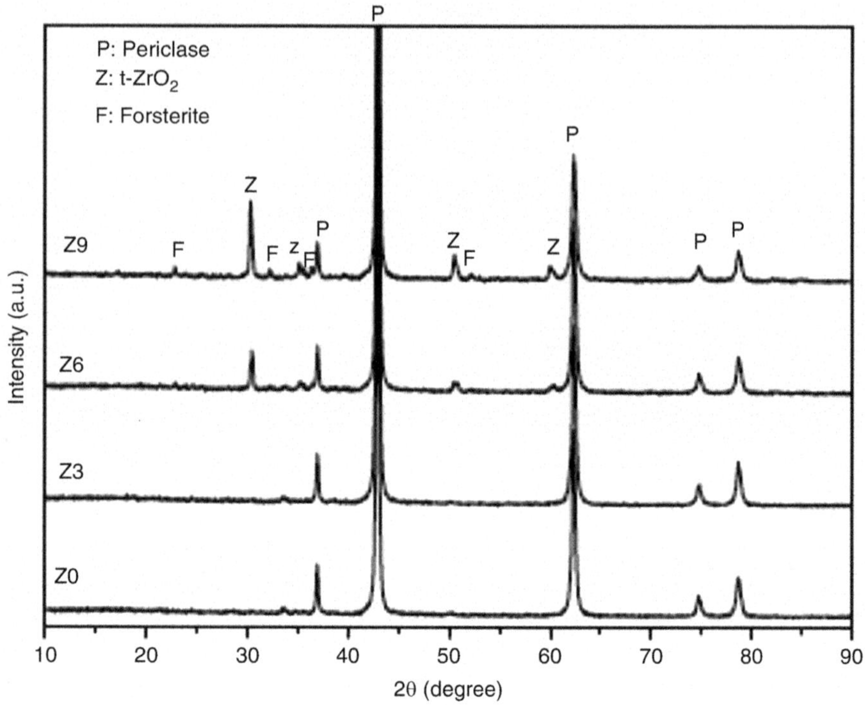

Fig. 1.18 The XRD of magnesia-zirconia refractory

References

1. Peijun Wang. Properties of Steel at elevated temperature. https://www.researchgate.net/publication/284995632_Properties_of_Steel_at_Elevated_Temperatures
2. ASTM standard – refractories carbon and graphite products, Volume-15.01, 1996
3. Technology of monolithic refractories. Plibrico Refractory Company, 1984
4. Bradt RC, Scott AN. Elastic properties of refractories – their role and characterization. Refractories Application and News, Vol-12, No-3, May–June 2007
5. Bradt RC. Fracture of refractories. In: Schacht C (ed) Handbook of refractories. Marcel Dekker Inc., New York
6. Hamuth H, Bradt RC. Investigation of refractory brittleness by fracture mechanical and fractographic methods. Refractories manual 2010, Interceram
7. Dahlem E. Characterization of refractory failure under combined influence of hydrostatic and shear loading at elevated temperature. Doctoral thesis, University of Limoges
8. Ru Zhou. Investigation of refractory brittleness by fracture mechanical and fractographic methods of refractories under combined mode I and mode II loading. Master thesis at the chair of Ceramics, Montan universitat, Leoben
9. Böhm A et al (2016) Thermal shock and thermo-mechanical behavior of carbon-reduced and carbon-free refractories. J Ceram Sci Tech 07(02):155–164
10. Prietl T et al (2005) The evaluation of refractory linings thermo-mechanical properties. In: Proceedings of EMC
11. Wear resistant engineering ceramics – Eugene Medvedovsky, wear(2001), 821–828, Elsevier

12. Ntakaburimvo N, Allaire C. Abrasion wear of Aluminosilicate refractories. CIREP-CRNF, Ecole Polytechniqe, Montreal
13. Pintaude G, Coseglio M (2009) Abrasive particle characterization following measurement of different shape factors. 20th international congress of mechanical engineering, Nov 15–20, 2009, Gramado, RS, Brazil
14. Malkin JT, Padgett GC (1979) Abrasion of refractories at elevated temperature. Interceram 28(Special Issue):228–233
15. Viljoen SD (2010) Reduced maintenance costs resulting from the use of wear resistant materials. J South Afr Inst Mining Metall 110
16. Xigao Peng et al. Research on abrasive resistance of several kinds of refractory products at elevated temperature. UNITECR 2009 proceedings
17. Dogan CP. Improved refractories for ICGC Power system. DOE/ARC/2002-014
18. Lee WE, Zhang S (2004) Direct and indirect slag corrosion of oxide and oxide-C refractories. In: VII international conference on molten slag, fluxes and salt. The South African Institute of Mining and Metallurgy
19. Garbers-Craig A. The study of molten liquid- refractory interactions – It is all about the phases. In: Advances in molten slags, fluxes and salts: proceedings of the 10th international conference on molten slags, fluxes and salts, molten −16, 2016
20. Ho Jung (2010) Overview of the application of thermodynamic database to steelmaking process. Computer coupling of phase diagram and thermochemistry 34. Elsvier
21. Shaobai S et al (2016) Improvement of gaseous corrosion resistance of refractories by pore structure design according to the seepage flow model. World Refractory Forum 8(3)
22. Serhat Baspinar M, Kara F (2009) Optimization of the corrosion behavior of mullite refractories against alkali vapor via $ZrSiO_4$ addition to the binder phase. Ceramics Silikaty 53(4)
23. Ballon M et al (2006) Design and maintenance of refractory linings in the reformer section of the ammonia plant. Ammonia Technical Manual
24. Brosnan DA Corrosion of refractories. In: Schacht CA (ed) A chapter in the handbook of refractories. Merkel Dresser Inc, New York, p 10016
25. Bost N, Poirier J, Ammar MT (2016) New data about the boudouard reaction in refractory materials: investigation by thermogravimetric measurements and in situ Raman spectroscopy. Eurogress, Aachen, Germany, 28–29th Sept, 2016
26. Poirier J et al (2017) Some good news: inhibition of boudouard reaction in refractory materials subjected to CO and H_2 reducing atmosphere. In: UNITECR proceedings

Chapter 2
Classification of Refractories

2.1 Classification of Refractories

There is no rigid norm for classification of refractories, and it can be classified in various ways. The basis of classification of refractories in different ways is discussed here.

2.2 Basis of Classification

Refractories are generally classified based on different parameters, e.g.:

(a) Basicity of oxides
(b) Form
(c) Manufacturing process
(d) Method of application
(e) Special chemistry
(f) Insulating property

Let us see the classification in details.

2.2.1 *By Basicity*

The oxide refractories can be classified according to the basicity of the oxides. Refractory oxide can be characterized by the electropositive character of the central atom in that oxide. The more electropositive is the central atom, the more basic is the oxide, and more electronegative is the central atom, and the more acidic is the

P. Sengupta, *Refractories for the Chemical Industries*,
https://doi.org/10.1007/978-3-030-61240-5_2

oxide. Electropositive character increases from right to left across the periodic table and increases down the column.

According to the position of the element in the periodic table, its oxide can be classified as acidic, basic, or amphoteric in nature. Accordingly, the refractories can be classified as acidic, basic, or amphoteric as shown in Table 2.1. This classification has significance in selecting refractories to be used in contact with molten slag with a known basicity. For example, basic refractories should be selected where the basicity of the slag, in contact with refractory, is more than 1.0 to get a reduced corrosion rate, although many other factors should be taken in consideration in selecting the right kind of refractories for a certain application.

2.2.2 *By Form*

The other classification of refractories has been made on the basis of its form manufacturing process and application. According to form, refractories are classified as shaped refractories, commonly known as refractory bricks, and the unshaped refractories as monolithic refractories. The shaped refractories are delivered in the form of definite geometrical shape, and unshaped refractories are supplied as loose powder, which is to be mixed with required amount of binder/water and applied at the site to take the shape of the area, where it is used. Figure 2.1 shows the scheme of the classification of refractories based on form.

2.2.2.1 Shaped Refractories

Shaped refractories are made in different geometrical shapes according to the requirement of the user. The shapes are designed depending on the geometry of the furnace or equipment which is to be lined with refractory. Shaped refractories are classified by the process of manufacturing adopted and are termed accordingly.

Pressed Shape

Different size fractions of different refractory raw materials are mixed with a liquid binder. The mixture is pressed in a mold at high pressure in mechanical press, followed by drying and firing to impart mechanical strength and volume stability at

Table 2.1 Classification of refractories based on basicity

Acidic refractories	Basic refractories	Amphoteric refractories
SiO_2 (silica)	MgO (magnesite)	Al_2O_3 (alumina)
ZrO_2 (zirconia)	CaO (lime and dolomite)	Cr_2O_3 (Chromia)
		Carbon

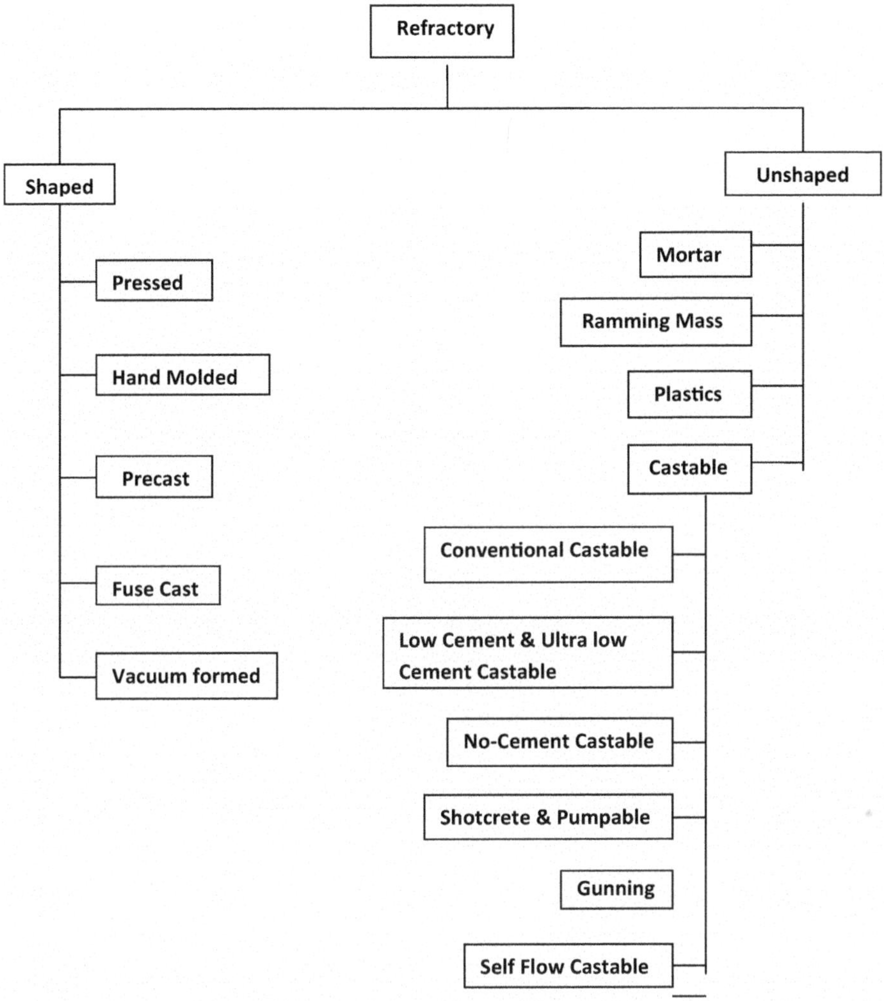

Fig. 2.1 Classification of refractories

high temperature to the body. These products are called pressed bricks. The bigger is the dimension of the brick, the higher is the capacity of the press needed. Product with complicated shape and geometry cannot be made by this process.

Hand-Molded Shapes

Hand-molded shapes are made by mixing different size fractions of different refractory raw materials with a liquid binder and to give shape by pneumatic ramming in a mold followed by drying and firing to impart mechanical strength and volume stability at high temperature. These products are having lower density and strength than pressed products. This method of production is much slower compared to production by pressing. Only very critical shaped items are made in this method.

Precast Shape

In this method different size fractions of refractory raw materials are mixed along with high alumina cement and other additives followed by mixing with required quantity of water. The mixture is then given shape in a mold by vibration casting or pouring. The cast body is then allowed to set and cured followed by demolding and drying to form the precast refractory. Sometimes the cast body is fired at high temperature to meet the desired properties. This is also a very slow process of production compared to pressing and used to produce complicated bigger shapes which cannot be made by pressing in mechanical press.

Fused Cast Shape

The mixture of raw materials is melted in a high temperature furnace, and the melt is poured in a graphite mold. The mold is cooled very slowly when the melt solidifies and takes the shape of the mold. These are called fused cast refractory shapes. Specialties of these products are almost zero porosity, and these are used mainly where the corrosion is the main factor to destroy the refractory, e.g., lining of the glass tank where the refractory remains in continuous contact with highly corrosive molten glass.

Vacuum Formed Shape

This method is mostly used for making ceramic fiber insulation boards or similar products with other shapes. In this process, the ceramic fiber along with the organic/inorganic binders are stirred in water at high speed to form a slurry, which is spread over a mold, fitted with fine wire mesh, and the water is sucked out under high vacuum pressure to give the shape of the body. The formed body can then be demolded and dried and heat cured to develop strength in it.

The advantages and disadvantages of using shaped refractories are:

1. The bricks made under high pressure attain very good properties like high density and low porosity.
2. The laying of shaped refractories does not demand very high skill of masons.
3. The laying process is normally manual and does not involve the use of expensive machines, although in case of rotary kiln lining the brick laying machines are being used very successfully.

4. The total time of refractory lining and preheating, i.e., starting of the installation to furnace ready-to -use duration, is less compared to lining with unshaped refractory.
5. Many different shapes are required in combination to line the equipment, and inventory cost is higher compared to unshaped refractories. On non-availability of even one shape, the total lining process may come to a halt.

2.2.2.2 Unshaped Refractories

Unshaped refractories are normally manufactured and supplied in bags as loose powder. The manufacturing of unshaped refractories is simpler than the shaped refractories. In its manufacturing, the different ingredients in different size fractions are mixed thoroughly in a mixer with required additives and bagged. During use, the bags are opened at the site; the powder is mixed with required quality and quantity of water or supplied liquid binder, in a mechanical mixer, and is rammed or cast or sprayed at desired place of installation. The unshaped refractories can be further classified based on its application method.

Mortars

Mortars are powder materials which are used for the purpose of joining the refractory bricks. The powder material is mixed with water or a liquid binder supplied along with the mortar and applied as glue on the brick surface to join two bricks. The mortar can set and harden either by chemical reaction (chemical setting mortar) at ambient temperature or by heating (heat setting mortar) the joined brick. Mortar can be of both basic and aluminosilicate type used to join either basic bricks or aluminosilicate bricks. Mortar joints in the brick lining act as a cushion to reduce the thermo-mechanical stresses generated during operation of the furnace.

Ramming Masses

Ramming masses are used in very less quantity and for very special purposes only, in cement industries. It is a mixture of coarse aggregates and fine powder with some additives, and it is mixed with water or liquid binder and rammed by hand or by pneumatic rammer at the desired area. It is normally applied to fix up the gaps which are difficult to be filled otherwise, for example, a small gap between the brick layers and brick layer and retainer plate. Ramming masses can be made in both basic and nonbasic composition, but in cement industry, the nonbasic ramming masses are used normally.

Plastic Refractories

Plastic refractories are ready-made unshaped refractories. These are supplied, in ready to use form, of small blocks, wrapped in polythene sheet. The material is unwrapped at site and put at the place where it is to be used and then stamped with wooden mallet or rammer with flat bottomed tool. Plastic refractories are made with aluminosilicate material only. This material is hardened fast after installation and can be dried and preheated faster, compared to castable. Because it is a ready mix material and binder is mixed already, it has got a low storage life, especially in hot climate of tropical countries. It has the advantage that it is a ready-mixed material, so the chance of mistakes due to wrong addition of binder or improper mixing of binder can be eliminated. It is a good material to use for lining the roof of a furnace. Ceramic anchors are used, similar to castable installation, to hold the material. Its installation is a slow process, but it can withstand fast heat up and does not develop crack.

Castables

Castables actually are the major products in the unshaped refractory category and are made up of aluminosilicate aggregate and calcium aluminate cement (also called high alumina cement) as binder, along with special additives to impart or modify certain special properties in it. The calcium aluminate cement, like Portland cement, has hydraulic property and sets to a hard mass in contact with water. In case of no cement castables, the liquid binders are supplied separately along with dry powder, and both are to be mixed during application. Of course, recently the no cement castable has also been developed in which the water can be used for casting instead of any separate liquid binder.

Castables are the material which can be cast after mixing with water or other liquid binder (in no cement castable), and it set hard at ambient temperature. Castables can be installed by different methods, e.g., by pouring, by vibration casting, by gunning, and by shotcreting. Two major steps in the installation processes are mixing of castables with water and the method of consolidation of the castable. Mixing quality is very important, and the aim should be to get maximum flow using the prescribed water or binder content. That is possible only when the mixing is very thorough and intimate, done in high shear mix.

The consolidation of the castable can be done by the following methods:

1. Pouring it as slurry or semi-slurry and to consolidate it with light vibration, called casting
2. Vibrating strongly a stiff thixotropic material when it starts flowing and the process called vibrocasting
3. Transporting and spraying of water mixed slurry called, shotcreting
4. Spraying a dry material which is mixed with water at the spray nozzle tip called gunning

Castables can have further classification depending upon its composition, method of application, and flow behavior. The purpose of this classification is to distinguish the difference between the products which are made out of the same basic ingredients but differ substantially in their rheological (flow behavior) character, drying behavior, etc. The relationships of different categories of castables are shown in Fig. 2.2.

Conventional castable with higher content of calcium aluminate cement requires high percentage of water to attain required flow character, and the mixing with water can be done even on floor with the help of shovel. The setting can be controlled and modified with the addition of special additives. The heat up curve to remove water is not very critical unless it is cured at a temperature below 10 °C.

In many occasions when a shutdown is taken at any plant, the refractory lining is first inspected to check for any damage in the refractory lining. Then small patching work is usually done to repair small areas of worn-out brick or castable lining by troweling or plastering the castable, mixed with water. Conventional castables are suitable for this purpose.

The low and ultralow cement castable, having almost the same ingredients as conventional castable, but with lower content of calcium aluminate cement. It can attain the desired flow character necessary for vibration casting, with almost half of the water as required in conventional castable. But it is to be borne in mind that very intense mixing is required to develop the desired flow property in the low cement castable and is to be mixed in a high shear mixer. Mixing of low cement castable

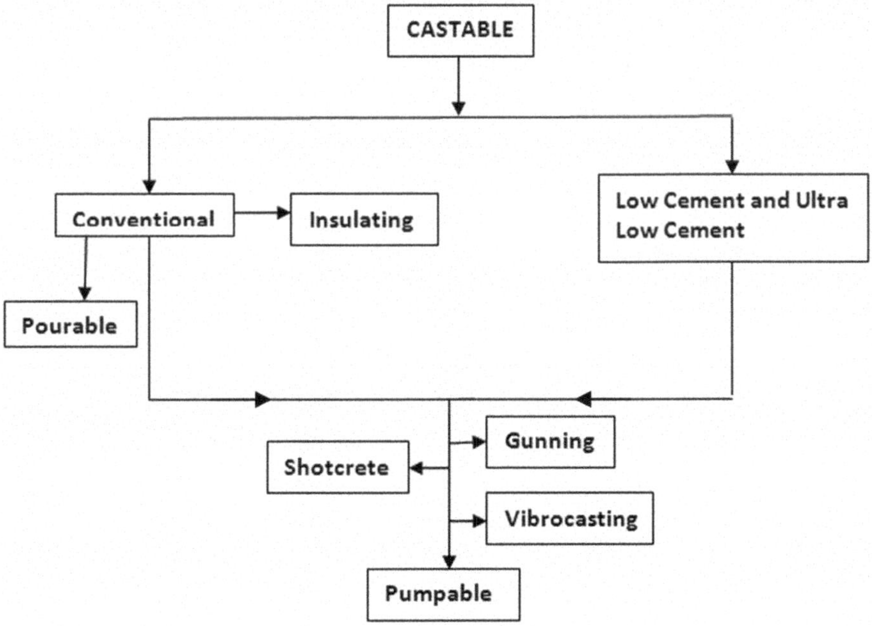

Fig. 2.2 Relationship between different castable products

and water in ordinary mixer or with shovels do not yield the desired flow property required for casting. This enhancement in the flow property, engineered by the particle size distribution and addition of certain special additives, which actually revolutionize the whole castable technology. Lower content of the calcium aluminate cement, as suggested by the name, imparts better properties to this type of castables at elevated temperature over the conventional castables because of lower content of CaO in the mix. Low cement and ultralow cement castables develop a thixotropic property on mixing with water in high shear mixer. The mix starts flowing only when a shear force is applied on the mix and requires strong vibration either by means of needle vibrator or form vibrator. This kind of castables forms a much denser body on casting and has lower permeability compared to conventional castable, and therefore its drying and preheating are very critical. The casting may crack on preheating unless proper care is taken.

A further development of low cement castable technology has made it self-flow type, which, when mixed with low percentage of water, can flow very easily and requires very mild or almost no shear force or vibration for its placement. It can be used at a place where the vibration casting cannot be done, and at the same time properties of low cement castables are required. This is actually an extension of the technology, developed for low and ultralow cement castable.

No cement castables either contains no calcium aluminate cement or a very low amount. Calcium oxide, in the cement, acts as a flux towards aluminosilicate raw materials and brings down the refractoriness of castable. Low and ultralow cement castables were developed to reduce the content of the cement and to overcome that draw back.

No cement castables were developed to completely eliminate the calcium aluminate cement and to replace it by alumina or silica sol as the binder. It is normally two pack system, and the powder component and the binder component (liquid) are supplied in two separate packets. The user is to mix both powder and liquid at the site and apply. It can be poured or cast similarly as calcium cement-based castable. It has an advantage over calcium aluminate cement-based castable that it can be preheated at a much faster rate than cement-based castable. A comparison of usual cement bonded and no-cement castables is given in Table 2.2. Now further

Table 2.2 Comparison of features of cement-based and no-cement castables

Feature	Castable with calcium aluminate cement	No cement castable
Raw material	Any aluminosilicate, or neutral raw material can be used	Any aluminosilicate, or neutral raw material can be used
Mixing agent	Water	Alumina or silica sol
Setting time	Can be controlled by additives	Can be controlled by additives
Mechanical strength	Comparable	Comparable
Drying and preheating	Slow	Fast
Storage life	Less	More

development has been made possible a one pack no-cement castable, in which the binder is used in the solid form and is available commercially.

Further progresses were made in castable technology to make it (pumpable castable) amenable for pumping as a water mixed slurry, for easy transportation. By this, the mixing activity of castable with water can be kept far off from the actual site, for ease of installation job. Pumpable can be used for casting a horizontal section because the slurry can rest of its own and set.

In the shotcreting process, the castable in similar way is pumped and transported through rubber hose and is sprayed through a nozzle with the help of compressed air on the target wall. It is very useful for fast installation of a large volume of refractory at a site, difficultly accessible, like preheater cyclone situated at a considerable height. Moreover, the fixing and dismantling of steel former, used for casting, involves manpower and takes time for fixing. The shotcreting method does not require any former and eliminates both time and cost. To enhance the setting of the slurry on the target wall and to convert it to a gel, so that it does not flow by its own weight, a gelling additive is added to the shotcrete material at spraying nozzle. This helps to install a thicker lining, and because a thick layer can be build up on a vertical wall, the requirement of former, as used in traditional casting process, can be eliminated.

Gunning material can be made both in aluminosilicate base and in magnesia base (basic) composition. Basic gunning materials are used generally for the hot repairing of the refractory lining in steel melting furnaces. There is no scope of use of basic gunning material and are never used in the cement industries. For cement industry, the gunning materials are made using formulation similar to castable but modified by adding special additives to suit the gunning process. In the process of gunning, dry material is conveyed pneumatically through the rubber hose of the gunning machine, to the nozzle. At the nozzle tip, water is added through a separate rubber pipe at high pressure and sprayed on the target area. It must set fast, gets hard quickly, and must not crack on fast heating. Gunning materials of varied specifications, based on aluminosilicate raw material and calcium aluminate cement mix, are used in cement industries. This method can be very effectively used for cold or hot repairing of an area inside the furnace without taking major shutdown. This method is not very effective for installation of large volume of material in short span of time. In gunning process there is always some loss of material due to rebound of the material when the gunning material hits the wall. The rebound percentage may vary depending upon the material or the operator, but it is always there.

The special features of the unshaped refractories are as follows:

(a) Although its manufacturing is easy, its installation requires very skilled and experienced hands with much attention. Installation of monolithic refractory requires less labor and can be automated to a large extent.
(b) The unshaped refractories are more environment-friendly material because it does not require any high-temperature operation for its production.
(c) It is a semifinished material. It is partly finished at the premises of supplier, and the final finishing is made at customer's premises. Therefore its quality of

installation, including consolidation, drying, and preheating, is very important, unlike shaped product.

(d) Normally, from beginning of the installation to ready-for-use, duration is much longer in case of castable refractories, because of longer time needed for (1) preparation (welding of steel anchors to hold the refractory) of the surface, over which the unshaped refractory will be applied, (2) preparation of the mix and consolidation of the refractory mass, (3) setting of the refractory to become hard, and (4) preheating the cast lining. The unshaped refractories comprise large group of products with varied application method and purpose to satisfy the particular technical requirement of a refractory lining. The unshaped refractories are further classified according to their purpose of application and method of application (Fig. 2.1), as discussed below.

2.2.3 *Insulating Property*

Refractories can be broadly classified by its purpose of use. As was discussed earlier, two main purposes of use of refractories are:

1. To contain the temperature and to withstand the abusive environment inside the furnace or the process equipment
2. To contain the thermal energy inside the furnace and not to allow it to flow out of the system

Most refractory linings are composite in nature and made of the combination of dense refractories, in front, to contain the high temperature and to withstand the harsh operating condition and insulating refractories, at back, to contain the heat and protect the energy loss from the system.

Insulating refractories, like dense refractories, are available in shaped and unshaped form. The main features of insulating refractories are the high porosity and low bulk density which leads to low thermal conductivity and low mechanical strength of the insulating refractory. The main purpose of use of insulating refractory is to preserve thermal energy. But the insulating refractories cannot withstand abusive environment, and they are used mostly as the backup layer along with the steel shell unless the furnace environment is very friendly.

Insulating refractories are made out of aluminosilicate range of materials because of their inherent lower thermal conductivity compared to many other materials. The insulation product can be designed for continuous use at a very high temperature (1600 °C). The use of insulating products is gradually increasing because of the demand of conservation of energy.

Besides normal insulating bricks and castables, there are other types of insulating refractories available made of ceramic fiber. Ceramic fibers are available with a wide range of properties. It is categorized mainly on its ability to withstand certain service temperature for long time.

The advantages of ceramic fiber lining are:

1. Ceramic fibers have very low bulk density. Ceramic fibers are virtually completely resistant to thermal shock.
2. Ceramic fiber lined furnaces can be heated up very fast and can be cooled also very fast.
3. Ceramic fiber can withstand thermo-mechanical stresses better than insulation bricks or castables.
4. Ceramic fiber lining is easier and cheaper to install.

Ceramic fibers are available in different forms like loose or bulk fiber, module, paper, rope, etc. for different applications.

2.2.4 Special Refractories

The refractories, which have very special chemistry, like carbides, nitrides, and borides (Table 1.1) are used for very special applications and not in bulk volume.

Chapter 3
Manufacturing and Properties of Refractories

3.1 Introduction

Study of the properties of refractory is very important for the purpose of its selection for various application. The properties of the refractory both at the ambient and high temperature are important to design the lining details of a furnace or kiln. Here we shall discuss the general manufacturing flow sheet and properties of different shaped and unshaped refractories.

3.2 Shaped Refractories

The manufacturing process of all shaped refractories is more or less similar. Refractory raw materials are normally available as hard lumpy materials, which are crushed first in primary crusher like jaw crusher (if the size of the lumps is above 50 mm) and then in the secondary crusher like hammer mill. The crushed materials are screened to several size fractions like 3–5 mm, 1–3 mm, 0–1 mm, and fines. The different size fractions are mixed together along with a binder (like molasses-water mixture) and pressed in a mold of desired size in high capacity press to form the brick. The brick is then dried to remove the moisture inside and fired at high temperature (1150–1750 °C) to impart the strength and desired properties to the refractory.

P. Sengupta, *Refractories for the Chemical Industries*,
https://doi.org/10.1007/978-3-030-61240-5_3

3.2.1 Silica Refractories

Silica refractories contain a minimum of 95% SiO_2. The common impurities are CaO, Fe_2O_3, and Al_2O_3. The basic raw material used, is quartzite available in different varieties, comprising of coarse crystalline uncemented variety to cemented fine crystalline variety. The complex behavior of polymorphism is most important to decide the suitability of the raw material for the production of silica refractories, but beside that the fracture behavior of the rock, microstructure, chemical composition, refractoriness, and porosity are all important parameters. The combined Al_2O_3 and TiO_2 should be as low as possible, but it should be below 2%, and alkali should be below 0.3% and for high-quality products it should be below 0.1%.

Quartz shows transformation to different crystalline forms when heat treated at different temperatures with some being reversible and some irreversible in nature as shown by schematic diagram, Fig. 3.1 [1]. The different forms are called polymorphic forms and the phenomenon as polymorphism.

The volume changes associated with polymorphic transformation of silica are shown in Table 3.1.

The horizontal arrows in Fig. 3.1 indicate the transformations which occur slowly, and the vertical arrows show the transformations that occur fast. The suitable quartzite for manufacturing of silica refractories must not degrade fast on heating. The degeneration can be measured by testing the sp. gravity of the quartzite after firing at 1460 °C from ambient temperature and holding for 1 h at peak temperature. Quartzite can be classified into four groups as shown in Table 3.2, and quartzite with average to extremely slow degeneration are used for manufacturing of silica refractories.

The flow sheet of manufacturing of silica refractory is shown in Fig. 3.2. The milk of lime (slurry of $Ca(OH)_2$ in water) is used as binder which provides 1–2.5% CaO, in the composition, that acts as a mineralizer and speeds up the polymorphic transformations in the horizontal direction as shown in Fig. 3.1. The addition of Fe_2O_3 also promotes the conversion of quartz to desirable phase.

The conversion depends upon temperature, the time temperature profile, the type of mineralizer, and also the brick shape [2]. CaO reacts with SiO_2 above 600 °C and forms β-2CaO. $SiO_2(C_2S)$ and $3CaO.SiO_2$ (C_3S), which play a significant role in the development of mechanical strength in silica refractory. But nowadays, silica refractory has been developed without CaO which gives superior performances under more abusive operational condition [3].

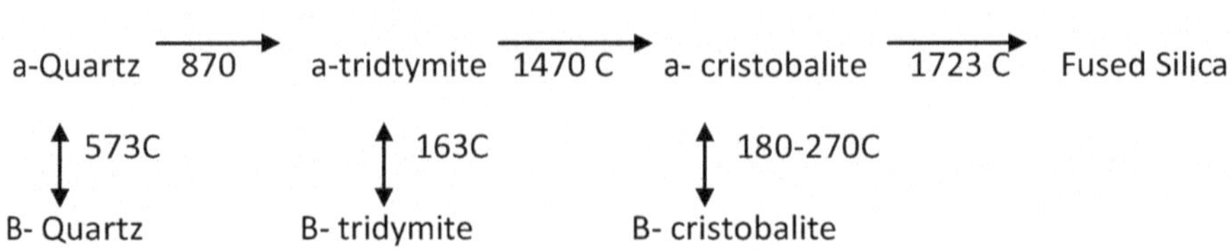

Fig. 3.1 Polymorphic transformation of quartz

Table 3.1 Volume changes associated with polymorphic changes in silica

Transformation	Temperature (°C)	Associated volume change (%)
β-Quartz to α-quartz	573	+0.82
γ-Tridymite to β-tridymite	117	+0.2
B-Cristobalite to α-cristobalite	250	+3.7
β-Tridymite to α-tridymite	163	+0.2
α-Quartz to α-tridymite	870	+16.0
α-Quartz to α-cristobalite	1000	+15.4
α-Quartz to silica glass	1723	+15.5
Silica glass to α-cristobalite	1000	−0.9

Table 3.2 Degeneration behavior of quartzite

Observation	Sp. gravity after firing
Rapidly degenerating	2.4
Average rate of degeneration	2.45
Slow degeneration	2.4–2.5
Extremely slow degeneration	2.5

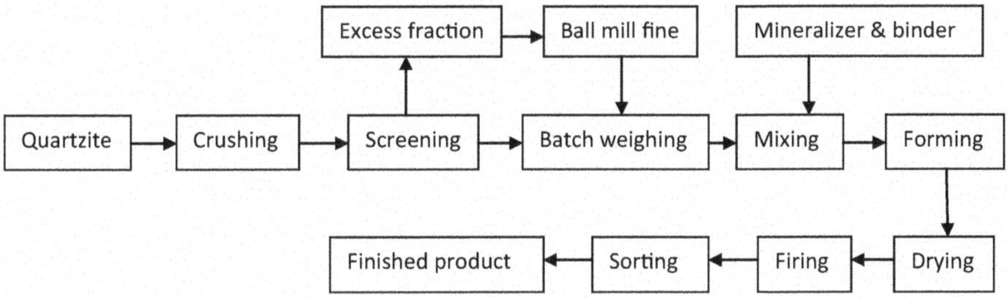

Fig. 3.2 Flow sheet of manufacturing of silica refractories

Tridymite is the most desirable phase in silica refractories although all the crystalline forms of silica exist in silica refractory. Residual or free quartz is the most undesirable phase in a silica refractory. The main problem associated with the existence of free quartz is its conversion to tridymite and cristobalite in service which is accompanied by 14% volume expansion and causes crack to the refractory. The presence of free quartz in a silica refractory can easily be recognized by its thermal expansion behavior with temperature (Fig. 3.3).

The determination of specific gravity also gives a good indication about the presence of free quartz in the refractory body. A silica brick without free quartz should have a sp. gravity below 2.34.

Silica refractories have very good thermal shock resistance above 600 °C and have extraordinary thermomechanical stability at high temperature up to 1600 °C and are being used widely in the construction of crown and superstructure of the glass tank furnace.

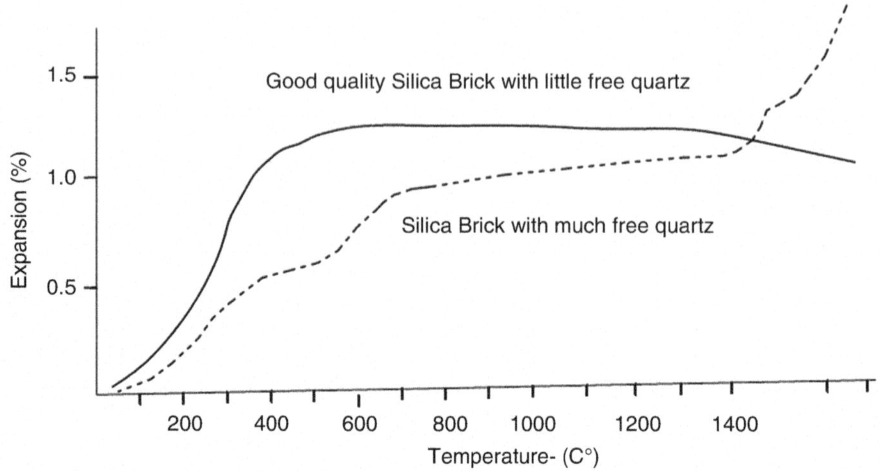

Fig. 3.3 Thermal expansion of silica refractories with free quartz

3.2.2 Aluminosilicate Refractories

The main chemical constituents of this category of refractories are alumina (Al_2O_3) and silica (SiO_2). If the Al_2O_3% is less than 50%, it is normally categorized as fireclay refractory, and refractory with more than 50% Al_2O_3 is called high alumina refractory, although this categorization has got no authentic base.

3.2.2.1 Fireclay Refractories

The main raw material used for fireclay refractories is calcined fireclay or chamotte, along with plastic fireclay. The flowchart given (Fig. 3.4) shows the manufacturing process of fireclay refractories.

Fireclay refractories have alumina and silica as main constituents. The other oxides which remain present as impurities are Fe_2O_3, TiO_2, CaO, MgO, K_2O, and Na_2O. All these oxides are acting as flux and bring down the softening temperature of the refractory. However, the presence of these oxides though undesirable but cannot be avoided because of their inherent presence in the naturally occurring raw materials. Microstructure analysis of the fireclay refractory reveals that it can contain up to 50% of glassy phase with embedded cristobalite and mullite crystals. The percentage of glassy phase increases with more fluxes and higher firing temperature. In a fireclay brick with 40% Al_2O_3, using raw materials with low amount of alkali and fluxes, the share of glassy phase can be reduced to 20% with mullite 55% and cristobalite 20–25%. The presence of glassy phase and mullite crystals and their sizes influence the thermomechanical behavior of the refractory. Fireclay refractories have good volume stability at high temperature, fairly good abrasion resistance, good thermal shock resistance, and very good resistance toward alkali attack at high

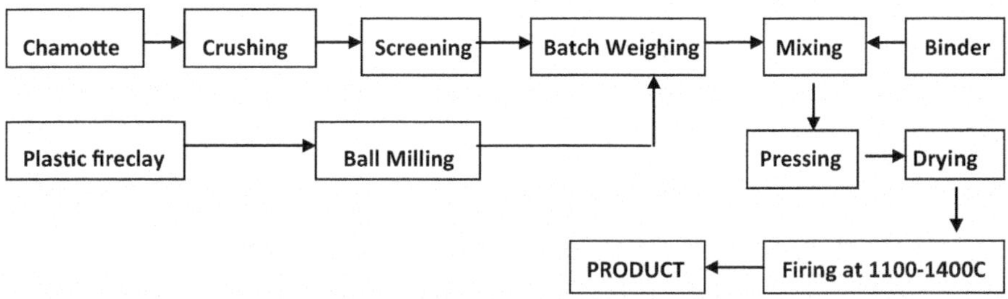

Fig. 3.4 Flow sheet of the process of manufacturing of fireclay-shaped refractory

temperature. Mullite is always a highly desirable phase in fireclay and high alumina refractories.

Addition of silicon carbide (SiC) in the fireclay refractory improves all those properties, to a great extent. In reducing atmosphere, e.g., in CO gas atmosphere, the presence of Fe_2O_3 catalyzes the decomposition of CO to deposit C in the brick structure, and it becomes fragile because of the expansion associated with the process. Fireclay refractories find a wide use in all types of chemical industries.

3.2.2.2 High Alumina Refractories

These are categorized as products containing 50–85% Al_2O_3. Different kinds of raw materials are used in high alumina refractories to achieve the target properties. The major raw materials used are sillimanite, andalusite, raw and calcined kyanite, calcined diaspore, calcined bauxite, mullite, brown and white fused alumina, tabular alumina, etc. The manufacturing flow sheet is same as that of fireclay refractory. But there are some qualities which are chemically bonded and do not require any high-temperature firing. The high alumina refractories can have varied properties depending upon its composition and raw material used. The refractory composition can be designed to attain particular properties required to perform well for a particular application. Refractories for certain application may demand high thermal shock resistance, and the other application may look for very high abrasion resistance or resistance to alkali attack at high temperature. In all those cases, the refractory composition and properties will be different. Refractory formulation is done looking to the operational conditions, where the refractory will be put in.

Mullite is one of the most important phases present in the high alumina refractories, and it is formed by the reaction of alumina and silica at high temperature (above 1100 °C). Mullite is chemically $3Al_2O_3.2SiO_2$. It contains 72% Al_2O_3. The presence of mullite improves the thermomechanical property and the thermal shock resistance of high alumina refractories, and the higher is the mullite content, the better are those properties. Mullite is also available as raw material which is made synthetically. Presence of higher percentage of corundum phase increases the abrasion resistance of the high alumina refractories. The corundum phase is contributed by calcined bauxite, fused alumina, or tabular alumina. High alumina refractories have poor alkali resistance at high temperature. It is found that the higher is the alumina content, the lesser is the alkali resistance. At high temperature, it reacts with alkali to form some feldspathic compounds like leucite, kaliophilite, etc. associated with high volume expansion which completely destroys the refractory [4–7].

$$Al_2O_3 + K_2O + SiO_2 = K_2O.Al_2O_3.2SiO_2 \tag{3.1}$$

$$3Al_2O_3.2SiO_2 + 3K_2O + 6SiO_2 = 3K_2O.Al_2O_3.2SiO_2 \tag{3.2}$$

$$Al_2O_3.2SiO_2 + 2SiO_2 + 2K + CO \qquad KAS_4\left(\text{Leucite}\right) + C \tag{3.3}$$

$$\begin{aligned} 3Al_2O_3.2SiO_2 + 6K + 3CO \rightarrow K_2O.Al_2O_3.2SiO_2\left(\text{Kaliophillite}\right) \\ + 2\left(K_2O.Al_2O_3\right) + 3C \end{aligned} \tag{3.4}$$

The properties of high alumina refractories depend upon the raw material base. The andalusite, sillimanite, and kyanite are converted to mullite easily at high temperature, and the bricks made out of those raw materials show high thermal shock resistance, high creep resistance at high temperature, and also good alkali resistance. Raw kyanite has got the property to expand at high temperature and has been used to impart a positive PLC in the refractory. The refractory made out of bauxite shows pyroplasticity (plasticity at higher temperature). By virtue of this property, bauxite-based high alumina refractories can deform under an applied stress at high temperature and can mitigate the effect of the stress. That is how these refractories show good performance in burning zone of rotary kiln with small to medium diameter. High alumina refractories in general and especially the bauxite-based refractories show a permanent linear expansion at high temperature, and this helps to keep the lining tight at high temperature during the service.

Phosphate-bonded high alumina refractories show increased resistance to attack by liquid phase in cement clinker and alkali and also have higher abrasion resistance. These refractories are normally fired at lower temperature during manufacturing.

The high alumina refractories are used for lining various equipment in almost all the chemical industries. In some quality of aluminosilicate refractories, special ingredients, e.g., SiC (silicon carbide), $ZrSiO_3$ (zircon), are added to impart special properties like anti-buildup behavior, thermal shock resistance, alkali resistance, etc.

3.2.2.3 Alumina Refractories

Refractories containing Al_2O_3 above 90% are used widely in the petrochemical industries. Different raw materials like tabular alumina, white fused alumina, and brown fused alumina are available as the major raw material for manufacturing of corundum refractories. Reactive alumina or good-quality kaolin clay is used as the binder. Alumina refractories are particularly useful in the presence of hydrogen atmosphere even at lower temperature because of reduction of SiO_2 in the presence of hydrogen gas. The flow sheet of the manufacturing scheme is shown in Fig. 3.5.

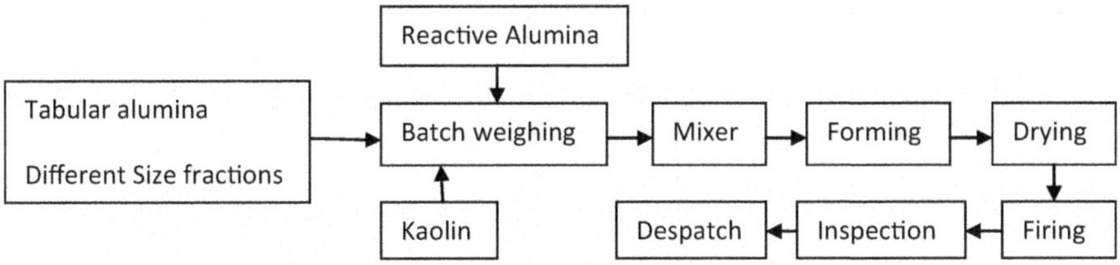

Fig. 3.5 Flow sheet of manufacturing of corundum refractory

The properties of the finished refractory product depend largely upon the raw material used besides the manufacturing process. Thermal shock resistance of different raw material is shown in Fig. 3.6 [8].

3.2.3 *Basic Refractories*

Basic refractories are used in lime kiln calcination, glass regenerator checker, and burning zone of cement rotary kilns with larger diameter. Earlier, cement rotary kiln burning zone was lined with high alumina refractories. But as the technology has changed over time, from wet to dry process and the kiln diameter got bigger and bigger, thermal load on the kiln burning zone increases, and as a result, the lining is changed from high alumina to basic refractory.

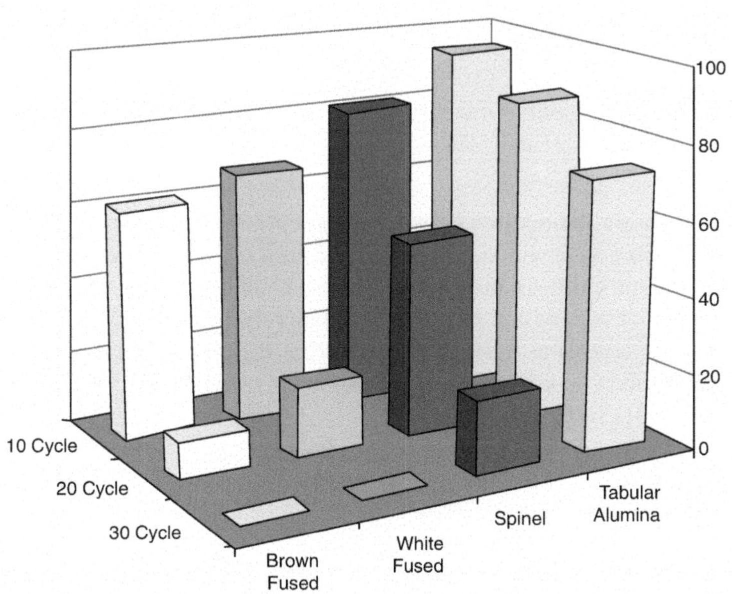

Fig. 3.6 Thermal shock resistance of different alumina raw material

The main raw material, for all the basic refractories, is dead burnt magnesia (MgO). It can be obtained by high-temperature calcinations of naturally occurring magnesite ($MgCO_3$) above 1700° C as per the reaction 3.5. The processed material is called dead burnt magnesite (DBM), because of its inertness toward hydration. The MgO formed is susceptible to react with atmospheric moisture and carbon dioxide to form magnesium hydroxide and magnesium carbonate as per the reactions 3.6 and 3.7. The higher is the calcination temperature, the more stable is the MgO formed.

$$MgCO_3 \rightarrow MgO + CO_2 \uparrow \tag{3.5}$$

$$MgO + H_2O \rightarrow Mg(OH)_2 \tag{3.6}$$

$$MgO + CO_2 \rightarrow MgCO_3 \tag{3.7}$$

MgO can also be synthesized from seawater. Seawater contains magnesium salts like $MgSO_4$ and $MgCl_2$, which, when treated with lime, $Mg(OH)_2$ is precipitated. The $Mg(OH)_2$ is then filtered, briquetted, and calcined above 1700° C to obtain DBM. In another method, the magnesite rock is fused by electric arc at very high temperature at about 3000° C and allowed to cool very slowly when MgO gets crystallized from the melt and forms fused magnesite with large crystals of periclase. It has been found that the properties of basic bricks depend on the properties of the magnesite raw material and its occurrence [9].

The important properties to characterize the magnesia raw material are as follows:

1. Bulk density and grain porosity
2. The impurities and their distribution
3. CaO/SiO_2 ratio
4. Boron content
5. Periclase crystal size

Table 3.3 shows the relation between the properties of magnesite raw material and their influence on the brick properties. The purity of magnesia raw material acts in different way in case of natural magnesia and synthetic magnesia [10–12], and a judicious blend of both will be the best. The higher the periclase (MgO) crystal size, the better is the corrosion resistance but lower is the coating formation in case of cement kiln burning zone use. Magnesite having higher Fe_2O_3 is found to have better plasticity [13] (Table 3.4).

3.2.3.1 Magnesite-Chrome Refractory

Magnesite-chrome refractory is used mostly in the burning zone of cement rotary kiln. The raw materials used in this product are high-purity dead burnt magnesite (DBM) or fused magnesite or their mixture and high-grade chrome ore. The flow

Table 3.3 Influence of magnesite on properties of basic refractories

Property of magnesia	Impact on the brick properties: Corrosion resistance to clinker liquid phases	Refractoriness	MOE	Coating formation
Chemical composition	H	H	–	H
CaO/SiO_2 ratio	L	M	L	–
Periclase crystal size	M	–	H	–
Grain density and porosity	H	–	L	L

H high, *M* moderately high, *L* low

Table 3.4 Effect of different spinels on MOE and wedge splitting test of basic refractory

Properties	Additions made to magnesite refractory: 5% hercynite	5% spinel	15% spinel	15% pleonaste
Dynamic E at RT (GP_a)	30.9	59.3	30.3	29.5
G_F (J/m^2)at 1100 °C	1088	631	741	1014
Rating based on wedge splitting test	Excellent	Moderate	Good	Very good

E modulus of elasticity, G_F fracture energy

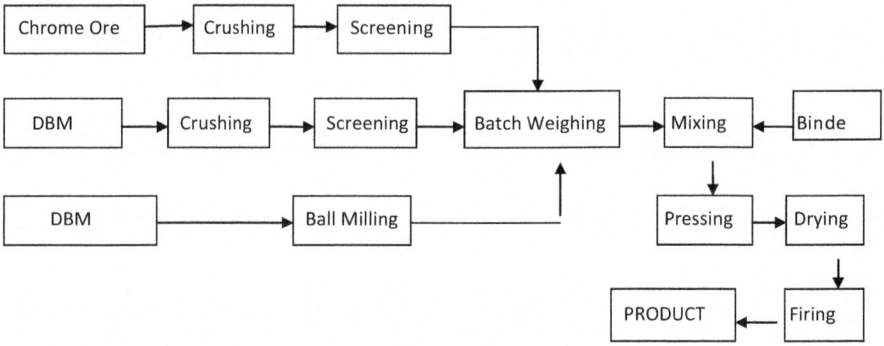

Fig. 3.7 Flow sheet of manufacturing process of magnesite-chrome refractories

sheet of its manufacturing process is shown in Fig. 3.7. Chrome ore contains MgO. Cr_2O_3 and $FeO.Cr_2O_3$ spinels. Addition of chrome ore reduces the modulus of elasticity [14] and improves the thermal shock resistance of magnesia refractory and its resistance to acidic slag. Magnesia chrome refractories are found to form a good coating of clinker on the brick lining surface in cement rotary kiln, which protects the brick lining from high temperature and thermal shock and enhances the campaign life. But many countries have put a ban on the use of magnesite-chrome refractories because of environmental pollution, caused by the disposal of the used magnesia-chromite brick lining after its use. The Cr_2O_3 in the refractory reacts with alkalis at high temperature, during its use, to form alkali chromates, containing water-soluble

Table 3.5 Basic refractories for cement rotary kiln

	Application area				
Properties	Transition zone	Burning zone	Burning zone	Burning zone	Burning zone
Refractory type	Magnesia-spinel	Magnesia-chrome	Magnesia-hercynite	Dolomite	Magnesia-zirconia
Bulk density (g/cc)	2.92–3.05	2.98–3.05	2.98–3.02	2.90	3.0
App. Porosity (%)	16–17	16–18	15–17	11–13	17
CCS (MPa)	50–60	50–60	80–100	>35	50
RUL (°C)	+1700	1650–1700	+1600	+1650	+1700
Thermal conductivity					
(W/mK) at 400 °C	2.90–3.00	2.60–2.80	2.40	–	4.2
1000 °C	2.40–2.50	2.10–2.30	3.00	2.6	
PLC(%) /1450 °C for 2 h	+0.3–0.5	+0.4–0.7	–	–	–
Chemical analysis (%)					
MgO	86–88	76–81	85–86	40–50	93
Cr_2O_3	Traces	6–8	–	–	–
Al_2O_3	9–11	2–5	3.0	–	–
Fe_2O_3	0.5–0.8	3–5	8.0	–	0.5
CaO	–	–	–	40–55	0.9
ZrO_2	–	–	–	–	5.0

hexavalent chrome salts, which are carcinogenic in nature. The rainwater can leach out the Cr^{+6} and can pollute the groundwater reserve to cause health hazards. The typical properties of this product are shown in Table 3.5.

3.2.3.2 Magnesia-Alumina Spinel Refractory

Magnesia-alumina spinel refractory is used in glass furnace regenerator and both in transition zones and burning zone of cement rotary kiln [15]. The main raw materials for this product are high-purity dead burnt magnesite and spinel. Spinel is a mineral phase having chemical formula $MgO.Al_2O_3$. Spinel is not available in nature and is made synthetically by high-temperature reaction between magnesia and alumina. The spinel has got a general formula AB_2O_4, where A is divalent atom and B is trivalent atom. Other mineral phases, having similar crystal structure as spinel, used in basic refractories for rotary kiln lining, are hercynite ($FeAl_2O_4$), galaxite ($MnAl_2O_4$), and pleonaste (Mg-Fe-Al spinel) [16].

Addition of spinel reduces the modulus of elasticity of magnesia refractories and makes it more flexible to withstand thermomechanical stresses. The flow sheet for manufacturing magnesia- spinel refractory product is given in Fig. 3.8. The spinel can be added as preformed spinel or it can be formed in situ or as a combination.

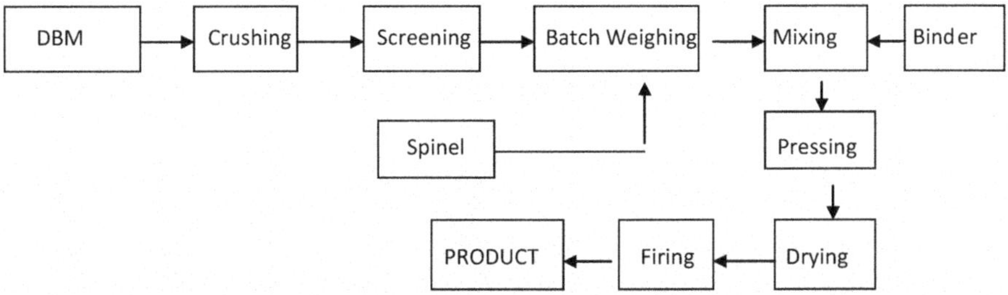

Fig. 3.8 Flow sheet of manufacturing of magnesia-spinel refractories

TiO_2 is added as catalyst for spinel formation at lower temperature [17]. Addition of SnO_2 is found to improve the thermodynamic stability of Al_2O_3-MgO spinel.

This product has got very high thermal shock resistance and performs very well in transition zone lining where the coating formation is less and the thermal shock is very high. It has also got a very high resistance to alkali attack and corrosion toward the cement liquid phase at high temperature. The typical properties of this product are given in Table 3.5.

3.2.3.3 Magnesia-Hercynite Refractory

Magnesia-hercynite refractory is made from DBM and hercynite. Hercynite is a synthetic mineral with spinel structure having the formula $FeO.Al_2O_3$. Addition of hercynite to magnesia refractories reduces its modulus of elasticity and enhances flexibility. It also reduces brittleness of magnesia refractories toward thermomechanical stresses. The magnesia-hercynite refractory has good ability to take up the clinker coating, and all the said properties have made it a suitable candidate for burning zone lining of cement rotary kiln. It has also got a very high resistance to alkali attack and corrosion toward the cement liquid phase. It has the added advantage of being free from Cr_2O_3 and therefore more environment friendly by nature. The flow sheet of manufacturing of this refractory is similar to magnesia-spinel refractory. The typical property of such refractory is given in Table 3.5.

The spinels are having lower thermal expansion than the periclase (MgO). As a part of the manufacturing process, these brick are fired at high temperature (1580–1650 °C). Due to mismatch of thermal expansion behavior, micro cracks developed in the refractory microstructure, surrounding the spinel grains, which makes the refractory more flexible and improves its capability to withstand more thermomechanical stresses. Magnesia and alumina may be added individually to form spinel in situ, and TiO_2 is reported to bring down the temperature, to synthesize mag-alumina spinel. Some additives are used to modify the properties, e.g., ZnO helps in densification, and SnO_2 improves the thermodynamic stability of spinel.

Figure 3.9 shows the typical thermomechanical behavior of a magnesia brick without spinel and a magnesia brick containing a spinel [18, 19]. A pure magnesia

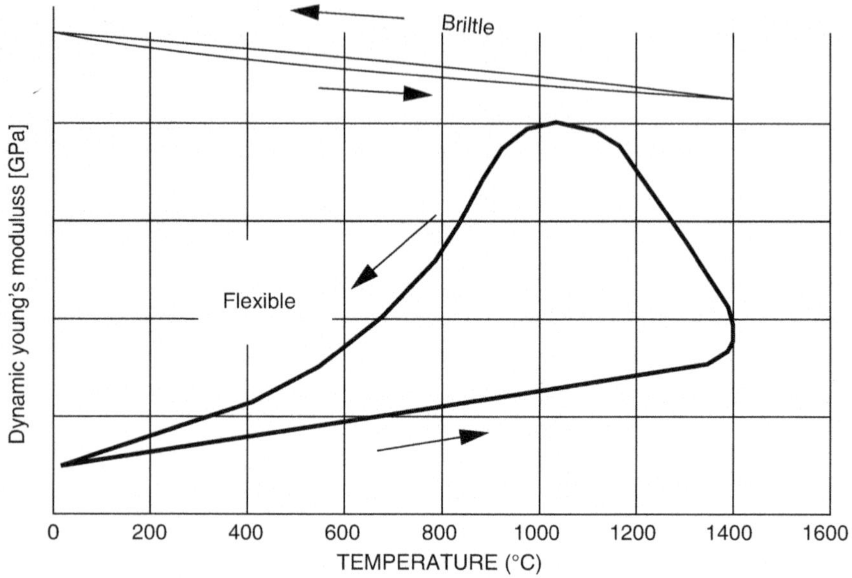

Fig. 3.9 Effect of the spinel on the dynamic Young's modulus

brick shows an almost constant Young's moduli at a high level, but with the inclusion of spinel, the Young's modulus starts at a low level, increases slowly during heating, and shows rapid increase in the first stage of cooling to 1000 °C. The flexibilization effect occurs during cooling at temperature below 1000 °C, and the Young's modulus is lowered significantly. The Young's modulus level provides an indication of the ability of an additive to flexibilize or elastify a magnesia brick. Spinels are also called as elastifiers, when added to basic refractories [20–25].

The effect of addition of different kinds of spinels on the basic refractory [18] is shown in Table 3.4, and it is clearly evident that the fracture toughness has been increased significantly by addition of the spinels. It has also been found that forsterite ($MgO.SiO_2$) phase, developed in the basic bricks, can also act as an elastifier [26]. But the chemistry and the microstructure of the brick are to be engineered properly to get the best effect.

3.2.3.4 Magnesia-Zirconia Refractory

Magnesia-zirconia refractory is made out of high-purity dead burnt magnesia or fused magnesia and monoclinic or stabilized zirconia [27–29]. This refractory body contains periclase and stabilized zirconium oxide (ZrO_2) or calcium zirconate ($CaZrO_3$). Uniform distribution of ZrO_2 in the brick structure improves thermal shock resistance due to micro crack formation. During the partial or total stabilization of ZrO_2, diffusion takes place between MgO and ZrO_2, and this helps to form direct bond and increases the hot strength which helps to withstand the mechanical stress at high temperature. Magnesia-zirconia bricks have high corrosion resistance against alkali and against basic slag at high temperature. The addition of ZrO_2 in magnesia-spinel bricks has been found to improve both coating formation and corrosion resistance in cement kiln burning zone application [30]. The simultaneous addition of $MgAl_2O_4$ and $CaZrO_3$ to small volume percentage (4%) is reported to improve its corrosion resistance and coating development property [31] for rotary cement kiln application. The flow sheet of manufacturing is same as magnesia-spinel brick. Typical property of magnesia-zirconia brick is shown in Table 3.5.

3.2.3.5 Dolomite Refractory

Dolomite refractory is made out of naturally occurring dolomite which is a solid solution of $MgCO_3$ and $CaCO_3$. The pure grade of dolomite is calcined at very high temperature to reduce its affinity to react with atmospheric moisture and to make it stable. The dead burnt dolomite is the basic raw material for manufacturing the dolomite refractories. There are a number of unique properties in dolomite brick that make it well suited for use as a refractory lining for burning zone of a cement rotary kiln [32, 33]. A high degree of refractoriness allows the dolomite brick to withstand the temperature and involved stress in *the* burning zone. The corrosion resistance to alkalis and a reducing atmosphere, as well as excellent ability for coating formation, makes it a good candidate for the lining of burning zone of cement rotary kiln.

In practice, the clinker coating on the refractory lining is not always present during campaign of the lining. Some or all coating loss may be caused by thermal shock. When the coating on the lining is destabilized, the bricks underneath are subject to a sudden temperature change, which will expose the bricks to severe thermal shock. For a long period of time, standard dolomite was limited to the area where stable coating existed over the refractory lining and shutdowns were minimal.

However, the thermal shock resistance of the dolomite brick has been improved significantly by the development of zirconia-enriched dolomite refractories [29–33]. Zirconia was chosen because it is relatively nonreactive with dolomite and addition of special grade of zirconia imparts high thermal shock resistance. Finally, the environmental safety of the used dolomite bricks is outstanding. One of the greatest disadvantages of dolomite refractory is its susceptibility to react with atmospheric moisture. The process is called hydration, in which the CaO of dolomite reacts with water vapor to form $Ca(OH)_2$ and causes destruction of the brick lining. In coating free area dolomite reacts with kiln gases to form $CaCO_3$, $CaSO_4$, or CaS [34, 35].

The phenomenon is more prominent in tropical countries with high level of humidity. Table 3.5 furnishes the typical properties and specification of dolomite refractories.

3.2.4 *Fused Cast Refractories*

The specialty of this kind of refractories is near zero porosity and very high abrasion resistance. Their manufacturing process is entirely different from the conventional brick-making processes. Fused cast refractories are available with different chemical and physical properties like β-alumina, αβ-alumina, alumina-zirconia-silica (AZS), alumina-silica, zirconia, magnesite-chrome, alumina-chrome, etc. The basic manufacturing process is same for all the qualities, but with variations in process parameters. As an example, for manufacturing of AZS fused cast blocks, a batch consisting of silica (SiO_2), alumina (Al_2O_3), zircon ($ZrO_2.SiO_2$) or zirconia (ZrO_2), and soda ash (Na_2O) and about 50% batch weight of recycled fused AZS are melted in arc furnace at very high temperature (2200–2400 °C).

The melt is treated with oxygen to oxidize zirconium oxycarbonitride or other reduced compounds, produced from the reaction with graphite electrode [36]. The molten material is then cast in specially prepared molds. The melt is allowed to solidify very slowly to avoid defects in the body for 7–10 days. The molds are either graphite mold or water-cooled metal molds. The total operation is shown in Fig. 3.10.

During the solidification process of the melt in the mold, cavity is formed in the cast body, due to shrinkage. Several techniques are applied to control the size and the location of the cavity with respect to the designated face of the block in use. Figure 3.11 shows the nature of the cavity actually formed in the block.

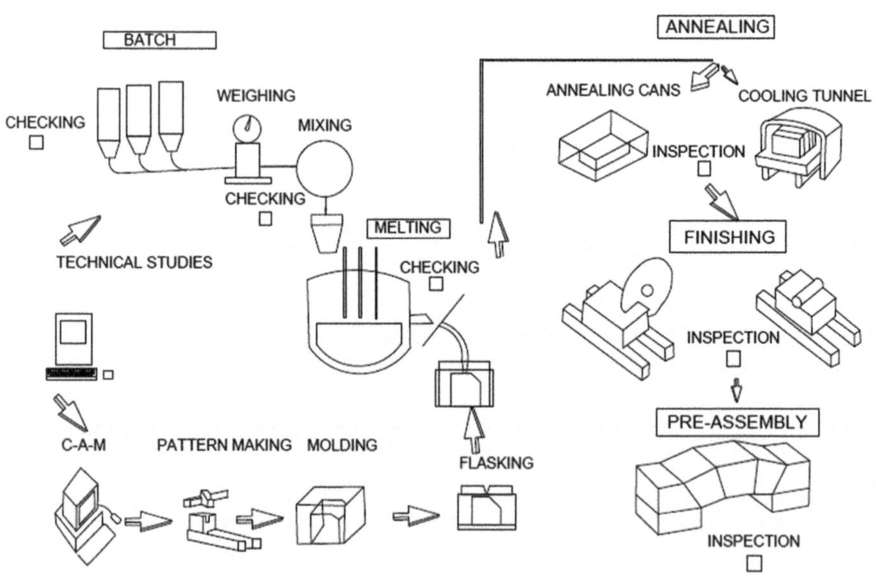

Fig. 3.10 Flow sheet of manufacturing process of fused cast refractories

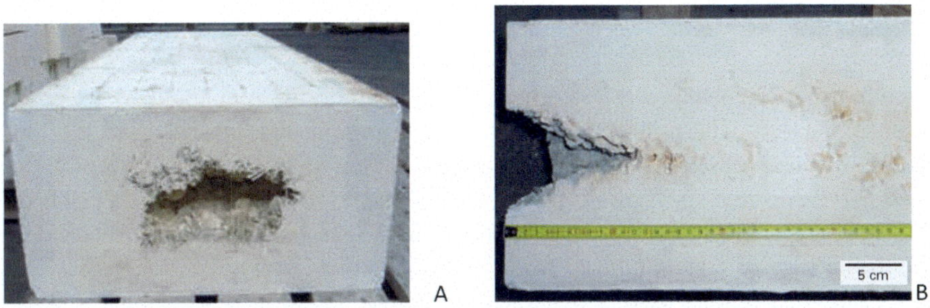

Fig. 3.11 (**a**) is the cavity in the block after the solidification of the melt and (**b**) the cross-sectional view of the block where the length of the cavity and the porosity formed are clearly visible

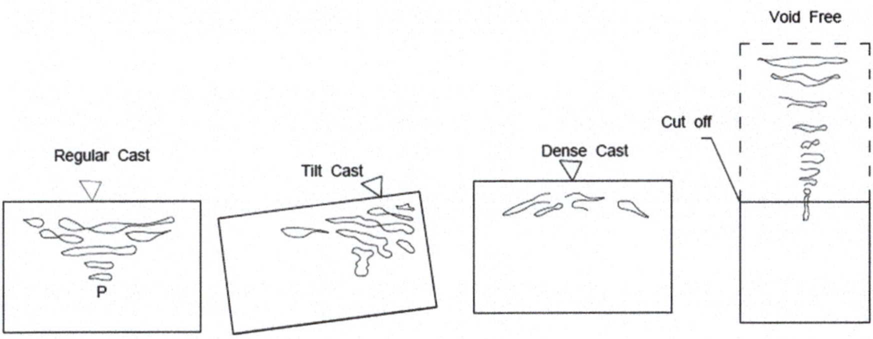

Fig. 3.12 Different types of fused cast refractory blocks

Four different types of fusion cast blocks are available from the supplier, namely, regular cast, tilt cast, dense cast, and void free shown in Fig. 3.12.

Regular cast-Here the normal casting is done, and for a block of size 300 × 450 × 600 mm, the casting face is 450 × 600 mm.
Tilt cast- Casting is done keeping the mold in a tilting position as shown in Fig. 3.12.
Dense cast- Here the casting is done by special method to reduce the blow holes. Casting face is 450 × 600 mm.
Void free- In this case, the blocks can be obtained by cutting off the section with blow holes.

The fused cast refractories have almost zero porosity, and that is why this has a very high corrosion resistance toward molten slag and glass. The fused cast blocks are used widely in the glass melting furnaces, both in glass contact area and noncontact area like superstructure. The fused cast refractories have different crystalline and glassy phases present in it. The quantity of glassy phase in the body influences the properties like glass exudation.

3.3 Unshaped Refractories

Various types of unshaped refractories are available for different applications in different industries. These include castables, ramming masses, plastics, and mortars. Out of all these, castables and mortars are the main types of unshaped refractories used in the different applications across the chemical industries. Castables can be of different types, meant for different applications. Castables are developed based on the technology of concrete, used for civil construction, with a difference that both aggregates and cement used in castable have high refractoriness. Castables are basically a dry mixture of refractory aggregates, calcium aluminate cement [37, 38], and some small but very important additives. The refractory aggregates can be chamotte, andalusite, sillimanite, calcined diaspore, calcined bauxite, brown fused alumina, white fused alumina, tabular alumina, silicon carbide, etc. of different sizes, varying normally, from 10 mm to superfine material. The different ingredients in different sizes and in different proportion as per the design of the formulation are mixed together along with high alumina cement as binder to make the castable ready for packing and dispatch.

3.3.1 Castables

The castables can be applied in different ways like pouring, vibration casting, gunning, shotcreting, ramming, troweling, etc. [39]. Depending upon the place of application and method of application of the castable, the quality and the percentages aggregates, of calcium aluminate cement and additives are decided to engineer its rheology or flow property, to make it suitable for a particular application method mentioned earlier [40].

The most important properties of castable, which are ignored, sometimes by the manufacturer, are its properties, related to installation, e.g., flow property, duration it remains workable, setting time, and also the scope of installation and preheating methods. If the installation is sound, half of the battle is won. Many a times desired campaign life is not obtained, because of improper installation. If the properties mentioned before, necessary for a good quality of installation, does not match with the actual properties of the castable, the installation quality will not be sound and it will not perform well. The refractory manufacturer must design the castable recipe taking into consideration the installation method to be employed and the condition prevails (e.g., ambient temperature, etc.), besides taking into consideration the operating condition. Similarly during the installation of the castable, the installer should have prior knowledge of the mentioned properties of the castable required for sound installation. The same castable, when installed in a country in Europe in winter and in a country in Middle East in summer, will not behave similarly because of different ambient conditions. Both the manufacturer of the castable and the user must

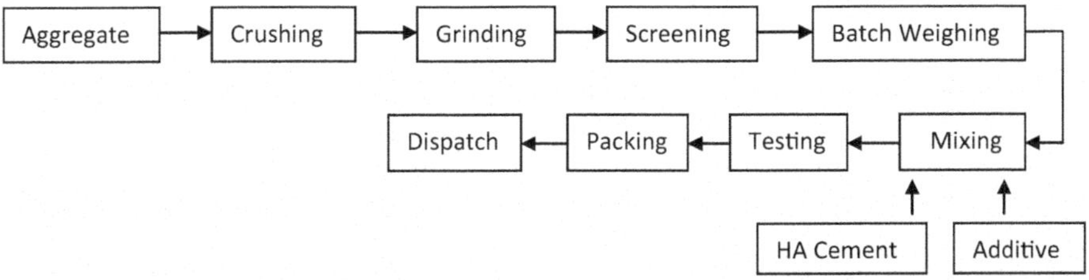

Fig. 3.13 Flows sheet of manufacturing of castable

bear this in mind during manufacturing and installation of castable refractories and proper attention should be paid.

Castables can be installed by different methods, similar to that of concretes used in civil construction, e.g., by pouring, by vibration casting, by gunning, and by shotcreting. Two major steps in the installation processes are mixing of castables with water and the method of consolidation of the castable. Mixing quality is very important, and the aim should be to get maximum flow using the prescribed water or binder content. That is possible only when the mixing is very thorough and intimate, done in high-shear mechanical mixer.

The general flow sheet for manufacturing of castable is given in Fig. 3.13. The characteristics of different types of castables are discussed below.

3.3.1.1 Conventional Castables

This is a mixture of refractory aggregates with 15–25% high alumina cement. These castables are required to mix with prescribed quantity of water and either cast, poured, or rammed. In castable, the aggregates used are normally aluminosilicate type, but in some special cases, SiC or zircon is also used. Conventional castables can be rammed, vibrocast, or poured. These are also good material for patching by troweling.

The special feature of these castables is the loss of strength, with rise of temperature, till it reaches a minimum value at a temperature range of 800–1000 °C, and then there is a rise of strength with increased temperature. The initial loss of strength with rise of temperature is because of the disintegration of the hydraulic bond formed, and later the increase of strength is due to sintering process. Typical specification of this type of castables used in cement plant is given in Table 3.6.

3.3.1.2 Low Cement Castable (LCC)

This is a mixture of refractory aggregates and 5–8% high alumina cement with other minor, but very important, additives which helps to get the desired flow property and enables installation by vibration casting with much lower water percent.

Table 3.6 Properties of different types of castables

	Targeted application feature				
Properties	Repairing and patching	Alkali and medium wear resistant	High wear resistant	High wear resistant	Fast curing and dry-up
Type of castable	Conventional	Low cement	Low cement	Ultralow cement	No cement
Al_2O_3(%)	40	45	60	85	75–85
Fe_2O_3(%)	1.0	1.0	1.0	1.5	1–2
CaO (%)	3.0	1.5	1.5	1.0	0.5
Grain size (mm)	(0–5)	(0–5)	(0–5)	(0–8)	(0–6)
Water for casting (%)	9.5–10.5	6–7	5–6	3–4	Silica sol 5–6%
Service temp. (°C)	1200	1400	1600	1700	1600
Bulk density (g/cc)	2.12	2.15	2.50	2.80	2.80–3.10
CCS (MPa)/110 °C	50	70	80	110	60–80
/1000 °C	35	80	100	120	100–110
/1300 °C	45	100	140	150	150
PLC (%) at 1100 °C	+/−0.5	+/−0.5	+/−0.5	+/−0.3	+/−0.5
Th. conductivity (W/mK)					
At 400 °C	0.90	1.1	1.7	2.0	2.0
At 1000 °C	1.0	1.2	2.0	2.2	2.1
Use	Preheaters, coolers			Cooler	Nose ring

Good flow property of the castable mix with low water content ensures better packing and lower porosity which leads to higher mechanical strength. The low cement castable should always be mixed with water in mechanical mixer and is to be installed by vibration casting. Ordinary mixing with water does not develop the desired flow property in the mix, and vibration casting with low water content is not possible. Addition of higher percentage of water will be disastrous and will impair setting and strength of the castable. Properties of some low cement castables are given in Table 3.6.

3.3.1.3 Ultralow Cement Castable (ULCC)

These products are very similar to the low cement castables except for further lower content (2–4%) of high alumina cement. Its features are similar to that of the LCC, but the high-temperature tensile strength is better than that of LCC with similar Al_2O_3 content because of lower flux content (CaO) due to the lower content of high alumina cement.

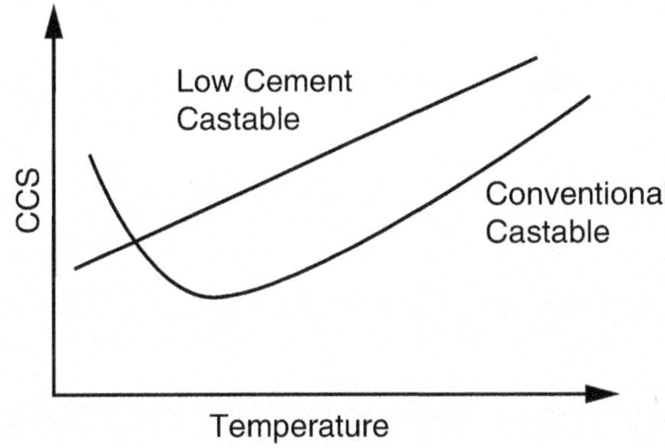

Fig. 3.14 Temperature vs. CCS relation – conventional and low cement castables

The very important differences of the behavior of the low cement and ultralow cement castables from that of conventional castables are as follows:

Obtaining required rheological property with addition of much lower amount of water, compared to conventional castable

Increased strength with increasing temperature (Fig. 3.14)

Having higher density, lower permeability, and high sensitivity to method of preheating because of the low water requirement for casting. Faster preheating may lead to explosive spalling and crack in the body, although the addition of special additives reduces its sensitivity toward heating rate.

3.3.1.4 No Cement Castable (NCC)

In these products, silica or alumina sols are used as binders, and a gelling agent (high alumina cement up to 1%) is used to set the castable.

These products are mostly two pack products. Refractory aggregates are supplied loose in bags, and the binder liquid (silica sol) is supplied in drum. The two components are mixed at site and are cast. But recently the solid powder is developed which forms sol when added to water, and one pack product of no cement castable can be produced using that binder. The special features of these products are as follows:

1. These products are to be applied by vibration casting.
2. The strength increases with temperature.
3. These can be preheated at a much faster rate compared to the cement-bonded castables (Fig. 3.15), and therefore total preheating time is much less which is a major advantage of chemically bonded castable [41].

Recommendations for mixer type, mixing time, and application methods for different types of castables are given in Table 3.7

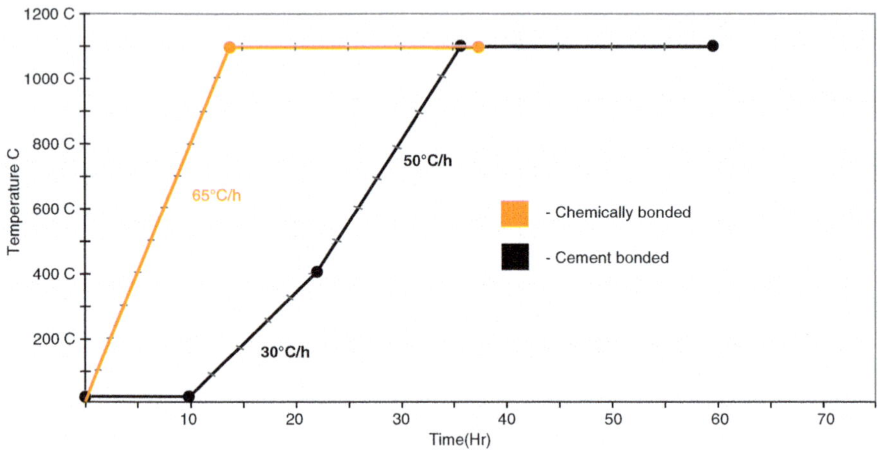

Fig. 3.15 Comparison of drying time of chemically bonded and cement-bonded castables

Table 3.7 Recommendation for preparation of castable mix (250 kg)

Mixer and mixing time	Insulating castables B.D < 1.2 g/cc	Insulating castables B.D > 1.2 g/cc	Conventional castable	LCC	ULCC	NCC
Ordinary mixer	Yes	Yes	Yes	No	No	No
High-speed paddle mixer	No	P	Yes	Yes	Yes	Yes
Mixing duration with slow water addition (min)	4	4	3	>3	>4	>4
Recommendation for application method						
Pouring/casting without compaction	Yes	Yes	Yes	No	No	No
Poking	Yes	Yes	Yes	No	No	No
Vibration	No	Possible	Yes	Yes	Yes	Yes
Self-flow	No	No	No	P	P	P

P possible

3.3.1.5 Gunning Mass

Gunning mass is supplied in powder form, and it is made out of aluminosilicate raw material and calcium aluminate cement mixed with other additives [42]. Gunning machine is used to spray the material on the place of installation. After charging the material in the hopper of the gunning machine, the material is transported through a long rubber hose, connected to a steel nozzle at the other end, by pneumatic pressure. The gunning material is mixed with water at the junction of rubber hose and steel nozzle, through a separately connected pipe, and ultimately sprayed on the application area.

It is used for both installation of new lining and repairing a worn-out lining. Gunning installation is a very specialized job, and the quality of the installation depends largely on the skill of the operator (gun man).

Portion of the gunning material sprayed rebounds from the target wall and falls down and is called rebound loss. The lesser is the rebound loss, the better is the quality of gunning. The rebound loss depends upon the material quality, the water percentage used for gunning (controlled by operator), distance of the gunning nozzle from the target surface, and the angle of nozzle with the wall to be gunned.

3.3.1.6 Shotcrete and Pumpable Material

Shotcreting is normally done in an installation where large quantity of refractory needs to be installed. It is very convenient to use this process when the installation is to be done at a height, for example, in preheater cyclone in Cement industry. In shotcrete installation, the shotcrete machine is kept at ground level. The castable material is mixed with recommended amount of water in the shotcrete machine to make slurry, and the slurry is then pumped through the rubber hose to the installation point and is sprayed on the target surface by compressed air.

Before the material comes out through the nozzle, a setting agent is mixed with the slurry to accelerate the setting. The rheological property of shotcrete is very important toward its successful use. The differences in the installation process of castables by gunning and shotcreting are given in Table 3.8.

As it was discussed in Sect. 3.3.1, basically, the castable with same chemical specification can be modified to install by different methods, e.g., by pumping the slurry, by self-flow technique, or by gunning or shotcreting depending upon the demand of the user. The basic ingredients of the castable remain the same, but the rheological or flow property is modified by addition of small amounts of additives [43] and control of particle size to achieve the desired flow property as the selected installation method demands. The properties of 50% alumina castable designed for installation by different methods are given in Table 3.9.

3.3.2 Phosphate-Bonded Castable

Most of the aluminosilicate monolithic refractories are based on the calcium aluminate binder (high alumina cement). But phosphate binders are also found to be very good to design aluminosilicate aggregate-based ramming masses, plastics, and castables. Mono aluminum phosphate (MAP), phosphoric acid, and different phosphate compounds are used as the binder [44–46].

When phosphoric acid or MAP and any other phosphate salt, used as the binder in the ramming masses or the plastics produced, are not hardened at ambient temperature

Table 3.8 Comparison of gunning and shotcrete materials and process

Gunning	Shotcreting
Castables with normal rheological property are used	Castables with deflocculating behavior are used
Dry transport of the powder material to the nozzle	Preparation of a slurry by mixing with water and pumped by double piston pump to nozzle
Water is mixed at the tip of the nozzle before spraying	A special solution is mixed at the nozzle tip to accelerate setting
Normal solidification by setting of high alumina cement	Solidification of the slurry by sol to gel transformation
Normally the applications are done at the same or at slightly elevated level	Can be done at much higher elevation, keeping the mixer and pump at ground level
Intensive dust formation	No dust formation
More loss of material due to strong rebound	Lesser loss of material due to slight rebound
Standard properties	Superior properties than normal gunning

Table 3.9 Properties of castables made out of the same base raw material but tuned to suit different installation methods

	Product type			
			Shotcrete	
Properties	Low cement self-flow castable	Pumpable castable	CaO accelerator	Na-silicate accelerator
Base raw material	Chamotte	Chamotte	Chamotte	Chamotte
Al_2O_3 %	50	50	50	50
BD (g/cc)/110 °C	2.34	2.24	2.08	2.34
/1000 °C	2.31	2.23	1.95	2.23
/1550 °C	2.26	2.29	2.02	2.23
AP (%)/110 °C	14.5	13.9	20.5	13.1
/1000 °C	17.0	17.6	27.8	16.0
/1550 °C	11.8	13.9	23.3	14.3
CCS (MPa)/110 °C	54.67	80.35	25.43	61.65
/1000 °C	89.37	114.6	14.1	100.6
/1550 °C	106.9	83.4	10.38	49.34

The results above indicate that lime as an accelerator does not work well in this composition

unless those are heated at about 100 °C. It is required to heat it beyond 350 °C to obtain a non-hygroscopic behavior of the material.

Alumina reacts with phosphoric acid to form Mono aluminumphosphate as per the reaction

$$Al_2O_3 + 6H_3PO_4 = 2Al(H_2PO_4)_3 + 6H_2O \quad (3.8)$$

$Al(H_2PO_4)_3$ is water soluble and sticky and this acts as a binder in phosphate-bonded castables. It decomposes to $AlPO_4$ at high temperature and is a refractory material [47].

In order to enhance the setting process, as required for a castable, based on phosphate bond, some accelerator is to be added. MgO, CaO, calcium aluminate, etc. can be used as accelerator to speed up the setting of the castable. Similarly some compounds, e.g., boron compounds, can be used to retard the setting process. The setting process is also influenced by the ambient temperature. The higher the ambient temperature, the faster is the setting process. The phosphate-bonded castable can be designed to achieve a certain setting time to suit the actual installation condition by addition of the said additives [48–51]. These castables are supplied in two pack system, where powder material and the liquid binders are supplied separately and are to be mixed in a definite proportion (prescribed by the manufacturer) at the site before application. The setting reaction is exothermic in nature and raises the mixture temperature. The higher is the generation of the heat, the faster is the setting, and this must be well controlled to get a desired working time.

Phosphate-bonded castables are available for installation both by casting and gunning method.

Phosphate-bonded castables have many advantages over the castables made using high alumina cement binder as given below [52].

1. The steam pressure generated during evaporation of the binder used in phosphate-bonded castable compared to water shows the steam pressure of earlier is 60% less at 80 °C and 20% lower at 100 °C (Fig. 3.15) which indicates that the phosphate-bonded castable can be preheated much faster compared to normal castables.
2. Its curing time is much faster than the alumina cement-bonded castables. The effect of shorter curing time and fast preheating time ultimately saves lot of time in the turnaround lining. One example [53]shows that installation of 120 MT of a phosphate-bonded castable of 200 mm lining thickness in fluidized coking environment saved 3.5 days
3. The effect of curing temperature is much lower compared to alumina cement-bonded castable.
4. The bonding of the phosphate-bonded castable with any refractory surface either brick or high alumina cement-bonded castable is very good and strong (Fig. 3.16).
5. Phosphate-bonded castables are much better to withstand the CO disintegration. The reason is that in the Boudouard reaction, free iron or iron oxides act as catalyst. But in phosphate-bonded castable, the Fe or iron oxide reacts with phosphoric acid to form iron phosphates as per the reaction 3.9 and 3.10:

$$2Fe + 6H_3PO_4 = Fe_3\left(HPO_4\right)_3 + 3H_2 \tag{3.9}$$

$$Fe_2O_3 + H_3PO_4 = Fe_3\left(HPO_4\right)_3 + 3H_2 \tag{3.10}$$

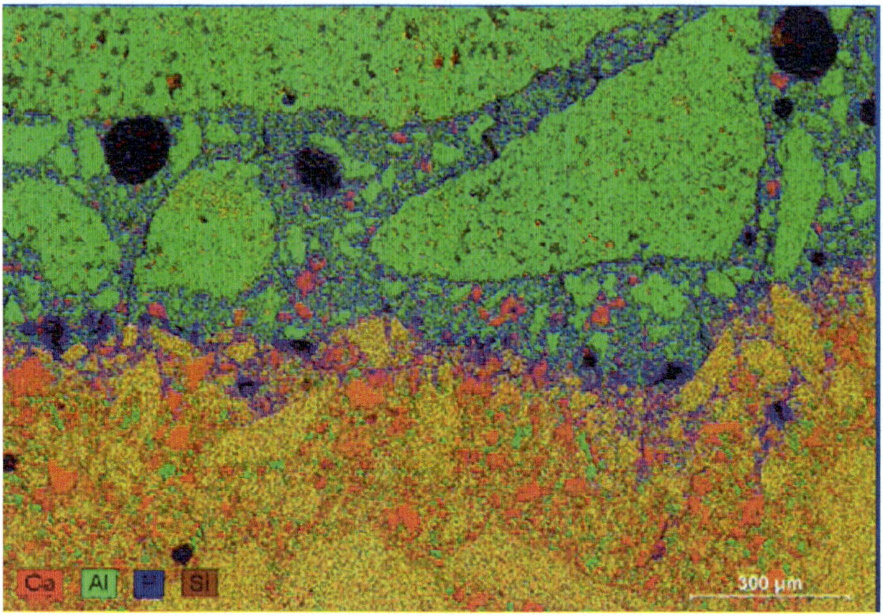

Fig. 3.16 The SEM photo of the adhesion of the phosphate-bonded castable (green) over the HA cement-bonded castable (yellow color)

Actual laboratory test of normal castable and phosphate-bonded castable supports it, and phosphate-bonded castable is found to have much better resistance to CO attack [53].

6. The phosphate-bonded castable when compared to HA cement-bonded castable with same alumina content and raw material base is found to have better abrasion resistance.
7. The PB castable has lower thermal expansion, higher thermal conductivity, and lower MOE compared to HA cement-bonded castable (Table 3.10) which contributes to its better thermal shock resistance over the other castables.

3.3.2.1 Steel Fiber Reinforcement

Steel fiber reinforcement is used widely both in castable (SFRC) and in precast blocks, like burner blocks. The addition of stainless steel fiber in all types of castables has been found to have a beneficial effect, especially where the castable lining or the precast shapes are exposed to high amount of mechanical abuse or temperature fluctuation. The addition of steel fiber to the extent of 2–3% by weight is found

Table 3.10 Comparison of properties of low cement and HA cement-bonded castable

	Andalusite-based low cement castable	Andalusite-based phosphate-bonded castable
Density (kg/m^3)	2.45	2.36
Th. expansion coefficient (K^{-1})	6.6×10^{-6}	5.5×10^{-6}
Th. conductivity (W/mK)	1.4	1.7
Poisson's ratio	0.3	0.3
Yield stress/20 °C (MPa)	14	5.7
/800 °C (MPa)	16	7.9
Strain (10^{-3})/20 °C	0.5	0.6
/800 °C	0.8	0.8
MOE (N/mm^2)/20 °C	28,000	9618
/800C	20,000	10,368

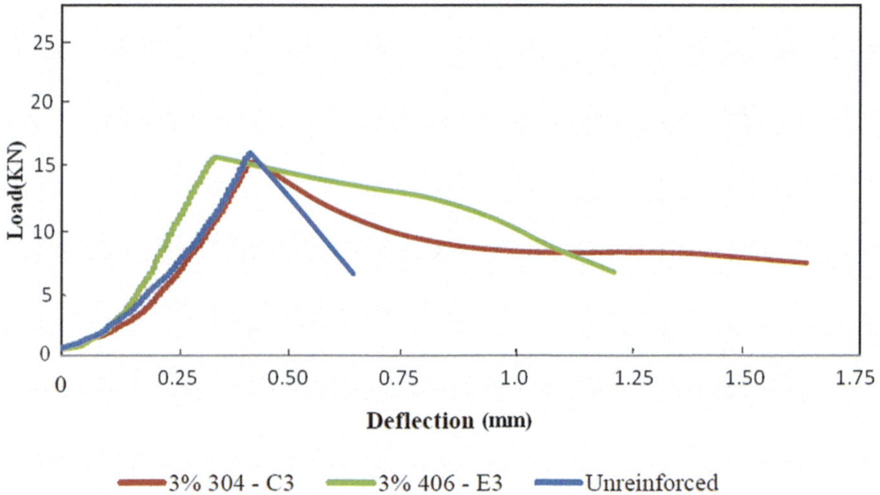

Fig. 3.17 Deflection pattern of the concrete, with and without steel fiber under tensile stress

to be enough to impart the beneficial effect. The addition of fiber improves its character of deflection under load and the toughness of the refractory concrete. Figure 3.17 shows that the addition of fibers in both the cases holds the sample better after crack initiation, showing higher ductility. The fiber chemistry, its diameter, length, aspect ratio, surface texture, etc. are all important parameters to decide the quality of steel fiber to be used [54–57]. The improvement in tensile strength of the concrete with addition of the fiber having different aspect ratio is shown in Fig. 3.18.

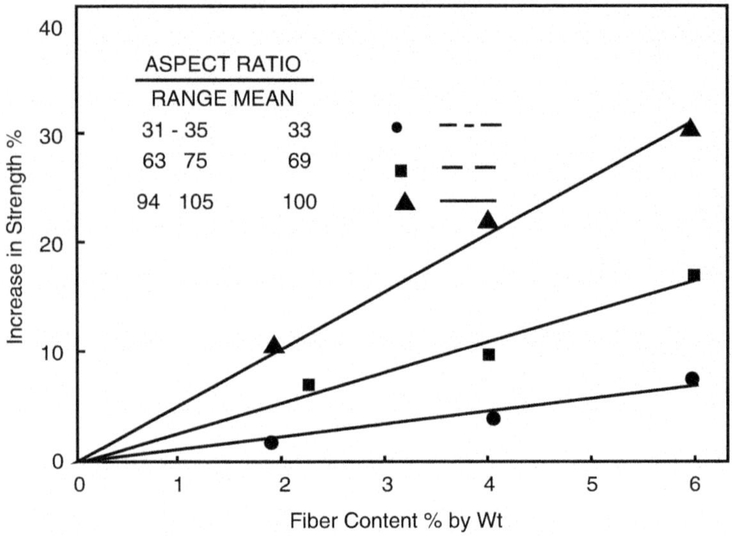

Fig. 3.18 Increase in strength of the concrete with addition of steel fiber

3.3.3 Ramming Mass

This is a combination of refractory aggregates and some additives as binders. It is supplied in powder form. The powder mass is mixed with prescribed quantity of water or any other specified liquid binder to get a semi wet consistency and then installed by pneumatic ramming or by hand ramming. These products are recommended for small repair jobs or to fill up some cavity or gaps, in the new brick lined construction in cement plant application. Typical property of a ramming mass is shown in Table 3.11.

3.3.4 Plastic Mass

This is manufactured and supplied in ready-to-use form, in small blocks, wrapped in polythene packet with stiff mud consistency. It is to be unpacked and to be installed either by ramming or tamping in the place of application. This is normally available in aluminosilicate range. The advantage of this product is its availability in ready condition and being time-saving. Quality consistency of the installed refractory lining is ensured because the water or the binder need not to be added at site. It gets set and hardens when it gets temperature. The preheating of this product can be done at a much faster rate compared to castables. Typical property of a plastic material is shown in Table 3.11.

Table 3.11 Properties of ramming mass and plastics

	Product type	
Properties	Ramming mix	Plastic mix
Base raw material	Calcined bauxite	Calcined bauxite
Binder used	Water	Ready mixed
Max. service temperature (°C)	1500	1500
Al_2O_3 %	80	85
CaO %	1.5	1.5
Fe_2O_3 %	2.5	2.5
Bulk density (g/cc)	2.68	2.75
Porosity (%)	20	21
CCS (MPa)	40	35
PLC/1200 °C (%)	±0.5	±0.5
Th. conductivity (W/mK)	1.20	1. 25

3.3.5 *Mortars*

Mortars are used to join the bricks in the refractory lining. Mortars used to have almost similar chemical composition as the bricks to be joined. Mortars are generally supplied as dry powder which is to be mixed with water or other specified liquid binders to form a paste which is trowelable and can be spread over the brick surface to act as glue to fix another brick. The mortars can be heat set, air set, or chemical set, depending upon the development of strength of the mortar on heating or in the presence of air or by its own. The heat set mortars develop its strength after heating the lining, whereas the chemically set or air set mortars develop its strength at ambient temperature itself, by chemical reaction. The mortar should have the property of good workability when mixed with water and should not get dried up fast, because of absorption of water in the brick pores, when applied on a brick surface.

3.4 Insulation Refractories

An efficient insulation allows the fuel source to be used more efficiently and saves money. Consumption of more fuel is uneconomic, and it has got a negative impact on environment for liberation of more greenhouse gases. Because of this reason, the use of insulation refractory is becoming more and more important. The thermal insulation can be provided by both shaped and unshaped refractories. The insulation refractories have high porosity, low strength, and low abrasion resistance, and due to this reason, the insulation refractories are used always as backup lining. The different insulation shaped materials have different properties and have different applications. The different types of insulation refractories are classified in different groups according to the ASTM standard as given in Table 3.12. The classification is based on the bulk density and the reheat change of the brick.

Table 3.12 ASTM classification of insulation refractories

Group no.	Reheat change <2% when treated at (°C)	Bulk density not greater than (g/cc)
16	845	0.54
20	1065	0.64
23	1230	0.77
26	1400	0.86
28	1510	0.96
30	1620	1.09
32	1730	1.52

The main features of the insulating refractories are high porosity and low bulk density. The pores are deliberately created in insulating refractories to entrap air, which provide very good thermal insulation and impart low thermal conductivity to the body.

All the thermal insulating materials are basically oxides, because compounds other than oxides have innate character of having higher thermal conductivity. The service temperature of insulating refractory and its volume stability at that temperature are also important to design a refractory lining. Alumina content of insulation products can vary from 20% to 95%.

3.4.1 *Insulation Shapes*

Insulation refractory shapes are manufactured from porous raw material, e.g., molar clay or exfoliated vermiculite, or from fireclay by introducing some organic combustibles in the mix which leave behind the pores on burn out. The properties of common fireclay insulation bricks are given in Table 3.13. Insulation shapes with very high alumina content can be manufactured from bubble alumina (Fig. 3.19) which are small hollow spheres made of alumina with 99% purity. Bulk density of bubble alumina is about 0.95–1.0 g/cc. This material is used widely to manufacture high alumina (+95% Al_2O_3) shaped and unshaped refractory products, used widely in chemical industries.

3.4.2 *Calcium Silicate*

This is available as boards of 25, 50, and 100 mm thickness and with different length and breadth. The shapes can be customized in curved section also to fit the contour of pipes and ducts. Calcium silicates are very light and efficient thermal insulating material and are used in large quantity in all types of furnaces and reactors. It is

Table 3.13 Properties of different types of insulation bricks

	Fireclay insulation	Fireclay insulation	Moler brick	Vermiculite brick
Service temperature (°C)	1260	1350	900	1100
Bulk density (kg/m^3)	600	850	570	475
Porosity (%)	–	–	76	80
CCS (MPa)	1.50	2.0	1.6	2.5
Sp. heat (KJ/kg)	–	–	0.7	0.94
PLC (%) (1000 °C/12 h)	1.50	0.5	1.0	1.0
Thermal cond. (W/mK)				
200 °C	–	–	0.11	0.14
400 °C	0.20	0.27	0.13	0.17
800 °C	0.25	0.30	0.15	0.19
Al_2O_3 %	35	40.0	8.0	7.0
SiO_2 %	–		72.0	47.0
Fe_2O_3 %	1.0	1.5	7.0	4.0
CaO %	–	–	6.5	2.0
MgO %	–	–	1.2	21.0
LOI %	–	–	1.0	7.0

produced synthetically by hydrothermal reaction between lime and silica. The calcium silicate blocks are available in two different grades suitable for application temperatures up to 800 and 1100 °C.

The thermal stability is based on the amount of calcium silicate hydrate shares (tobermorite $5CaO.\ 6SiO_2.5H_2O$ and xonotlite $6CaO.6SiO_2.H_2O$), due to their different shrinkage behavior and removal of bonded water. The calcium silicate blocks can be sawed and drilled as per the need. The properties of these materials are given in Table 3.14. Calcium silicate insulation blocks can be used as backup lining for both brick lining and castable lining.

3.4.3 Insulating Castables

Insulating castables are made from porous refractory aggregates and mixed with calcium aluminate cement as binder. The content of high alumina cement used in insulating castables is normally much higher than dense castables to get the desired mechanical strength. The insulating property of the castable depends upon the type of aggregate used and the composition of the castable. The properties of two insulating castables are shown in Table 3.15 as an example. Insulating castables requires much higher percentage of water to attain the consistency required for its placement.

Fig. 3.19 Bubble alumina

Table 3.14 Properties of calcium silicate block insulation

	Grade	
Properties	I	II
Service temperature (°C)	1000	1100
Bulk density (g/cc)	0.225	0.245
CCS (MPa)	2.6	2.7
MOR (MPa)	1.9	1.8
Sp. heat (KJ/kg)	0.84	0.84
PLC(%), when heated for 12 h at 50 °C below service temperature	1.0	1.5
Thermal conductivity (W/mK)		
200 °C	0.08	0.09
400 °C	0.10	0.11
600 °C	0.12	0.14
SiO_2 %	45	47
CaO %	45	45
LOI %	8	6

Table 3.15 Properties of insulating castables

Properties	Insulating castables	
Type of castable	High cement	Medium cement
Al_2O_3 %	30	33
Fe_2O_3 %	7	1.6
CaO %	15	5–6
Grain size (mm)	(0–3)	(0–6)
Water for casting (%)	45–50	28–32
Service temperature (°C)	1100	1350
Bulk density (g/cc)	1.00	1.28
CCS (MPa)/110 °C	3.5	5.3
/1000 °C	3.0	3.5
PLC (%) at 1000 °C	−0.6	−0.15
Th. conductivity (W/mK)		
At 400 °C	0.25	0.34
At 1000 °C	0.27	0.37

3.4.4 Ceramic Fiber Insulation

Ceramic fibers are artificially made and consist mainly of alumina and silica. Because of its inherent advantages (refer to Sect. 2.2.3), it is widely used to line different types of furnaces, reactors, ducts, etc. in almost all kinds of industries. Ceramic fibers can be divided into two broad categories, i.e., vitreous CF, for use below 1200 °C, and crystalline CF for use above 1200 °C. Figure 3.20 shows the effectiveness in energy saving of various insulation materials.

Ceramic fiber of vitreous quality is manufactured by melting alumina, silica, kaolin, etc. at very high temperature above 2000 °C in electric arc furnace, and the pouring stream of the melt is blown at, by jet of high pressure steam, when the molten aluminosilicate material will form fluffy fibrous material. It is also manufactured by pouring the melt on a very-high-speed rotor and is converted to fiber by the high centrifugal force (Fig. 3.21) [58]. But neither of the process can convert 100% material into fiber. Some of the material that is not converted into fiber is called "shots." In terms of mass, the fiber and shots are almost equal. The content of shots is measured by fiber index, which is the ratio of mass of fiber and shots. The higher the fiber index, the more is the content of fiber. Shot content in the fiber is not desirable as it increases the thermal conductivity of the fiber and reduces its efficiency as an insulation material.

The shots are removed from the fiber by washing with water. Traditional method of production of ceramic fiber by melting and blowing and washing produces fibers with fiber index of 60–80%. These fibers are woven or processed to form ceramic blanket, ceramic papers and ropes, etc. These fibers have amorphous structure and are thermodynamically unstable. These fibers tend to attain crystalline structure at favorable condition at higher temperature associated with shrinkage and increased thermal conductivity.

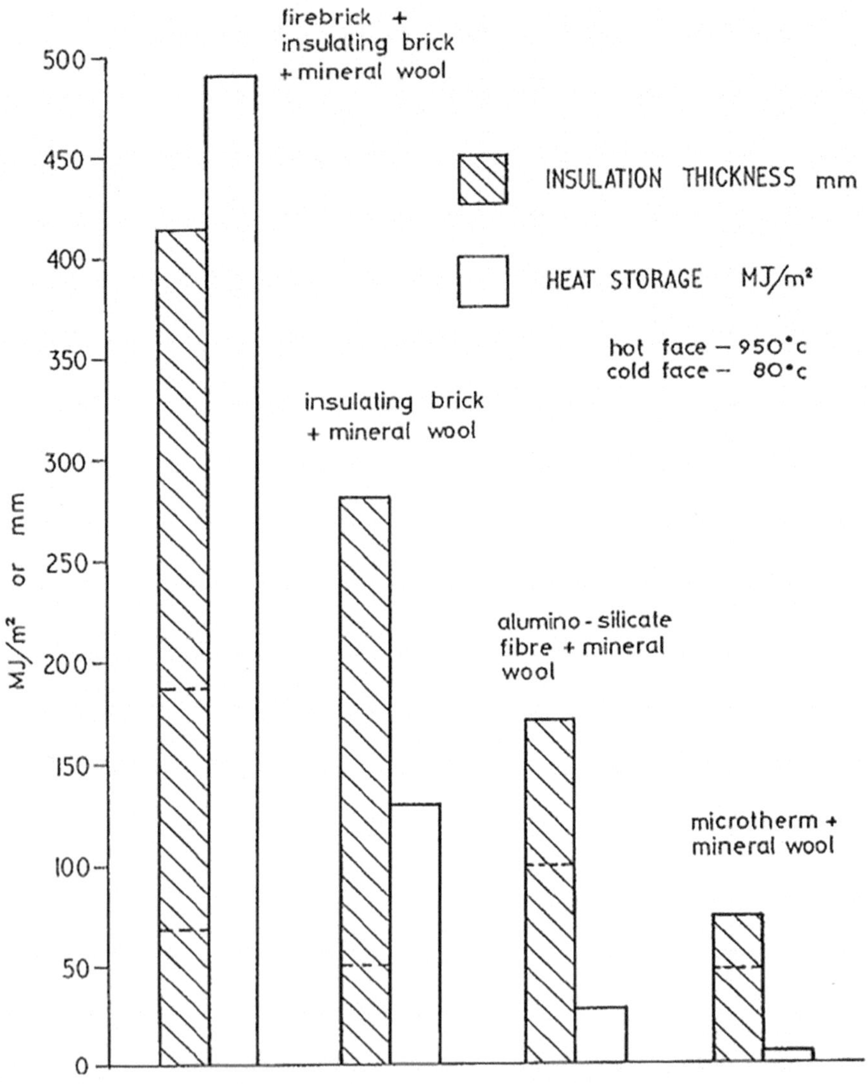

Fig. 3.20 Comparison of the furnace lining thickness and heat storage with same hot face and cold face temperature

The relation of the extent of recrystallization with temperature and duration of exposure at that temperature is shown in Fig. 3.22a [58]. Thermal conductivity of ceramic fiber is controlled by the fiber diameter, shot content, and the bulk density. Ceramic fiber products have lower thermal conductivity when the fiber diameter is small, shot content is less, and bulk density is low. Thermal conductivity of ceramic fiber products is dominated by radiation in the service temperature, and therefore the fiber surface area controls the thermal conductivity. The relation of thermal

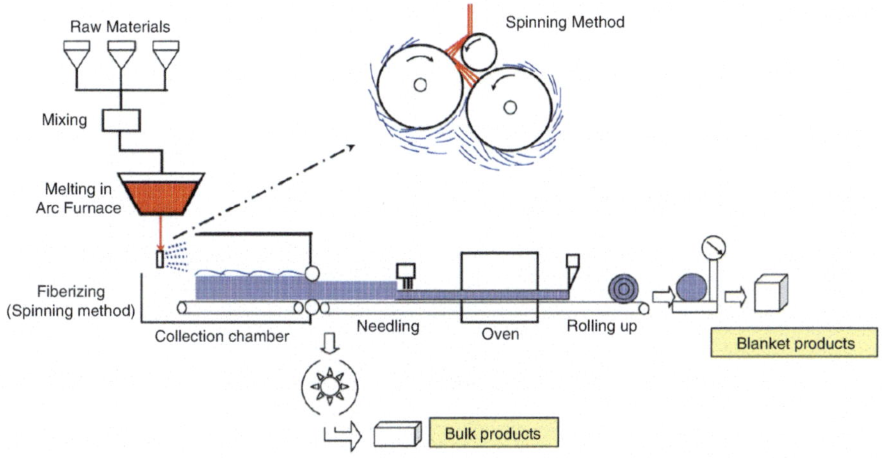

Fig. 3.21 Flow sheet of manufacturing of vitreous ceramic fiber

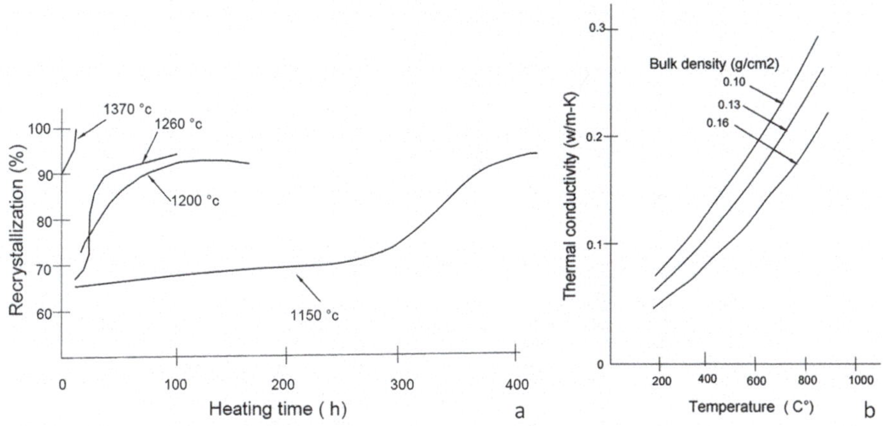

Fig. 3.22 (**a**) Effect of heating duration on recrystallization of ceramic fiber; (**b**) temperature vs. thermal conductivity of ceramic fiber

conductivity of ceramic fiber with temperature is shown in Fig. 3.22b [59]. Characteristics of some aluminosilicate ceramic fibers are shown in Table 3.16.

In case of crystalline fiber manufacturing, as the raw mix contains higher percentage of alumina, the same process of melting cannot be adopted here. Therefore fiber manufacturing route through sol-gel process is adopted here which is shown in Fig. 3.23 [60]. In this process, the sol-gel precursor is spun at room temperature, and the fibers formed are heat-treated above 1000 °C to form the ceramic fibers.

Three different types of fibers are available depending upon their method of manufacturing and composition. The refractory ceramic fiber (RCF) is the most common grade of ceramic fiber. But these fibers are categorized as hazardous

Table 3.16 Characteristics of ceramic fibers

	Fiber A	Fiber B	Fiber C
Service temperature (°C)	–	1500–1700	1600
True sp. gravity	3.1	3.6	2.90
Fiber diameter (μm)	4	3	2–4
Fiber length (mm)	50–100	–	<50
Al_2O_3%	72	95	72
SiO_2%	28	5	28
Crystal phase	Mullite	Corundum, mullite	Mullite

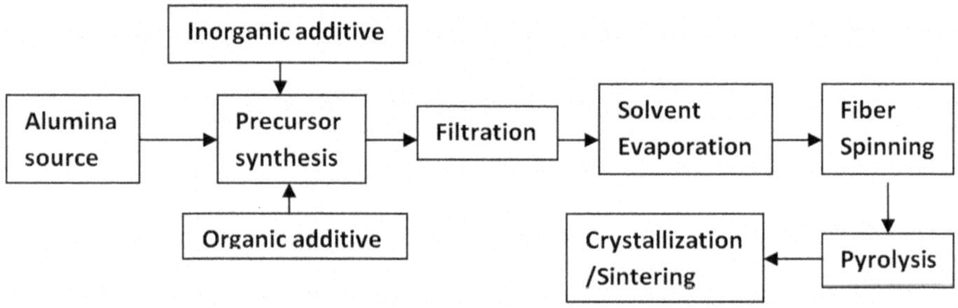

Fig. 3.23 Flow sheet of manufacturing process of ceramic fiber by sol-gel process

material because long exposure to this kind of fibers may cause respiratory problems, and it may cause carcinoma. The other grade of ceramic fiber is alkaline earth silicate (AES) type. This kind of fiber has much lesser quantity of alumina in it (Table 3.17). This is also branded by superwool by some companies.

These fibers are found to have 20% better insulation performance over RCF-type fibers and have 20% lesser weight. These fibers are exonerated and are not categorized as carcinogenic by IARC or under any national regulations on an international basis. The PCW fibers are perfect for chemically aggressive and high-temperature applications. These fibers are produced by sol-gel technology through processes, explicitly designed to produce fibers of defined dimensions. These fibers are classified up to 1600 °C application. Table 3.17 shows the properties of these three different types of fibers. Sometimes ZrO_2 and Cr_2O_3 are added in the fibers to enhance the service temperature and other specific properties.

All these different grades of fibers can be used to produce modules, veneers, vacuum-formed boards, papers, ropes, etc. and are used widely as insulation material in different equipment during the refractory lining. The ceramic fiber products are characterized by their continuous temperature of use, tensile strength, high temperature shrinkage, and thermal conductivity values. Properties of some of the qualities of ceramic blanket products are shown in Table 3.17.

Table 3.17 Properties of ceramic fiber modules based on different types of fibers

	Product		
Properties	RCF	AES	PCW
Al_2O_3 %	44–45	–	80–95
SiO_2 %	50–56	62–68	5–20
CaO + MgO	–	29–39	–
$Na_2O + K_2O$ %	0.10	Others <1	–
Classification temperature (°C)	1200–1300	1260–1300	1600
Density (kg/m^3)	64–160	64–160	96–128
Shrinkage (%) after 24 h	1.5 at 1000 °C	0.2–1.5 at 1000 °C	1.0 < at 1370 °C
	3.0 at 1200 °C	<1 at 1200 °C	1.5 < at 1600 °C
Th. conductivity (W/mK)			
200 °C	0.07	0.04–0.06	0.05
600 °C	0.20	0.11–0.18	0.11
1000 °C	0.42	0.23–0.42	0.30

Chemical Stability-Ceramic fibers are largely resistant to sulfur oxides and sulfuric acid. A reaction with sulfuric acid occurs at the hot face forming aluminum sulfate, but it has no negative impact on the performance of ceramic fiber, but rather it forms a protective coating on the surface and prevents the further attack of the acid and ingress of other gases.

Alkali metals affect the stability of ceramic fiber which may cause high shrinkage, fiber embrittlement, and sometimes destruction of the lining. The oxides of most common alkali metals like sodium and potassium are strong fluxes, and their reaction with fibers brings down the melting point and causes sintering of fibers at lower temperature which results in high shrinkage. The extent of the fluxing reaction is a function of the concentration of the alkali metal present and the temperature. As a general guideline, the concentration of the alkali metals should be kept below 0.2% to have a longer campaign life of the fiber [61, 62].

Oxides common to most slags, e.g., CaO, MgO, Fe_2O_3, V_2O_5, Na_2O, and K_2O, are all highly volatile, and they react with the ceramic fibers at higher temperatures.

Environmental Issue- The impact of the ceramic fibers has been studied and has found that the fibers must be low bio-persistent (LBP), and that is a drawback with RCF, because of which its use is banned in many countries, including European countries who have kept a norm of presence of RCF fibers in workplace below 0.3 fibers/ml. Low bio-persistent material means any fiber which is breathed in the lungs is rapidly removed by the body's natural mechanism, before it gets accumulated and causes harm to the body. Disposal of RCF also is a problem and requires special landfill area. Morgan has reported to have developed and has established the performance of a new fiber, with the name Superwool Extra, which does not produce any crystalline phase and has been exonerated from any carcinogenic classification under Nota Q of Directive 97/69EC [63].

3.4.5 Microporous Insulation Product

Microporous insulation refractories are the most efficient form of insulation. The original theory of microporous insulation was developed in 1940, and the first microporous insulation was used in nuclear reactor in 1950 and was subsequently used in aerospace and other industries. Its use in chemical reactor is more recent [64]. It has been found three to four times more efficient than the conventional refractory insulations.

The base material of microporous refractory is pyrogenic silica. Pyrogenic silica is made synthetically. It is produced by the chemical reaction 3.11 given below:

$$SiCl_4 + 2H_2 + O_2 = SiO_2 + 4HCl \tag{3.11}$$

The $SiCl_4$ is vaporized and mixed with hydrogen and air and fed into a combustion chamber. A hydrogen flame containing $SiCl_4$ burns to produce water vapor that reacts with $SiCl_4$ to form SiO_2 and hydrochloric acid gas. The silica formed is amorphous in nature as shown by its X-ray diffraction pattern compared to quartz as shown in Fig. 3.24 [65], and that is why it is also called synthetic amorphous silica (SAS). It has also got a low-density, high surface area and a structure which helps to get a very low thermal conductivity.

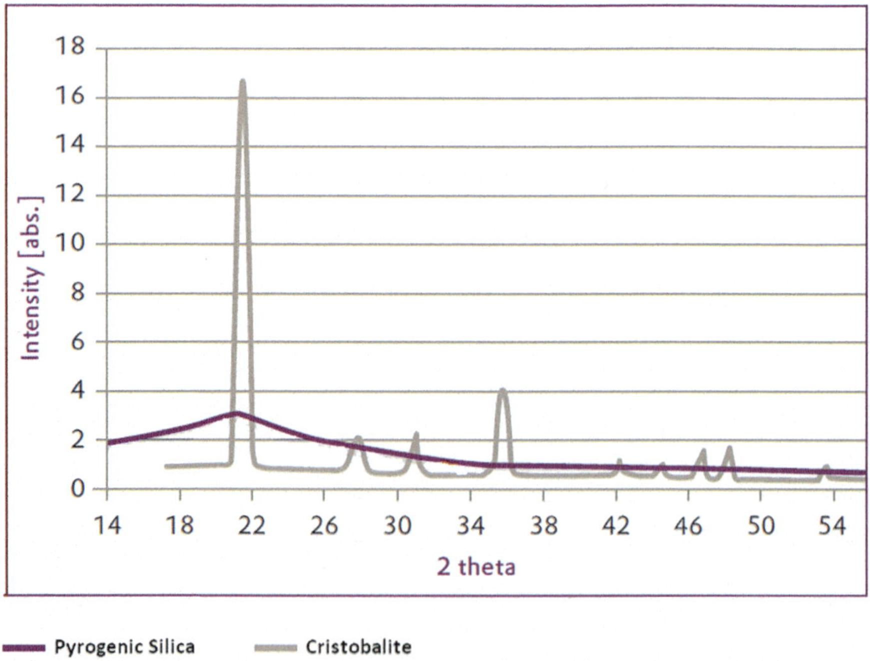

Fig. 3.24 The X-ray diffraction pattern of pyrogenic silica and cristobalite

Basically, microporous insulation is a collection of fine compacted powders of pyrogenic silica, opacifiers and fibers. It is manufactured in boards, panels, flexible panels, and preformed shapes. Microporous insulation is usually completely inorganic and noncombustible, though some manufacturers add organic content to improve the machinability of the product.

Microporous insulation achieves its low thermal conductivity by resisting all forms of heat transfer. Heat transfer through materials takes place by a number of different modes. An efficient and effective insulator must address each of these modes in order to achieve the lowest possible thermal conductivity. The most important modes of heat transfer in insulation materials are gaseous conduction, solid conduction, and radiation.

Insulations usually contain significant volumes of air in voids. The air in the voids conducts heat by collisions between molecules, allowing transfer of energy from fast-moving "hot" molecules to slow-moving "cold" molecules. In microporous insulation, the void volume is 90% of the total volume of the material, but because the voids are so small, the collisions between gas molecules are eliminated. Effectively, each air molecule is trapped in a void, smaller than the mean free path of the air molecule. This effectively blocks the ability of transferring energy to a neighboring air molecule. Air under these conditions has a far lower thermal conductivity than free air. This is known as the microporous effect and is the primary reason why a microporous material has a thermal conductivity lower than that of still air (Fig. 3.25).

Microporous insulation minimizes solid conduction in several ways. The most important is the low density of the material, with a high ratio of gas to solid. Furthermore, microporous insulation is largely composed of amorphous particles with a low intrinsic thermal conductivity compared to most solids. Finally, the particles are very small and randomly packed, which results in long heat paths through the material along which solid conduction can take place. Heat flux by conduction is inversely proportional to the distance along which the heat has to travel; thus, long heat paths reduce heat transfer.

At higher temperature above 550 °C, radiation is the dominant mode of heat transfer (Fig. 3.26). Radiation has a major impact at elevated temperature because it is proportional to fourth power of temperature differential between the two surfaces.

The second base material in the microporous insulation is special opacifier with controlled particle size and distribution. These opacifiers embedded in the core of the material scatter 95% of the infrared radiation and thus reduce the heat transmission to the lowest possible level. The effect of the opacifier can be demonstrated by comparing the thermal conductivity of pyrogenic silica with and without opacifiers in it (Fig. 3.27) [66].

The benefits of the microporous insulations are as follows:

1. These are resistant to most of the chemicals but not to liquids.
2. Very low thermal conductivity.
3. High strength and low bulk density.

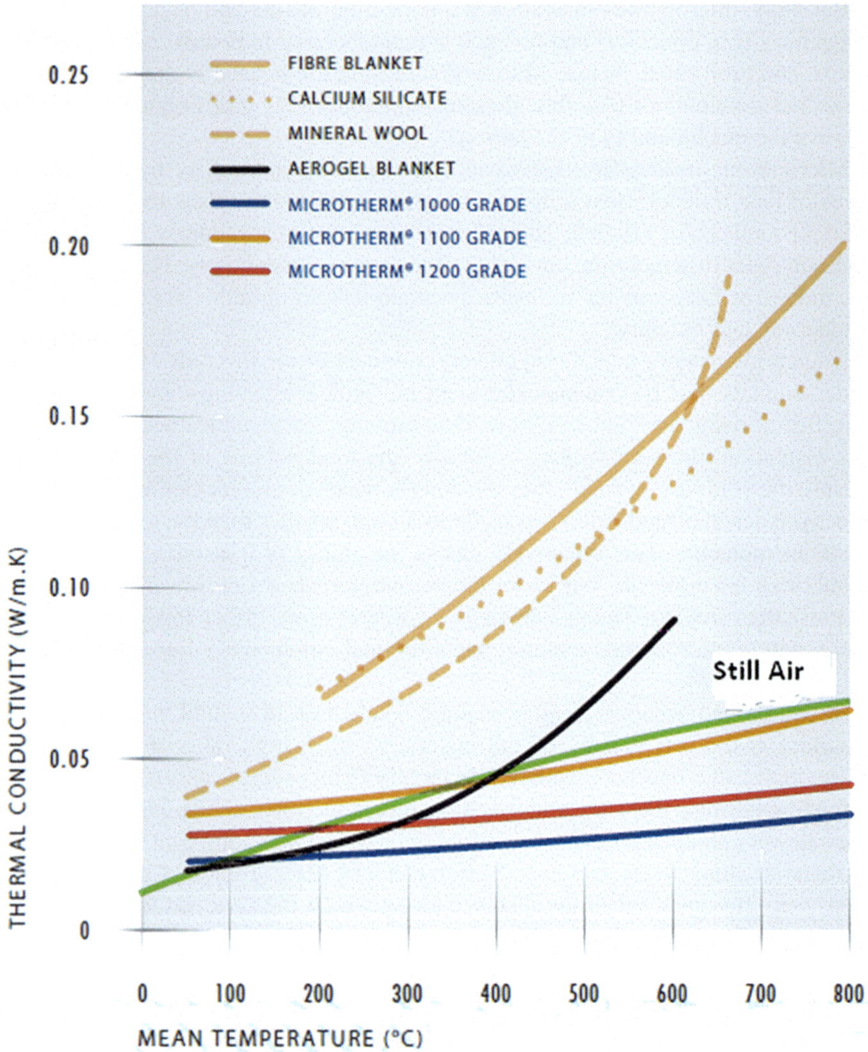

Fig. 3.25 Thermal conductivities of different thermal insulating materials. Microtherms are the microporous insulation materials. (Courtesy: Promat High Temperature Insulation catalogue, Promat International NV, Belgium)

4. Can be custom-made and flexible (Fig. 3.28) and so can be obtained directly as pipe sections.
5. Clean and easy to install.
6. Environment friendly.

Selection of insulation refractory in some cases becomes very important and crucial especially for vessels of bigger size, for example, in petrochemical industry.

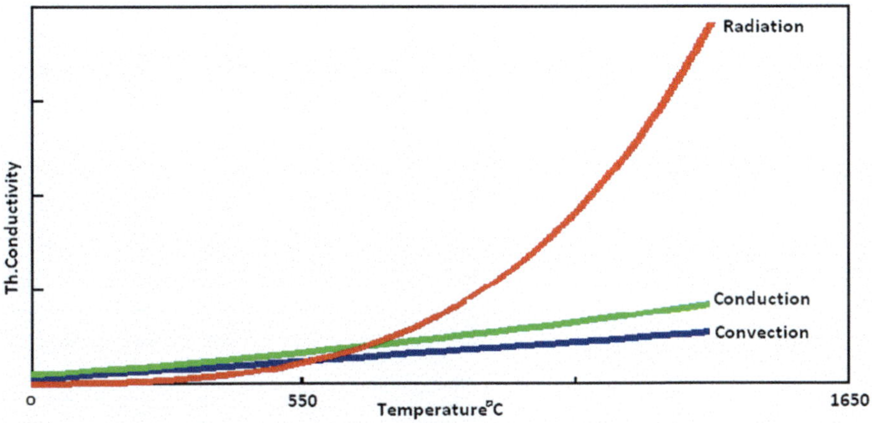

Fig. 3.26 Impact of different modes of heat transfer on thermal conductivity, as a function of temperature. (Courtesy: Unifrax Ceramic Fiber Technical Bulletin, July 18, 2011)

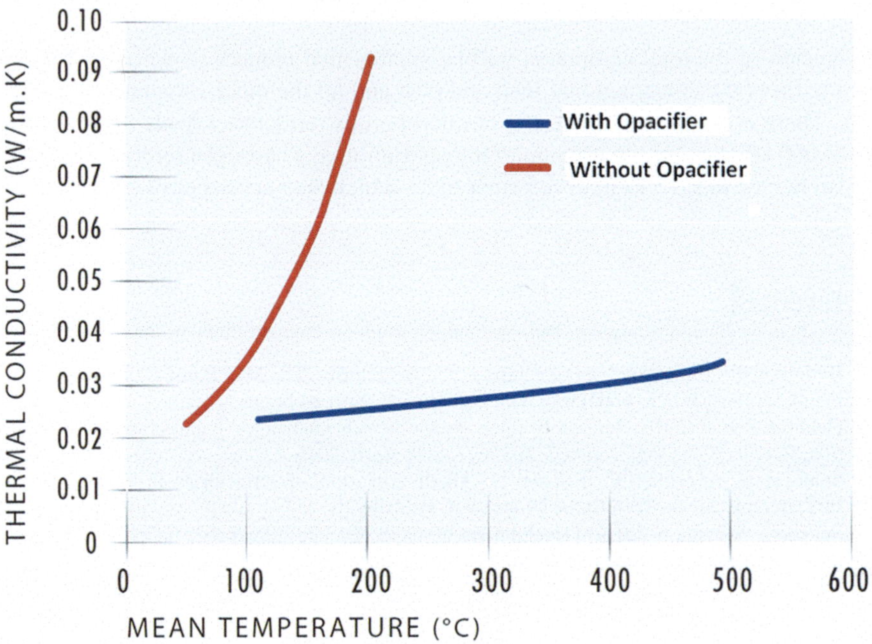

Fig. 3.27 The thermal conductivity of pyrogenic silica with and without opacifier

Apart from the recurring saving of energy on a long-term basis, it saves substantial cost on manufacturing, erection and transport of a reactor or furnace. When the insulation material is very efficient, the thickness of the insulation will get reduced for the same designed shell temperature. As the insulation thickness reduces, the

Fig. 3.28 Microporous insulation in flexible form. (Courtesy: Promat High Temperature Insulation catalogue, Promat International NV, Belgium)

dimension of the total equipment will be smaller that reduces its weight which in turn reduces the constructional load and cost and all the other associated cost factors. There is evidence that the use of microporous refractories leads to the saving of 40,000 USD, alone, on account of transportation of a hydrogen reformer in 2008. It also helped to save 17 m of reformer tubes which are very expensive [64].

References

1. Budnikov PP. The technology of ceramics and refractories. The MIT Press, Cambridge
2. Brunk F (2001) Silica refractories. CN Refractories, Special Issues 5
3. Heidrich R et al (2016) Refractory solutions for the new challenges in glass furnace construction. ResearchGate Publication 272085677, 20th April, 2016
4. Schlegel E, Aneziris CG, Fischer U. Alkali corrosion of refractories in cement kiln – PPT. Bergakademie, Technische Universitat, Freiberg
5. Sengupta P, Gupta N (2010) Development of alkali resistant refractories for cement industry. AUCBM conference, proceedings
6. Demirsan V, Buyukcayir I (2013) Study of refractory bricks with improved alkali resistance for the cement industry. Refractory World Forum s(2013)1
7. Bemdiek K, Klischat H-J (2011) Installation and performance of Monolithic materials with graded alkali resistance. UNITECR proceedings 2011, 2-B2-1
8. Buchel G et al (2007) Review of tabular alumina as a high performance material. Interceram, Refractories manual
9. Goorman F et al (2013) The performance of high quality Magnesia raw materials in cement applications. UNITECR proceedings, 2013, pp 177–180
10. Lampropoulou et al (2004) Mineralogical study and properties of Magnesia refractories derived from Evian Magnesite. Bulletin of Geological Survey of Greece XXXVI

11. Sodje J, Klischat HJ (2012) Magnesia an essential raw material for cement kiln refractories. Refractory World Forum 4(2)
12. Steinkelner P. Basic bricks for rotary kilns – cement. PPT, RHI
13. Wirsing H et al (2015) Magnesia brick containing iron spinel –troubleshooters for thermo-mechanically stressed kilns. UNITECR proceedings, 2015, Vienna
14. Jastrzebska I et al (2015) Elastic behavior of magnesite – chrome refractories at elevated temperature. Mater Technol 496 913–918.
15. Kitaguchi DY et al (2011) New chrome free bricks for the burning zone of cement rotary kiln. UNITECR proceedings, 2011, 2-B2-3
16. Geith M, Jorg S, Krischanitz R (2017) Influence of flexibiliser on basic cement rotary kiln brick properties. RHI bulletin, 1
17. Ghanbarnezhad D et al (2013) The effect of TiO_2 on properties and microstructure of chrome free basic bricks. UNITECR proceedings, pp 171–176
18. Geith M et al (2017) Influence of flexibilizers on basic cement rotary kiln bricks. UNITECR proceedings
19. Jorg S, Krischanitz R (2006) First in service results of the new Magnesia Galaxite Brick. RHI Bulletin 3:7–10
20. Jong S et al (2007) Ankral ZE, outstanding performances at the Caravate Plant, Colacem s.p.A Italy. RHI Bulletin 2:31–34
21. Jorg S et al (2010) Ankral QF – a new brick to combat tyre stress. RHI Bulletin 1:8–12
22. Jorg S (2010) RHI's Ankral Q2 Solution. The Refractories Engineers
23. Sodje J et al (2013) Aspects of elastification reactions in basic cement kiln bricks. Refractories World Forum 5:4.
24. Klischat H et al (2017) Magnesia Pleonaste bricks for essential requirements in rotary kilns. UNITECR proceedings
25. Klischat HJ, Wirsing H. Practical application of mineralogical variations for cement kiln refractories. Refratechnik Cement GmbH, Germany
26. Klischat HJ, Wirsing H (2013) Higher thermochemical resistance by installation of Magnesia Forsterite bricks. UNITECR proceedings, pp 181–186
27. Kozuka H et al (1995) Further improvements of MgO-CaO-ZrO_2 bricks for burning zone of rotary cement kiln. UNITECR proceedings, pp 256–263
28. Radovanovic S (1997) Reaction behavior of Spinel Zirconia and Monocalcium Zirconate under working conditions of cement kiln. UNITECR proceedings, pp 1613–1630
29. Kozuka H et al (2005) Minimizing structural spalling of MgO-CaO-ZrO_2 refractories for cement rotary kiln using large amount of wastes. UNITECR proceedings
30. Ohno M et al (2017) Magnesia Spinel bricks with good coating adhesion and high resistance to corrosion and spalling for cement rotary kiln. UNITECR proceedings
31. Sniezek E, Szczerba J (2018) Magnesia Zirconia refractory materials with Aluminate phases. Refractories World Forum 10:1
32. Bongers U et al (1997) Experience with a new generation of Chrome free Magnesia enriched dolomite bricks in cement kiln. UNITECR proceedings, pp 1605–1612
33. Heartenstein J, Krischanitz R (2018) Benefits of Modern Doloma-Magnesia linings in modern cement kilns. Journal of Refractory Innovation, RHI-Magnesita
34. Prange R et al (1995) Present status and future trends in the use of basic refractories in cement and lime kiln. UNITECR proceedings, pp 248–255
35. Bartha P (2004) The cement rotary kiln and its refractory lining. Interceram, Refractories manual
36. Canaguier J (2016) How to reduce glass defects generated by Fused cast AZS. 16th conference on Electrical Glass Melting, Sept 14, 2016, Prague, Czech Republic
37. Parr C et al (1998) Calcium aluminate cements for refractory gunning application. 3rd international symposium on Refractories, Beijing, China

38. Parr C, Roesky R, Wohrmeyer C (2004) New developments in Calcium Aluminate Cement containing castables for steelmaking application. Teheran international conference on Refractories, 4–6th May, 2004
39. Ninsheng Z et al (2004) Advances in modern refractory castables. Proceedings, Tehran international conference on refractories, May, 2004
40. Lee WE et al (2001) Castable refractory concrete. Int Mater Rev, 46, 3.
41. Weidhaas M. Development in monolithic refractories for the Portland Cement Industry. Refratechnik North America Inc
42. Hundere A et al (1997) Free flowing castables – a pre-requisite for wet gunning of refractory castables. Proceedings, VIIth international metallurgical conference in Ustron, Poland
43. Fukami N et al (1995) Effect of additives on the pumpability of the low cement castables. UNITECR proceedings, Vienna
44. Kingery D (1950) Phosphate bonding in refractories. Thesis paper to Doctorate of Science, MIT
45. Fundamental investigation on Phosphate bonding. Georgia Institute of Technology, School of Ceramic Engineering, December, 1983
46. Chung DDL (2003) Review, acid aluminium phosphate for the binding and coating of materials. J Mater Sci 38:2785–2791
47. Giscow R et al (2004) The variety of phosphates for refractory and technical applications by the example of aluminium phosphates. cfi/ber.DKG 81(5)
48. Luz AP et al (2015) High alumina phosphate bonded refractory castables $Al(OH)_3$ sources and their effects. Ceram Int 41:9041–9050
49. Luz AP et al (2018) High alumina chemically bonded Refractory castables containing liquid and powder binders. Refractories World Forum 10(2)
50. Lopes SJS (2017) Self flowing high alumina phosphate bonded refractory castable. Ceram Int 43:6239–6249
51. Luz AP et al (2016) Monoaluminium phosphate bonded refractory castable for petrochemical application. Ceram Int
52. Decker J (2010) Advantages of liquid phosphate bonded refractories. Refractory World Forum 2(1)
53. Decker J, Hagberg TS (2008) Better, stronger, faster and lasts longer. Hydrocarbon Engineering
54. Meddah AA (2015) Effect of the fiber geometry on the flexural properties of steel fiber reinforced refractory concrete. Proc Eng 108:185–192
55. Peterson RA (2014) Engineered design of fiber reinforced refractory precast material for Taconite processing furnace. Thesis submitted to Graduate School, University of Minnesota, USA
56. Bernhart G et al (2008) Fiber reinforced refractory castable for SPF tooling. Mater Sci Eng Technol 39(4–5) Wiley
57. Krasnikovs A et al. Crack propagation in short steel fiber reinforced concrete. Modeling and experimental investigation. Concrete Mechanics Laboratory, Riga Technical University, LV 1048, Riga, Latavia, Email: akrasn@latnet.lv
58. Ceramic fiber and insulating technology – Nippon Steel technical report, No-98, July
59. Refractories handbook – The technical Association of Refractories Japan, 1998
60. Ceramic fibers and coatings. Advanced material for the twenty first century. National Academic Press, p 42. ISBN:978-0-309-05996-1. https://doi.org/10.17226/6042
61. Klinn AF (2008) Microporous insulation for energy efficiency. Industrial Heating
62. Unifrax, Ceramic Fiber Technical Bulletin, July, 18, 2011
63. Maity M et al (2013) Deterioration of refractory ceramic fiber lining in an ethylene cracking furnace – a case study. UNITECR
64. Jubb G (2020) A viable replacement of ceramic fibre. The Refractories Engineer
65. Aerosil Fumed Silica – technical overview, Evonik Industries, Germany
66. High temperature microporous insulation – Promat International, NV, Belbium

Chapter 4
Design, Installation, and Maintenance of Refractories

4.1 Introduction

Performance of refractory lining depends largely on both the design and installation of the refractory lining. The kiln, furnaces, or the reactors used in modern cement, lime calciner, petrochemical, or glass industries are huge and massive in size and require thousand of tons of different refractories of different physical and chemical characteristics. At high temperature, many physical and chemical changes happen, and unless the design and the installation are sound, the refractory lining will not last long. A cement rotary kiln, for example, of 6 meter diameter and 60 meter length, weighs about 1100 MT, including refractory lining and feed material. When such a huge structure, with huge mass, rotates on two or three supports, different typed of thermo-mechanical stresses are generated in the refractory lining, and therefore lining design and perfection in installation is very much needed to reduce the effect of those stresses. Designing of the refractory lining in any equipment needs the consideration of the following.

4.2 Refractory Selection Criteria

Selection of right kind of refractories for a particular application depends upon the following information:

(a) The process that will be carried out in the furnace. The chemical reactions that will be carried out and physical parameters (temperature, pressure, etc.) that need to be maintained.

P. Sengupta, *Refractories for the Chemical Industries*,
https://doi.org/10.1007/978-3-030-61240-5_4

(b) Design and type of the furnace/reactor. The lining pattern of cylindrical and rectangular vessel will be different. It also differs from horizontal to vertical cylindrical vessel.
(c) Fuel used in the furnace or the reactor. The combustion products and the characteristics of the ash, in case of solid and liquid fuel, are important. The possibility of reaction of combustion products and fuel ash with refractory at operational temperature must be taken into consideration.
(d) The operation of the furnace, whether continuous or intermittent, is an important factor. The frequent heating and cooling imposes the condition of high thermal shock.
(e) Furnace atmosphere, whether oxidizing or reducing or inter changing, is important because all refractory materials are not stable in reducing or oxidizing conditions.
(f) The state of the matter that will be treated in the furnace or reactor is important factor because when the solid material is treated in any furnace or reactor, its loading and movement inside the furnace causes abrasion on the refractory lining, whereas if the reactants or charge to the furnace or reactor, is liquid the corrosion may play the dominant role towards the damage of the Refractory.
(g) Surrounding condition like temperature, wind velocity, and humidity of ambient atmosphere are important because the heat transfer from the furnace or shell surfaces are influenced by those factors.
(h) The shell temperature of the furnace/reactor, as per design parameter, is important to design both the type of insulation to be used and the thickness of the composite lining.
(i) Storage facility and the total quantity of the Refractory to be installed taken into consideration.
(j) Selection of refractories also depends, sometimes, upon the position where the refractory application is done. For example, if a refractory installation is to be done at a substantial height above the ground level, then instead of shaped refractory, installation of castable by shotcreting method can be considered which is much faster. On the other hand, if the space inside the reactor or the furnace is not sufficient, required for installation of unshaped refractories, the shaped refractories will be a better choice.

4.3 Shaped Refractories

Refractory bricks for a lining can be designed in any shape but the following points are to be borne in mind during designing the shapes of the bricks used for a particular lining:

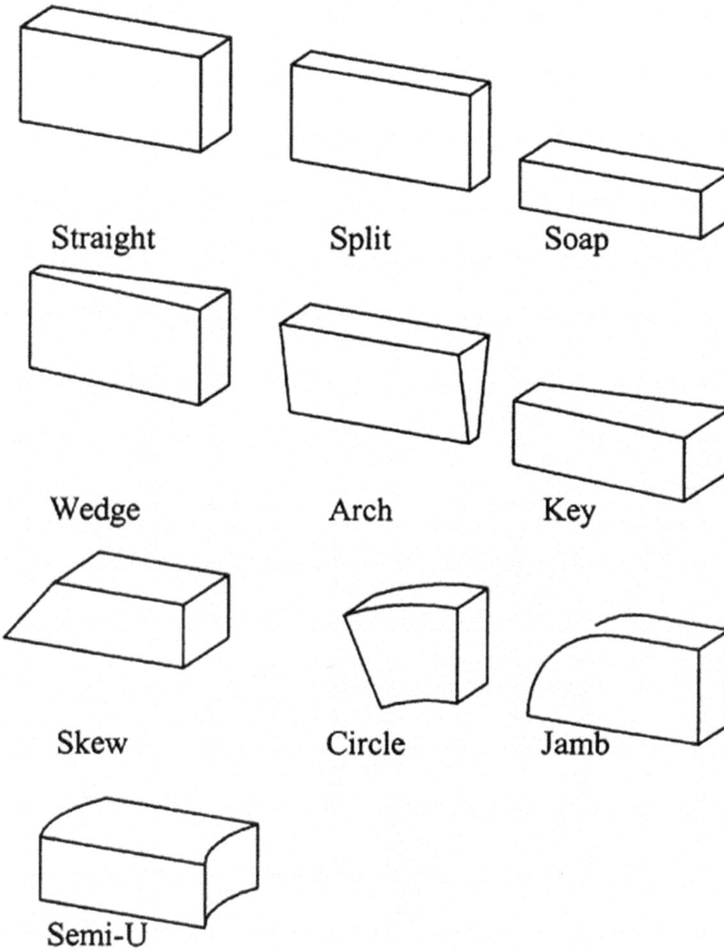

Fig. 4.1 Different shapes of the refractory commonly used

(a) It is better always to choose the bricks out of the standard shapes as given in Fig. 4.1.
(b) If the furnace/reactor design is of complicated nature and the lining needs special shapes, those can always be used, but it is always preferable to use lesser number of shapes for any furnace or reactor lining.
(c) The larger is the size of the brick, the lesser is the number of joints, but the stress developed in the individual bricks due to the existence of a temperature

profile, with steeper gradient, will be higher. Therefore optimum dimension of the bricks are chosen. Moreover it is always more difficult to manufacture bricks of bigger shape with the same properties as bricks with smaller dimension.

(d) The shapes to be used depend upon the geometry of the furnace, for example, the brick shape will be different when the shape of the furnace or reactor is rectangular or cylindrical.

4.3.1 Features of Lining with Shaped Refractories

Advantages of using shaped refractories:

(a) Shaped refractories are normally denser and thermally more stable as they are already heat treated at higher temperature.
(b) The installation of shaped refractories takes more time in laying but takes less time in preheating.
(c) The installation of furnace lining with shaped refractories is more forgiving in nature.

Disadvantages of using the shaped refractory lining:

(a) To carry a big inventory of many different shapes especially when the furnace geometry is complicated.
(b) Lining of the shaped refractories is more labor-intensive and is also difficult to introduce automation in the lining.
(c) Joints in the shaped refractory lining are weak points and lead to penetration of slag and gases which corrodes the furnace or reactor shell.
(d) Shaped refractories cannot be used for repairing or patching the worn-out refractory lining of the furnace or reactor in hot condition and need complete cooldown of the furnace, removal of worn-out refractory, and installation of new refractory shape there.

4.4 Installation of Shaped Refractories

Refractories can be a substantial investment for any chemical industry, either in new construction or in repair work. Installation of refractories may comprise a small part of the entire budget, but it protects the equipment, protects the people, and protects the production, and their value is greater than cost. Refractory performance depends on the quality of installation to a great extent. The best quality of refractory does not ensure the best performance, unless the Installation quality is sound. Installation techniques are different for different types of refractories. Installation methods of shaped and unshaped products are quite different. Complete knowledge on the character and properties of the refractory products, under installation, is required for good installation quality.

4.4.1 Basic Rules of Refractory Lining

Certain basic rules must be followed to get a sound lining:

(a) Refractory bricks must be laid horizontally unless the design of the lining demands an inclined lining as in the case of arch construction.
(b) The construction dimensions in the design drawing must be followed taking in consideration the dimensional tolerances. The first layer must be installed with extreme care, and then "go ahead" must be given after checking dimensions and alignment.
(c) Brick joints must be made with prescribed material, and joints must not be more than 2 mm thick.
(d) If due to size tolerance of the brick the prescribed joint thickness cannot be maintained, the decision should be made to deviate from the prescribed thickness or to sort out the bricks.
(e) If the steel wall of the furnace is not level, then the brickwork must be adjusted to the shape of the steel wall in case the deformation is not too big.
(f) The brickwork must be kept clean specially the expansion joints.
(g) Already laid bricks can only be aligned in the direction of the bed or vertical joint.
(h) Readjustment is not possible if the mortar has started to become hard; bricks not laid correctly must be removed, cleaned, and installed again with fresh mortar.
(i) Bricks with cracks can only be used if the cracks are insignificant or it is in the rear side of the brick which will remain on the cold face of the lining.
(j) The brickwork must be designed and made in such a way that no gaps or hollow spaces are formed, because the deposition of dust in the hollow space can develop an uncontrollable buildup of pressure resulting into the destruction of the lining.

4.4.2 Vertical Wall

In case of vertical lining of wall, in a straight or circular wall, steel retainer plate is put after every 15–20 layers, so that the load is distributed on the retainer plates. In that case, ceramic blanket is used to fill up the gap between the topmost brick layer and the retainer ring to accommodate the expansion as shown in Fig. 4.2.

In any refractory lining, a temperature profile exists across the refractory lining between hot face and cold face. In case of cylindrical lining, when the refractory lining is heated up, it starts expanding, and initial stresses are nullified by the re-adjustment of lining material and compression of mortars, in case of the brick lining with mortar joints. At still higher temperature, stress starts to build up, because there is no further space to expand for the brick. In refractory, fortunately, like many other materials, there is a transition from the brittle to plastic behavior with rise of temperature, which mitigates the effect of stress buildup.

In case of flat wall, when it receives the heat at one face of the wall, the refractory bends towards the heat, as a result of the thermal gradient through the lining. Due to the higher operating temperature, the hot face will thermally expand (elongate) more than the cold face, causing the wall to curve or bend towards the heat. The result is a tendency for refractory linings to pull away from backup linings and steel shells, creating voids and, ultimately, leading to lining failure (Fig. 4.3).

To prevent this bulging of wall, flat brick walls must be tied back to the steel shell, to resist the pull away from the shell. For brick linings, this is normally accomplished by utilizing "anchor brick" that are secured to the steel shell by a stainless steel anchor (Fig. 4.4).

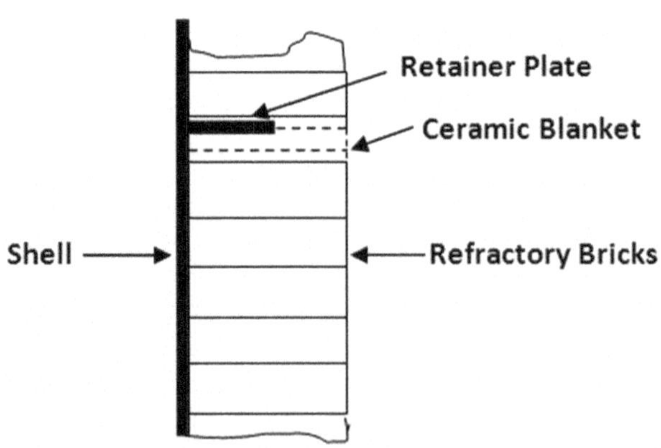

Fig. 4.2 Arrangement of special joint between brick lining and retainer plate

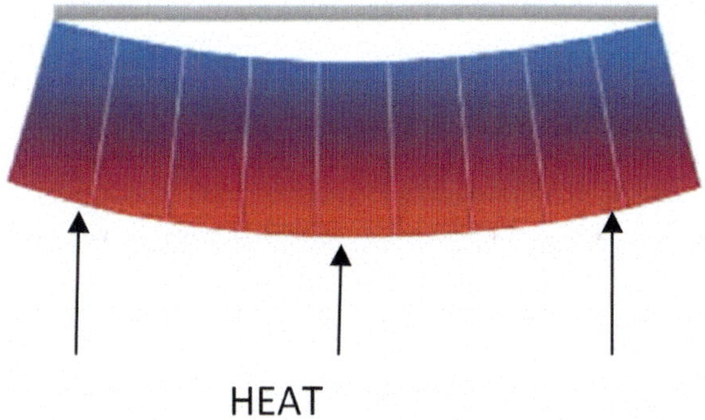

Fig. 4.3 Bending of the lining due to thermal gradient, across the lining thickness

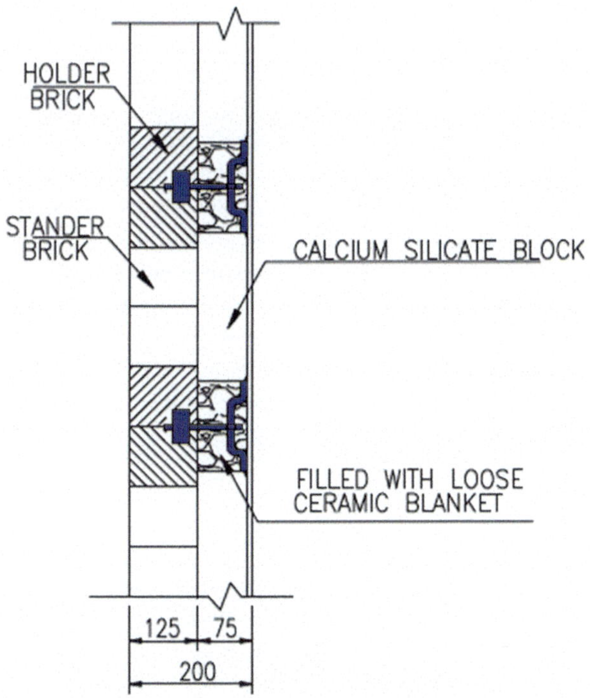

Fig. 4.4 The arrangement of holder bricks anchored to the wall

4.4.3 Expansion Joints

This is a very important aspect of all refractory installations. Refractory expands when gets heated up in service. A provision is to be kept to accommodate the thermal expansion and to relieve the stress, without which the lining may collapse. It is to be borne in mind that during the operation of a kiln or furnace, the whole refractory lining maintains a temperature profile across its thickness. Therefore the expansion will be more at working face and minimum at the cold face. Moreover, the refractories attain plasticity at high temperature and mitigate the compressive stress generated due to restricted thermal expansion. The provision for expansion can be kept in different ways. For example, in case of constructing a ring with refractory bricks in a cylindrical vessel, the total expansion of hot face is calculated considering the circumference of the vessel and operating temperature. Then the expansion per brick is calculated, and the cardboard of required thickness is put after every 5 or 10 bricks.

Calculation of expansion joint – if the diameter is D, thermal expansion coefficient is α, operating temperature is T, and number of bricks per ring is N. Then the expansion per brick

$$E_b = \frac{3.14\, D\alpha T}{N} \tag{4.1}$$

If the brick interval, for providing expansion joint, considered is 5, then the thickness of cardboard to be used to provide expansion joint should have thickness $5 \times 3.14 D\alpha T/N = 15.7\, D\alpha T/N$.

In cement rotary kiln, no expansion joint is provided in the rings to keep them tight during operation. During the closure of a ring in the lining, the 1–2 steel shims are struck in the joints, and during high-temperature operation of the kiln those steel shims soften or melt and reduces the thermo- mechanical stress due to expansion. The expansion joints are more important in axial direction because of larger length provided in the axial direction of the lining after every 1–2 meters length. The thermal expansion joint is, especially, very important for the lining of basic refractories, which have high thermal expansion.

4.4.3.1 Example

Let us consider the lining of basic refractory over a length of 15 meter. Basic bricks have thermal expansion of about 1.5% at 1000 °C. Steel has the thermal expansion coefficient of 12×10^{-6}/°C. If we consider the lining working face temperature is 1000C and the steel shell temperature as 350C, then the relative expansion of the lining hot face compared to steel shell will be $15{,}000 \times [0.02 - (350\text{–}30) \times 0.000012] = 242.4$ mm. The ambient temperature is considered 30C. That means the lining hot face will expand 242.4 mm on a length of 15 meter of the lining compared to the steel shell. If the required space is not allowed to expand, the tremendous compressive stress generated can be calculated easily (Eq. 4.11) which can crush the top portion of the bricks at the plane, parallel to kiln axis. Figure 4.5 shows the damage of 70% Al_2O_3 refractory bricks in a kiln due to insufficient expansion joints given in the lining in the axial direction. The phenomenon becomes more

Fig. 4.5 Damage in the 70% Al_2O_3 brick lining due to insufficient expansion joints

prominent if the preheating is made faster when the stress does not find time for relaxation (Eq. 4.12).

But fortunately, in actual practice, a temperature profile exists across the lining thickness, and the temperature of the cold face of the brick remains at a much lower temperature compared to working face, and the effect of thermal expansion is minimized.

The expansion joint in wall and corner of the wall is shown in Fig. 4.6a, b.

4.4.4 Roof design

Furnace roof is a critical area in the furnace construction. The roof is to be built on the peripheral walls, and no supports are available in between. Two very common designs of roofs are arch/dome construction and flat suspended roof.

Arch is a very old and common form of design used in building construction, which is adopted in many types of furnace construction. For rectangular furnace, it is called arch or crown, and for a furnace with circular cross section, it is called dome. Figure 4.7a shows the different parameters of arch design.

In the arch design, the load of the arch is transmitted on the vertical wall through skew back brick (Fig. 4.6b). It has been found when the ratio of the rise/span = 1:8, the stability of the roof is best [1]. During the arch construction, a wooden former is used over which the bricks are laid, and after the brick laying is over, the wooden former is removed, and the brick structure forms an arch and remains stable of its own. The total load of the roof is transmitted to the peripheral wall through skewback

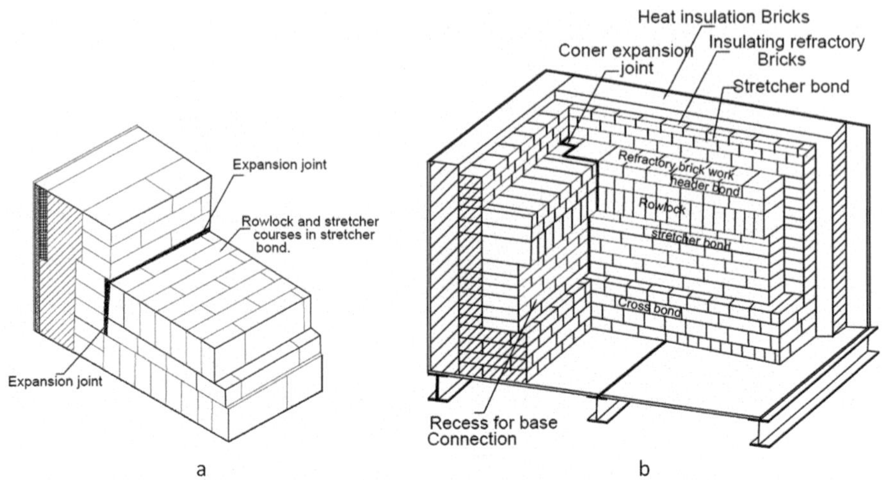

Fig. 4.6 Expansion joint, in (**a**) wall construction (**b**) at corner of two walls at 90 degree

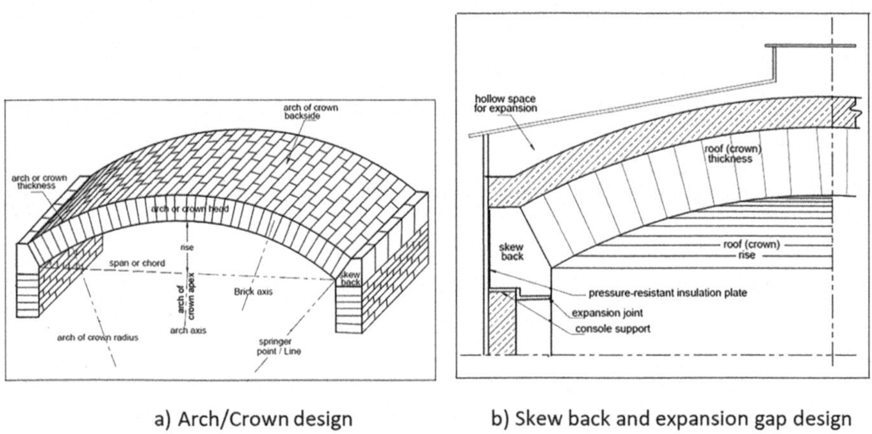

Fig. 4.7 (**a**) Arch/crown design. (**b**) Skew back and expansion gap design

bricks. During laying of refractory, care is to be taken to ensure that the brick axis must point to the arch center. The bricks used are normally side arch shape, and two types of bricks with two different tapers are used for lining. The interrelation between the parameters shown in Fig. 4.8 is given in Table 4.1 [1].

Suspended Roof Suspended roofs are constructed flat, and unlike the arch construction, the roof stability does not depend upon the brick thickness. The roof thickness can be anything in this construction. The roofs are constructed either by rider bricks with intermediate bricks or only of rider bricks (Fig. 4.9). Rider bricks in suspended roofs are fixed by metallic anchors to roof beams or cast rails. The roof beams or rails in turn are suspended from main roof beams. The rails and beams are

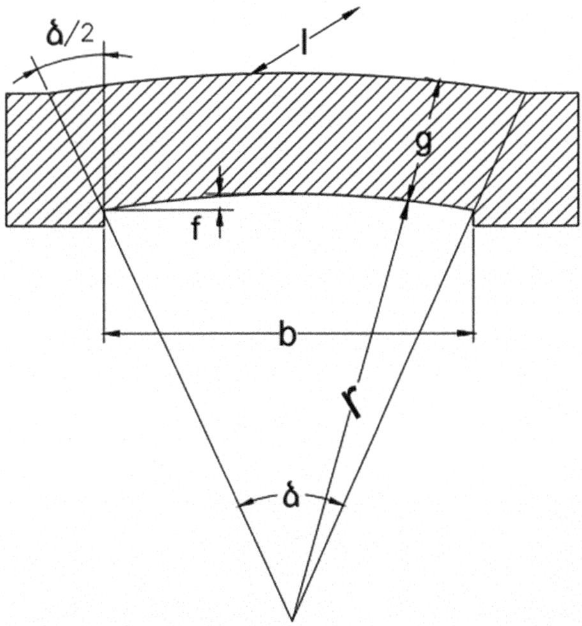

Fig. 4.8 Different parameters of arch construction
where b = span or width, f = crown rise, l = crown length, s = crow, r = radius, α = arch angle

Table 4.1 Interrelation of different parameters in an arch design

	Selected parameter		
Dependent parameter	f	r	α
f=	f	$r - \frac{1}{2}\sqrt{(4r^2 - b^2)}$	$\frac{b}{2}\tan\frac{\alpha}{4}$
r=	$\frac{b^2}{8f} + \frac{f}{2}$	r	$\frac{b}{2\sin\frac{\alpha}{2}}$
α=	$4\text{arc}\tan\frac{2f}{b}$	$2\text{arc}\sin\frac{b}{2r}$	α

arranged in such way that in service, the bricks get space to expand without any hindrance.

The roof can be constructed with castable also where the refractory anchor bricks are used to hold the suspended roof (Fig. 4.10).

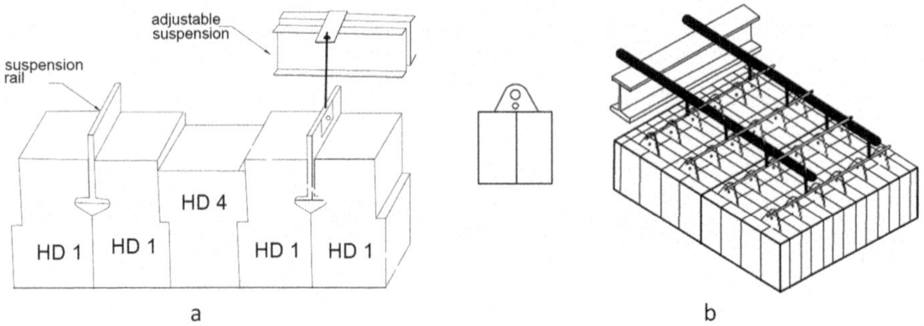

Fig. 4.9 Suspended roof brick shapes. (**a**) Rider and intermediate bricks combination. (**b**) Only rider bricks

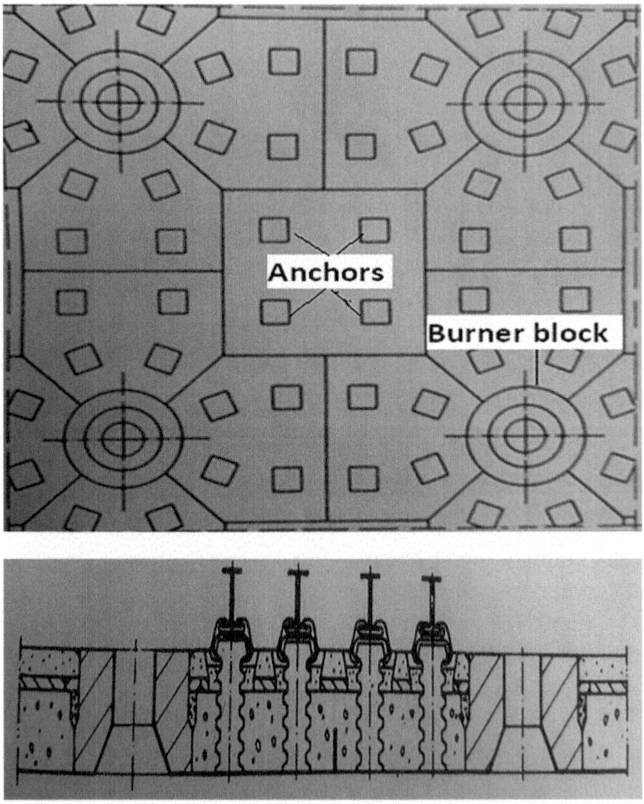

Fig. 4.10 Suspended castable roof supported by refractory anchors

4.5 Installation of Monolithic Refractories

Installation of monolithic refractories encompasses a large category of products, namely, conventional castables, low cement castables, gunning and shotcreting material, ramming masses, and mortars. Monolithic refractories are semifinished refractories, part of which is manufactured in the premises of supplier, and the final finishing is done at application site. The monolithic refractories are, mostly, a combination of coarse aggregates and fine powders, supplied as dry powder to the user. The loose powder is to be mixed with water or a specified liquid binder at site before application. Its consolidation and compaction, drying, and preheating ultimately influence its properties to a great extent. The application of monolithic products, of different categories, e.g., conventional castable, low cement castable, gunning, shotcreting, etc., is quite different. The accuracy of its installation largely influences the performance of these products. The mixing of castables must always be done in mechanical mixer, and the blades and scrappers must be checked that they are proper and not worn-out. The gap between the mixer bottom and paddles and scrappers should not be more than 5–6 mm. For the sound installation, the small tools and tackles are very important and must be arranged before the start of the job of casting. The water must be added by measuring properly the water quantity as per the recommendation of the supplier of the material. The vibrators must be checked that those work properly before the work starts.

The mixer must be made clean before the start of the job. The old hardened castable should not be adhered to mixer, scrapper, and sidewall because its presence in newly mixed castable alters its property and makes the setting faster. Before the start of the job, installer must know the workability and the setting characteristics of the castable.

4.5.1 Conventional Castables

Conventional castables are high cement containing castables. These products contain relatively higher percentage of high alumina cement and require higher percentage of water, compared to low cement castable. The mixing of castable with water, preferably, is to be done in a mechanical mixer to ensure proper homogeneous mixing. These castables can be installed by casting, pouring, or ramming. The water, to be mixed, depends upon the method of installation. Least percentage of water will be required in case of ramming. Conventional castables are also used largely for patch repairing of existing worn-out lining by troweling. During installation of fresh castable over an insulating castable as backup lining, the backup layer must be sufficiently wetted beforehand or a waterproof coating must be made on it, to prevent the backup layer from absorbing moisture from the freshly applied castable. Similarly in case of application of castable over an insulating board like calcium silicate, the board is to be quoted with a waterproof paint like bitumen, before installation of castable.

4.5.2 Low Cement/Ultralow Cement/No Cement Castables

These castables contain high alumina cement up to 6%. Those are to be mixed with water, only, in a high shear mechanical mixer. Without a high shear mixing, the desired flow property will not be developed in these castables, with low water percentage. The quality of mixer is very important in determining the water content required to get a desired level of flow. The castable mix can be installed only by vibration casting and the intensity of required vibration may vary depending on the flowability of the castable.

Hydration characteristics of Calcium Aluminate The setting and strength developed in castable is due to the hydration reaction of water with high alumina cement, which basically is calcium aluminate. When water is added to the castable, the calcium cement it first gets dissolved in water, and then different hydrated phases get separated from the mix. The hydrated phases are CAH_{10} ($CaO.Al_2O_3.10H_2O$), C_2AH_8, C_3AH_6, and AH_3 gel, and gibbsite which are recognized as the hydrates of calcium aluminate [2].

The formation of the calcium aluminate hydrates is influenced by the temperature during hydration. The hydrates formed at different temperatures are as follows:

Up to 20 C: CAH_{10}
20–35 C: C_2AH_8, C_3AH_6 and AH_3
More than 35C: C_3AH_6 and AH_3

Actual experience suggests that CAH_{10} is likely to be formed when the ambient temperature is low as in winter and C_2AH_8 and/or C_3AH_6 is formed during installations in summer. When dried at about 100C, the product is exclusively C_3AH_6. Properties of the hydrates are shown in Table 4.2.

The chemical reactions associated with the hydration of the calcium aluminate and the conversion of the hydrates are as follows:

$$CA + 10H \rightarrow CAH_{10} \tag{4.2}$$

$$2\,CAH_{10} \rightarrow C_2AH_8 + AH_3 \tag{4.3}$$

$$2\,CA + 11H \rightarrow C_2AH_8 + AH_3 \tag{4.4}$$

Table 4.2 Properties of different calcium aluminate hydrates

	Chemical compositions				
Properties hydrate	CaO	Al_2O_3	H_2O	Crystal system	Sp. Gravity
CAH_{10}	16.6	30.1	53.5	Hexagonal	1.72
C_2AH_8	31.3	28.4	40.3	Hexagonal	1.95
C_3AH_6	44.4	27.0	28.6	Cubic	2.52
AH_3	–	65.4	34.6	Hexagonal	2.42

$$3CAH_{10} \rightarrow C_3AH_6 + 2\,AH_3 + 18H \tag{4.5}$$

$$3\,CA + 12\,H \rightarrow C_3AH_6 + 2\,AH_3 \tag{4.6}$$

$$C_2AH_8 \rightarrow 2\,C_3AH_6 + AH_3 + 9H \tag{4.7}$$

The setting of high alumina cement is influenced by the temperature during hydration reaction (curing stage) and is longest around 30C. Setting time rapidly decreases above 30C (Fig. 4.11).

Curing temperature has also got a profound effect on the strength of the hardened cement. Figure 4.12 shows the relation between strength and the hydrates formed at two different curing temperatures 20 °C and 40 °C in a high purity (70% Al_2O_3) high alumina cement. It is clear from the figure that increased formation of CAH_{10} and C_2AH_8 increases the strength of the hydrated cement. Strength decreases as CAH_{10} content decreases due to its conversion to C_3AH_6 at higher curing temperature. C_3AH_6 or AH3 has higher sp. gravity than CAH_{10}, and therefore the porosity is created around the cement hydrate during conversion, and it causes the decrease in strength. Figure 4.13 shows how the strength gets reduced with degree of conversion. The degree of conversion can be expressed as

$$\text{Degree of conversion} = \left(\text{Weight of } AH_3\right) \times 100 / (\text{Weight of } AH_3 + \text{Weight of } CAH_{10}) \tag{4.8}$$

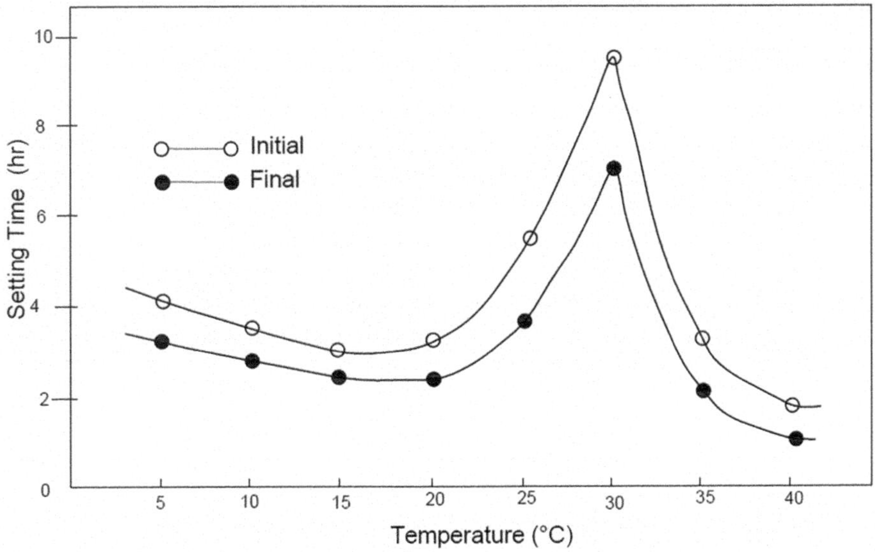

Fig. 4.11 Influence of temperature on the setting time on high alumina cement

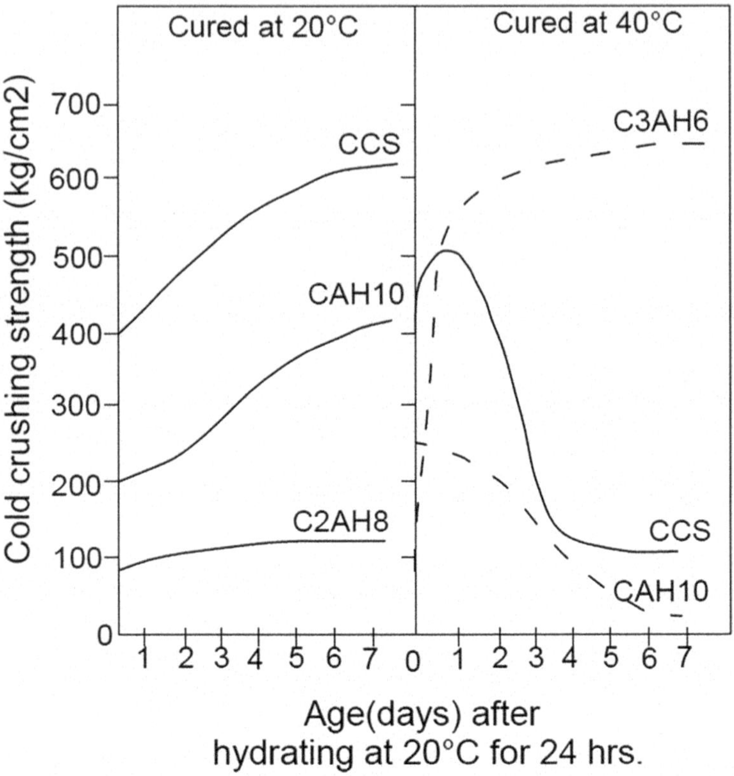

Fig. 4.12 Influence of curing temperature of the high alumina cement on its strength

Precautions to be taken during installation to get best performance are:

(i) Castables are to be stored in cool dry area. Castables are to be used on first in first out basis.
(ii) Mixers and tools must be clean and free of any old castable residues, cement, or foreign material, adhered to it, that may contaminate the mix and may affect the setting time and the ultimate strength of the castable.
(iii) Only clean drinkable water is to be used for mixing castables. The dissolved contamination in water may affect the flow and setting of the castable. In cold weather, warm water can be used to raise the temperature of the mix to 15–20 °C. In hot weather, the water or castable should be cooled, so the temperature of the mix does not exceed $30^{\circ}C$. Cold water can retard setting time and warm water accelerates the set. After placement the temperature of the casting should be maintained between 5 and 35 °C. To control the mix temperature, it is more important to control the temperature of castable powder rather than the temperature of the water which is evident from Table 4.3.

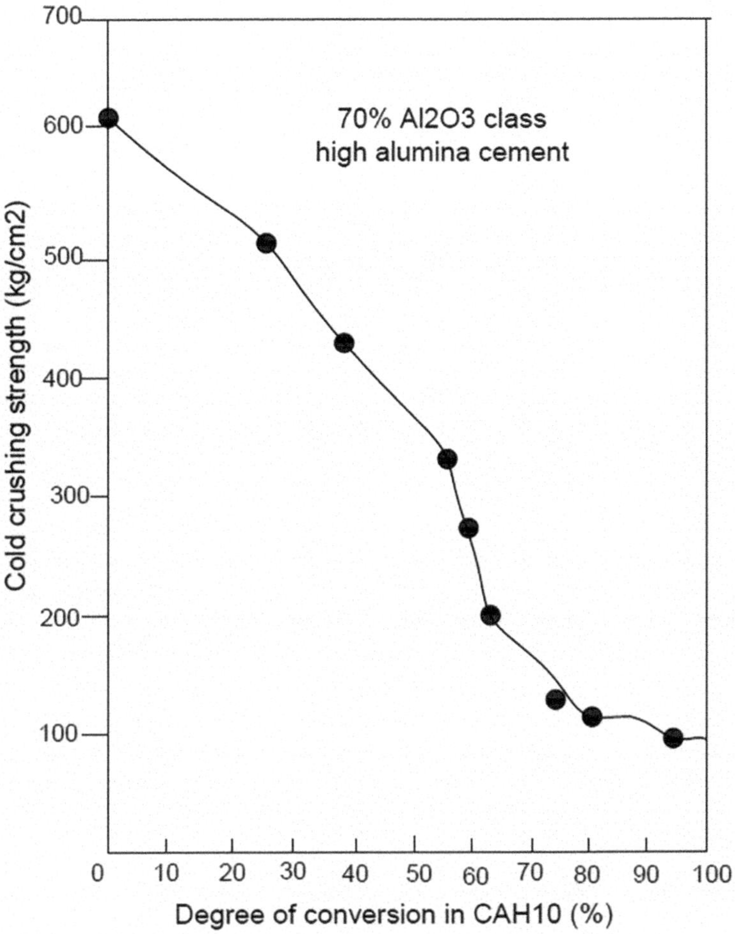

Fig. 4.13 Influence of degree of conversion on the castable strength

Table 4.3 Effect of castable and water temperature on the mix temperature

	Weight %		Temperature °C		
Castable type	Castable	Water	Castable	Water	Mix
Conventional	90	10	40	15	34
			22	35	25
Low cement	95	5	40	15	36.6
			22	35	19.2

Courtesy – Calderys Refractory Ltd., India

Mixing temperature also has a bearing on the working time of the mix which means the duration for which the castable water mix have sufficient flow property for placing and compaction. In actual installation, the higher is the working time for certain percentage of water, the better it is. Figure 4.14 shows the relation between the mix temperature and workability with varied water addition.

(iv) Use of excess water, than required, reduces the strength of the castable lining. The water quality is important factor to influence the flow behavior of the mix, the setting behavior, and the final strength of the cast body. The impact of water variation is more prominent in case of low cement castable compared to conventional castables with higher cement content (Fig. 4.15). The recommended quality of the water (Table 4.4) should always be used to get best results [2]. Castable should be mixed only long enough to produce an uniform mix (approximately 3–5 min). Mixing for longer time generates heat and speeds up the setting time.

When a castable has been mixed to the desired consistency, it is to be placed in the work area as rapidly as possible. The recommended consistency for cast installations is the ball and hand test. A ball of properly mixed castable, when tossed 6–12 inches, should adapt to the shape of the hand when caught, but should not flow through the fingers.

(v) The forms used for casting should be coated with oil or grease or liquid sodium silicate, to prevent moisture loss from the castable and to aid easy removal of the forms. A form vibrator or pencil vibrator of rating 10,000 to 12,000 vibrations per minute should be used to consolidate the castable mix. The vibration removes air pockets and air bubbles and increases the density of the castable.

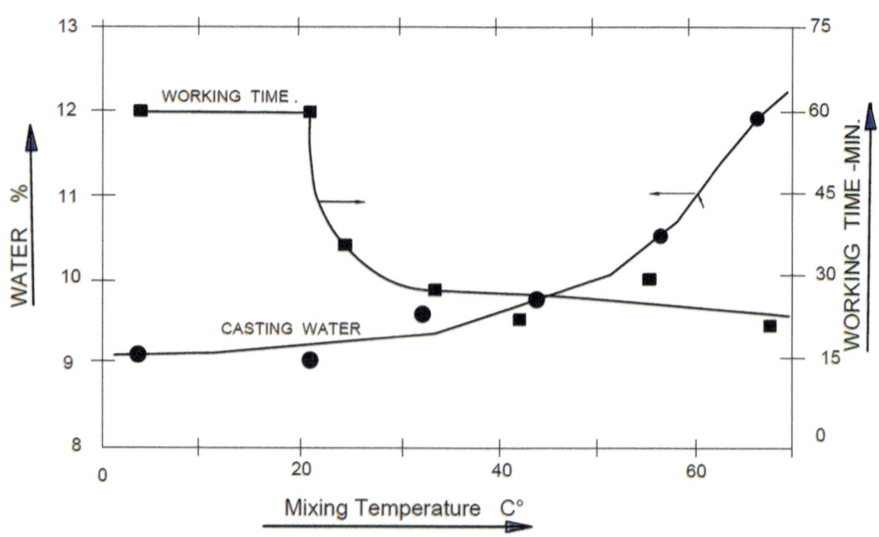

Fig. 4.14 Effect of mixing temperature on workability of the castable

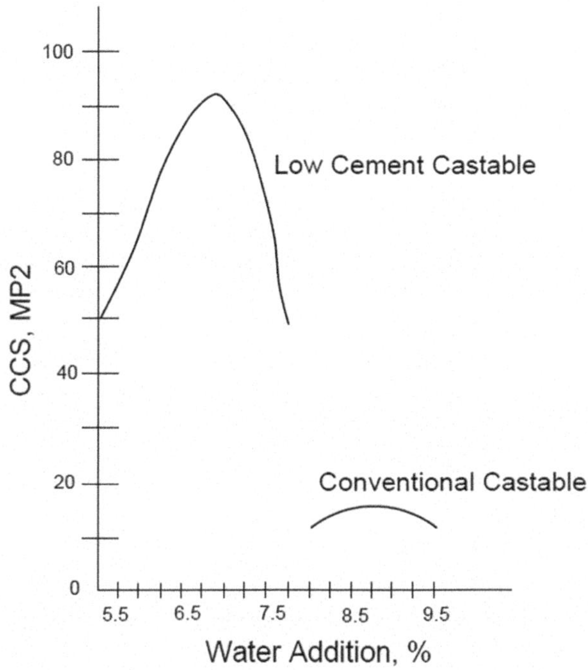

Fig. 4.15 Impact of water addition on the strength of castables

Table 4.4 Recommendation of water quality for casting

Factors	Tolerable limit
pH	6 - 8
Ca^{+2}	< 300 ppm
Mg^{+2}	< 300 ppm
Residue after evaporation	< 750 ppm
Cl^{-}	< 200 ppm
Fe^{+3}	< 500 ppm
SO_3^{-2}	< 200 ppm

Over vibration causes segregation of the castable mix and weakens the casting. The outer casing can be drilled with 2–3 mm diameter hoes (weep holes) to aid in dissipation of moisture during the drying and firing process.

(vi) After placement and vibrating, the castable should be cured for 18–24 h. All calcium aluminate-bonded castables generate heat while curing. They should be covered with polyethylene sheets, damp sacks, or sprayed with cold water periodically to prevent moisture loss and promote uniform curing. After the above procedure is completed, it is to be uncovered for air drying for 12–24 h. during the casting, curing, or drying process, the castable should never reach freezing temperatures. In addition, at no time during the curing process, should the casting be moved or shaken, vibration will interrupt the bonding process, thus greatly reducing the ultimate strength of the casting.

(vii) The performance of castable of any type depends upon the soundness of installation. Soundness of installation depends on the ease of placement, compaction, drying, and preheating. Different types of castables have different sensitivities towards the soundness of installation; some are very sensitive and some are forgiving type (Fig. 4.16). During selection of a castable for a particular application, one must take this point into consideration, and the selection of the type of castable should depend upon the probable installation quality. The castable type which is very sensitive to the installation soundness must not be suggested for an installation where proper facility, maybe in terms of machinery or trained manpower, is not available to carry out a sound installation.

4.5.3 Shotcreting

Shotcreting is most modern technique of installation of a castable, although the conventional castable recipe is to be modified to get the desired flow and setting property. Initially, this technique was developed and was largely used for the civil engineering construction of roadways and railways tunneling through mountainous regions. This method is, specially, useful for an installation of refractories, to be made at a height. The basic principle of the installation of this type of product is to

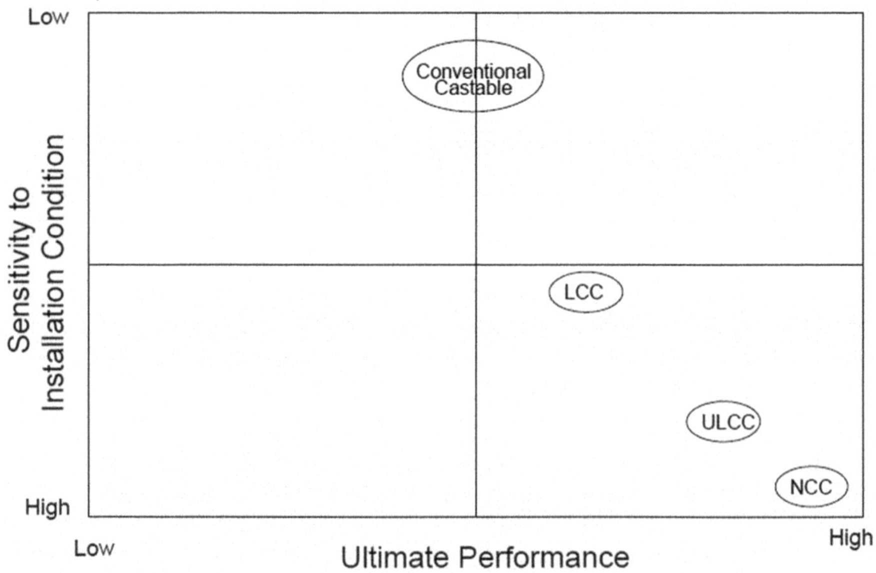

Fig. 4.16 Impact of soundness of installation on performance

mix it with water to obtain a pumpable slurry, which is pumped by high capacity pump at the desired height, where the installation is to be made. The slurry is conveyed through steel pipe and rubber hose connected at the end with a steel nozzle (Fig. 4.17). At the nozzle, high pressure air line is connected which sprays it to the target to be lined with castable. The nozzle design is very important part of the machine.

The nozzle diameter is generally 50 mm and the nozzle tip is made of special rubber. Shotcrete installation requires compressed air of sufficient quantity with sufficient pressure. A typical value is 300 kg per minute or 5–10 tons per hour at a pressure of 4.5–5.5 kg/cm^2 at an ejection velocity of 45–60 meter per second required for proper compaction [3–5].

Accelerator is introduced to the mixing ring by a high-pressure (7 kg/cm^2) low-volume dosing pump. The pump may be electric or air powered and typically capable of metering the additive at a rate of 100 Liter per hour. The additive type and amount may vary from product to product. The additive gels set the castable within few second [6, 7]. The common additives used are lime, sodium silicate, aluminum sulfate, or sodium aluminate dilute solution. The extent of dilution is important to maintain proper dispersion and atomization of the additive. Mixing of the castable below temperature 5 C may affect the gelling property of the castable and may need higher percent of additive. The adjustment of the dosage of accelerator is very important. The lower addition will cause slumping, and the higher addition will reduce working time and may form lamination. During the installation, the nozzle distance from the gunning surface is kept 2–4 ft.

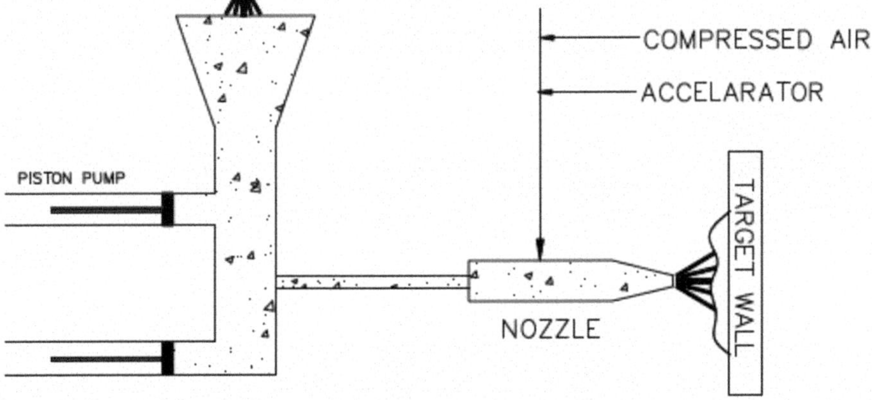

Fig. 4.17 Typical layout of different equipment used in shotcreting

The nozzle should be kept in constant circular motion during installation. The gunning should not be done on an area that has set for more than 5 min. If the flow of the castable gets interrupted, the flow of dispersant is to be stopped immediately.

The installation of castable by shotcreting on a vertical wall does not require any shuttering as is used in vibration casting or pouring of castable mixes. Thus it is a very fast and cost-effective method for installation and saves lot of manpower, although the machine is very expensive. This method is suitable for large installation only and is not suitable for small patch repair, etc. It is a convenient method to line in cyclone preheaters in cement plants, which are at a height, because transporting and storing refractory material to that height is very difficult and sometime unsafe.

4.5.4 Gunning

This method is used normally for small installation and cold or hot repairing. In this process the gunning material is poured dry in the hopper of the gunning machine, and the material is conveyed dry by the pneumatic pressure through a rubber hose to a nozzle. High-pressure water is added at the nozzle tip, where the water gets mixed with the castable before it gets sprayed on the desired target (Fig. 4.18). The gunning machine can be of batch type or continuous type machine. The air pressure needed is 4–5 Kg/cm^2 for a 100 ft. gun hose. For additional 50 ft. length, the pressure is to be increased by 0.5 Kg/cm^2.The water booster pump must be able to supply 20 liters of water per minute at a pressure of 7 kg/cm^2 [8, 9]. The ideal placement temperature is 15–35C. Predampening of the castable, before feeding the gunning machine, helps in better installation. Quality of the installed lining depends on the distance, angle of nozzle with the target, adjustment of water content, and finally the skill of the operator. Depending upon the machine capacity, it can place 1–100Kg material per minute at a nozzle velocity of 30 meter per second [10].

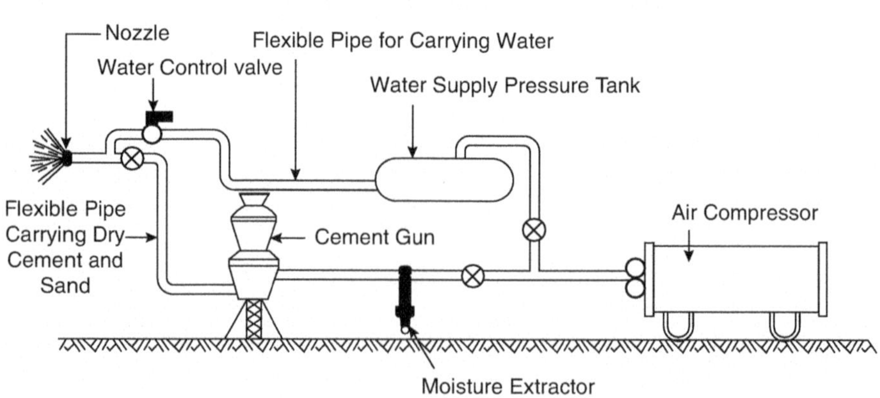

Fig. 4.18 Typical arrangement of the equipment for gunning installation of castable

4.5.5 *Pumping*

Pumping the castable mixed with water can be done for placement of large quantity of castables [11]. The equipment required are a paddle-type mixer with capacity if one ton. The high pressure piston pump is used to pump the castable slurry. The slurry is pumped through 65–75 mm diameter pipeline. A combination of hard pipe and rubber hose is used. Rubber hose should be used for the final section to distribute the slurry, because rubber hose significantly increases the resistance to flow.

The line diameter reduction should be done at the pump outlet with reducer section as long as possible to minimize the resistance to flow and pipe jam. The elbows used must have larger radii to reduce to minimize the resistance. Adequate personnel must be engaged to handle the hose filled with material, which is very heavy and weighs 40–70 Kg per 3 meter of 75 mm diameter pipe.

4.5.6 *Monitoring and Safety*

For job safety, installation efficiency, and quality of job, an audio system should be used for communication between the pump operator and placement crew. Prior to mixing and pumping the castable, the pipeline must be lubricated by pumping slurry of bentonite or any good plastic clay. The first batch of castable, to be pumped, is generally mixed slightly "wetter" than optimum. This ensures that it will pass through the pipeline with no leaks or restrictions. Once this is established, subsequent batches are mixed and pumped at the desired placement consistency, at normal pump pressure. Once pumping of castable mix begins, it is important that the pumping operation be as continuous as possible. Refractory castable, even those designed especially for pumping, can have a relatively short working time. Exposure of pipeline, conveying slurry, to the sun or hot ambient condition, can cause accelerated stiffening or setting. If a significant work interruption occurs or the pumping pressure increases (indicating the castable is stiffening in the line), the pipeline and pump hopper should be purged and cleaned.

4.5.7 *Joints*

Different kinds of joints need to be provided for the sound lining of castables. The joints are provided to take care of expansions or shrinkages of the refractory lining during its use.

4.5.7.1 Expansion Joints

Joints must be provided to accommodate the thermal expansion in monolithic refractories; those are not supported by anchors. The size of the joint is usually computed by multiplying the average temperature of inside and outside of the refractory lining with the coefficient of thermal expansion of the refractory and the dimensions. From this value, the shrinkage of the refractory due to drying and the PLC of the refractory at the operating temperature are subtracted. If the value is negative, no expansion joint is required.

Separation Joints

Monolithic refractory usually shrink during post installation drying or when cooled after heating. To protect the lining from damage due to shrinkage, joints are provided at appropriate intervals which can absorb the shrinkage. These joints are made by sectional installation of the lining or by inserting thin sheets of millboard or ceramic paper into the lining during installation. Cuts made halfway through the thickness of lining, called scoring, can also be used as a separation joint. The following methods are used to provide expansion joints:

(a) Castable refractory is poured in alternating segments between the expansion joints. After the castable sets, foam styrene is fixed to both joint surfaces of the lining, and castable water mixture is poured in remaining segments.
(b) Foam styrene of 3 mm thickness is fixed on veneer boards, and the boards are fixed firmly on the form at the required spacing so that it will not be displaced during installation. Castable is poured on both the sides of the partition to neutralize the pressure of the castable.

4.5.8 *Choice of Installation Method*

For a furnace or equipment construction if it is decided to use castable refractory, then the question arises that which method of installation will be the best. There are essentially four different methods by which the castable refractory can be installed, namely:

(a) Vibrocasting
(b) Gunning
(c) Shotcreting
(d) Pump casting

But the decision of the method to be used depends on the following factors:

- Location and site condition
- Environmental condition and equipment

- Volume required to be installed
- Ease and speed of installation
- Storage conditions
- Skill of the installation team
- Curing and drying requirements
- Budget

Refractories are not commodities, and the selection of proper installation method needs a careful evaluation of the detail project requirements. Any installation method may work, but the best will be the easiest and accurate one. The decision can be made considering each of the points mentioned before, e.g.,

Location and Site Condition The important points are:

- Accessibility of the location, where the installation is to be done inside the equipment
- Height of that place from ground level
- The obstacles on the way to reach that place
- The space around for the movement of people and equipment

If the place is at a height, then it will be easier to use shotcreting or pump casting where there is no need to carry the material at that height. In that case, the machine and the material can be kept at zero level.

Environmental Condition and Equipment The points important here are:

- How clean does the environment need to be, around the application area?
- What kinds of safety measures are required?

Volume of the Refractory Need to Be Installed The important considerations are:

- How much materials are to be installed in what time. If a large volume of castable is to be installed in a short time, then shotcreting or pump casing should be the choice. If a small quantity is to be installed, then casting or gunning should be better method.

Ease and Speed of Installation Some methods of installation are much faster, for example, shotcreting installation is much faster because it does not require any former to install as is required in case of casting. Therefore the time required for fixing of the formers, greasing it, and removal of the formers can be saved. A comparison of typical time requirement for installation of castable by various techniques [12, 13] is shown in Fig. 4.19.

Storage Condition If there is no storage space available around the area of installation to keep the mixer or required equipment, then instead of vibrocasting, shotcreting will be a better method, because the shotcreting equipment can be placed far off from installation point. The material to be installed, also, will be stored, naturally,

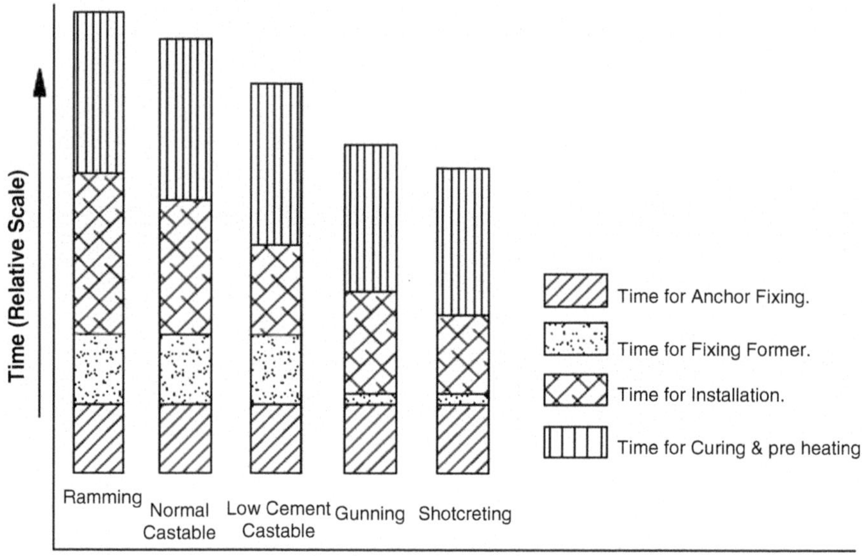

Fig. 4.19 Comparison of time consumption for different lining methods

near the shotcrete machine, eliminating the requirement of space around the installation area.

Skill of Installation Team Guniting, shotcreting, and pump casting require more trained and experienced crew members than that by vibrocasting.

Budget Many a times, though, the cost of the materials (considering equal properties) are the same, but the size of the project can make the difference because of wastage. In pumping or shotcreting application of 100 ton of castable, the wastage due to material in the hose pipes may not be significant but will be a very significant cost factor for installation of 5 ton material for which gunning is a better choice. The main cost differentiating factors are equipment cost and labor cost.

4.5.9 Anchor

For a complete monolithic lining of any equipment, anchor system is an integral part of refractory design and largely responsible for achieving desired campaign life of the lining. The anchor system includes the selection of anchor construction material, anchor design, anchor spacing, and the fixing of anchors. The anchors can have multitude of sizes and shapes.

4.5.9.1 Anchor Construction Material

Metal anchors are used where the service temperature does not exceed 1200°*C*. At high temperature, the mechanical strength of the metal decreases, and it is corroded fast. Different grades of alloy steels are used for making anchors, and the choice of material of construction depends upon the actual temperature the anchor is to withstand. In selecting the anchor material, the temperature of the tip of the anchor embedded in the lining must be taken into consideration. This can be predicted from the temperature profile of lining. Table 4.5 shows the steel grades and alloys, suitable for different temperatures [14–16].

Anchors are made normally of steel rod and plates. Above 1200°*C*, ceramic anchors are used, which are basically, refractory shapes of aluminosilicate quality, with Al_2O_3 content varying from 30% to 80% depending upon the application temperature. During selection of the anchors, the temperature profile across the lining is a very important factor to be taken into consideration [17].

Higher operating temperature can lead to carbide precipitation in the metal that changes the original properties of the metal, causing rapid oxidation and premature failure of the anchor and lining. The atmosphere prevailing in the furnace can change the temperature profile in the lining. It has become more important today because of more and more use of alternate fuels. The high nickel content (without any special additive in the alloy) makes the steel highly sensitive to sulfidation [18].

Reducing sulfurous and nitriding environment can severely affect the metallic anchor, and care must be taken in the choice of metal alloy. This led to the development of new alloys that contained especially added components. Additional corrosion resistance is provided by the rare earth elements (e.g., cerium, lanthanum). In the environment of a cement plant, the first micro alloyed commercialized steel grade 253MA is found to work better than SS 304 and SS 309. The composition of 253MA is given in Table 4.6 [14, 15].

Anchor Shape and Size Design Flat anchors, made from steel plates, are cheap, simple to manufacture and readily available. Therefore these anchors are often used, but those have some major disadvantages. The surface area of flat anchor is more, compared to rod anchor with similar mechanical strength. Because of higher surface area, corrosion rate is higher in flat anchors [15]. The steel rod anchors are mostly in the shape of "V" with the angle 60 degree. The arms of the "V" are sometime twisted to provide an extra surface area and better holding. The arm lengths of the

Table 4.5 Recommendation of steel grades of anchor at various temperatures

Maximum service temperature (°C)	Steel grades
450	Carbon steel
900	304 SS
1000	309 SS
1100	310 SS
1150	253 MA
1200	Inconel 601

Table 4.6 Composition of alloy 253MA

Element	C	Mn	P	Cr	Ni	N	Ce
%	<0.03	<0.8	<0.04	21.0	11.0	0.17	0.08

Table 4.7 Recommendation of steel rod diameters of anchors for different applications

Roof	Rotary kiln	Static parts	Burner
8–10 mm	8–10 mm	6–8 mm	4–5 mm

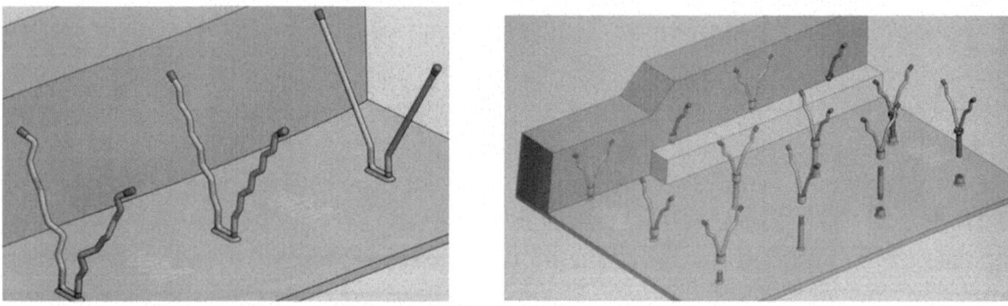

Fig. 4.20 Different designs of metal anchors

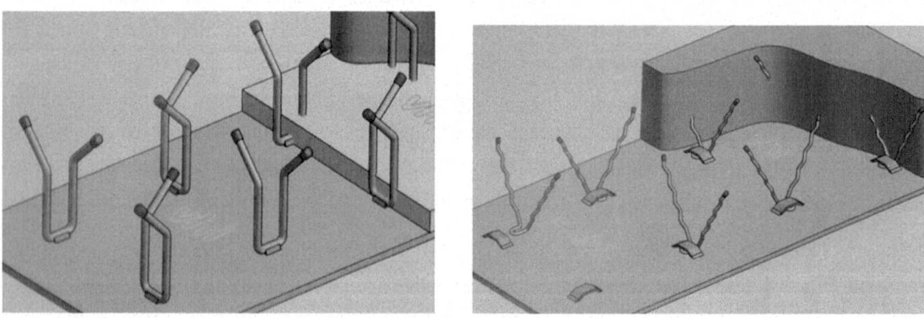

Fig. 4.21 Different designs of metal anchors

"V" are normally 25 mm less than the lining thickness. Sometimes the arms of the "V" are of different lengths to prevent spalling in layers. The selection criteria of diameter of the anchors are shown in Table 4.7.

The said anchors are suitable for lining up to 230 mm thickness. Different designs of anchors are shown in Figs. 4.20 and 4.21. During use at high temperature, the metallic anchors will expand, and if the space is not provided for expansion, the lining will develop crack. The plastic caps are put on the tips of the "V" arms, and the bitumen paints are brushed on the anchor rods. At high temperature, these will burn out, leaving a space for the expansion of the anchors.

For dense monolithic linings with thick cross sections (greater than 230 mm), use of pre-fired refractory anchor is the preferred method of anchoring the structure.

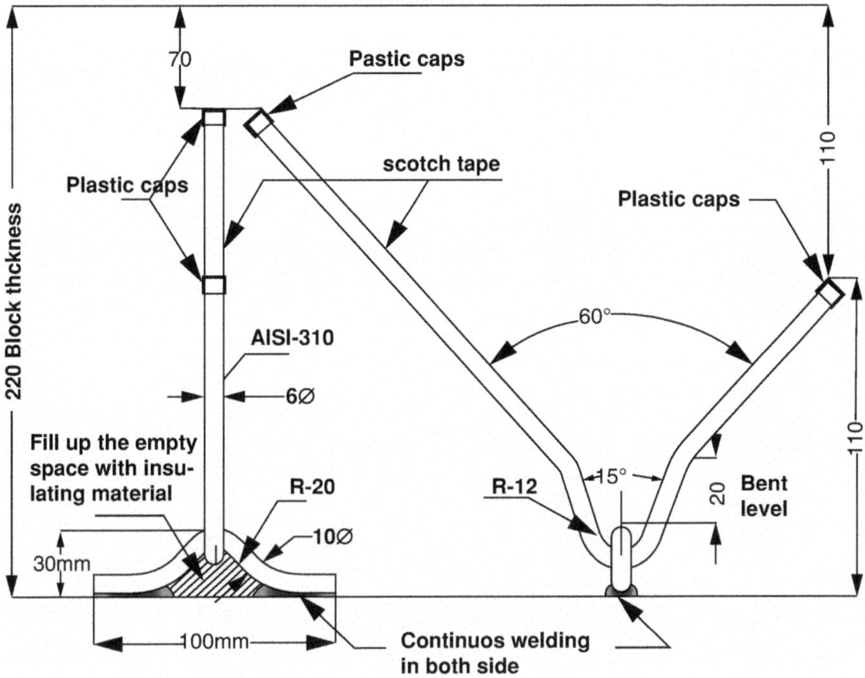

Fig. 4.22 Design of floating anchor

Ceramic anchors have several advantages over other types of anchoring systems. They have more holding power than metal anchors due to their design and greater surface area. They also extend through to the hot face providing extra retention of the lining. Also being ceramic material, those can withstand much greater temperatures and tougher atmospheric conditions than standard metallic anchors.

Fixing of Metal Anchors

Quality of fixing of anchors on metal shell of the furnace or equipment determines the soundness of installation. The whole load of monolithic refractory is transferred on the anchors. The anchors must be able to hold that load for long duration at a high temperature. Therefore, the welding joints of the anchors on metal shell of the furnace must be very sound, and it is to be tested by hammering the anchors after welding.

In some design, the anchors are not fixed rigidly on the metal shell but are fixed through a support (Fig. 4.22) so that it remains free floating and can have movement in three dimensions to take care of any movement of refractory lining at high temperature. The anchor support should be in the shape of inverted U with tips bent 90 degree and welded upon the metal shell.

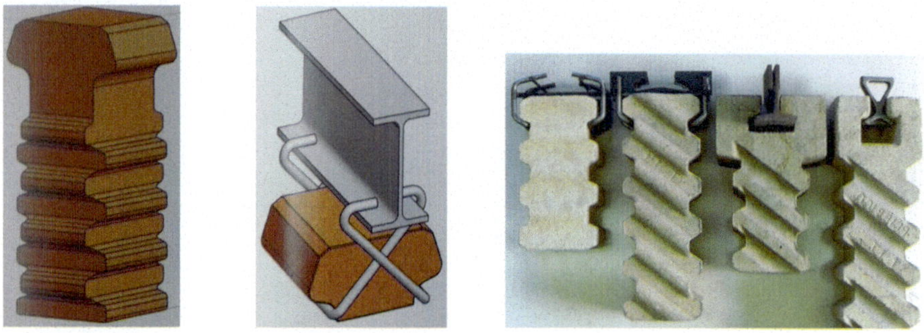

A B C

Fig. 4.23 Design of some ceramic anchors and their holding arrangement

The support diameter should be 2 mm more than the anchor diameter. V anchor should be positioned at the middle of the support and just tack welded to the support. The welding should not be rigid. It should be such that welding can be broken by hand. After fixing the anchor arms should be coated with bitumen paints or plastic tape to create provision for expansion.

Fixing of Ceramic Anchors

Ceramic anchors cannot be welded like metal anchors and are fixed to the vertical metal shell or roof of the furnace or equipment through metal holders of different designs. The metal holders and clips must be able to withstand the load of the lining. Some of the designs of the holders and anchor bricks are shown in Fig. 4.23. The metal holders hang from the steel I-section beam as shown in Fig. 4.23B. In case of fixing ceramic anchors on vertical wall, C-clips are mostly used (Fig. 4.23C).

4.5.10 Anchor Spacing

The spacing and orientation of the anchors are also very important. The following factors determine the spacing between anchors:

1. Strength of anchor
2. Strength of refractory
3. Thickness of the castable
4. Furnace operating temperature
5. Position of the lining (vertical, horizontal, inclined)
6. Lining shape
7. Existence of pressure

8. Method of installation
9. Mechanical movement of the furnace or related equipment and the external force on furnace lining

The spacings between the anchors can be determined in two different ways:

1. Considering the tensile strength of the support or anchor
2. Considering the strength of the refractory

4.5.10.1 Support Spacing Based on Strength of Anchors

Spacing between the anchors can be calculated on the basis of the strength of anchor at the operating temperature and the weight of the refractory that are supported by the anchors.

1. Horizontal lining (ceiling) in single layer

If t_1 = the ceiling thickness, l_a = the spacing between the anchors, d_1 = the bulk density of the refractory lining, a = the area of cross section of the support, and σ_m = the permissible tensile stress of the support material at operating temperature, then

$$l_a \leq \sqrt{\frac{a.\sigma_m}{t_1.\gamma_1}} \tag{4.9}$$

2. Support spacing in a three layer refractory lining

If the first, second, and third layer thicknesses are t_1, t_2, and t_3 and their bulk densities are d_1, d_2, and d_3, then

$$l_a \leq \sqrt{\frac{a.\sigma_m}{\left(t_1.\gamma_1 + t_{2.}.\gamma_2 + t_3.\gamma_3\right)}} \tag{4.10}$$

4.5.10.2 Support Spacing Based on the Strength of the Castable

1. Horizontal lining (Ceiling)

If t_1 = the lining thickness, l_r = support spacing, γ_m = the bulk density of the refractory lining, and σ_r = bending strength of the refractory at the operating temperature, then the permissible distance between the supports are

$$l_r = \sqrt{\frac{\sigma_r.t_1}{3\gamma_1}} \tag{4.11}$$

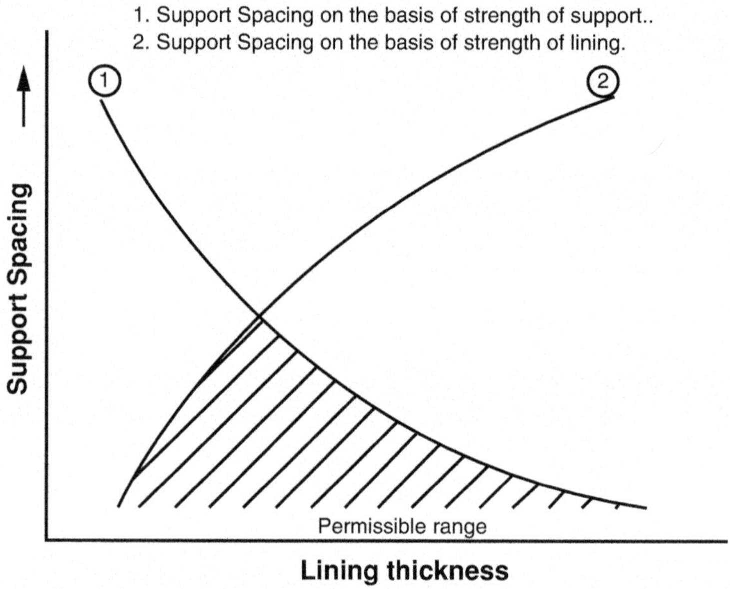

Fig. 4.24 Relation between lining thickness and support spacing

2. Multilayer lining

Let us consider t_1, t_2, and t_3 are the lining thicknesses of the refractories in the first, second, and third layer. Their bulk densities are γ_1, γ_2, and γ_3, respectively. Here the layers 1, 2, and 3 are considered from hot face to cold face. Normally the refractory layers 1 and 2, used next to shell, are insulating type and having very low strength. Therefore the strength of the dense refractory layer is to be considered for calculation purpose, which is σ_r, then

$$l_r \leq \sqrt{\frac{\sigma_r . t_1^2}{3(t_1.\gamma_1 + t_2.\gamma_2 + t_3.\gamma_3)}} \quad (4.12)$$

The support spacing calculated separately based on the support strength and the anchor strength are related to each other as shown in Fig. 4.24 [2, 19]. Table 4.8 gives the general norm of selection of anchors and spacing.

The anchors positions are kept in a regular manner, and the orientation of the anchors are made in such a way that two adjacent anchors in one row forms an equilateral triangle with the nearest anchor in the other row (Fig. 4.25).

Table 4.8 Suggested anchor spacing

		Anchor types	
		Wire and rod	Ceramic anchor
Location	Lining thickness (mm)	Suggested anchor centers (mm)	Suggested anchor centers
Wall and slopes	50–100	150	–
	100–200	225	–
	200–300	300	450
	300–400	–	450
	>400	–	900
Roofs and bullnose	50–100	150	–
	>200	225	300
Floors	50–100	300	–
	100–225	450	–
	>225	500	–

4.6 Drying and Preheating

This is one of the most important steps in refractory installation. The part of the water used during installation of castable remains within the lining. In brick lining, also some physically absorbed moisture remains, and the extra moisture comes from the mortar used. The internal moisture inside the lining will get evaporated and removed during the operation. But the main idea of preheating is to remove the physically and chemically bound moisture in the castable or in the brick lining slowly and in a controlled manner. Fast removal of water from the system may develop crack in the refractory lining and may cause explosion in castable lining. Besides the moisture removal, the other major function of preheating is to achieve a thermal equilibrium to stabilize the whole system, including refractory lining. In case of rotary kiln, especially, because it is a moving structure, the bricks, of the lining, rearrange themselves during the preheating period accompanied by its movement, and the lining attains stability. In a large industry like cement plant, where several thousand tons of refractories are used to line different equipment, this thermal equilibrium is to be attained slowly. The plant runs smoothly once it is established. The hot air generated for drying purpose must have sufficient volume to create turbulent flow for more efficient heat transfer. The additional burners very often are required to be installed in kiln inlet or preheaters to increase drying efficiency. Refractory lining, very often, consists of various qualities with different drying characteristics. During the drying of such lining, the heating curve, for the material with lowest permissible heating up curve, is to be chosen. It is widely accepted that truly effective means of drying out refractories is to deploy some form of convective heat source, passing large volume of hot gases over the refractory surfaces.

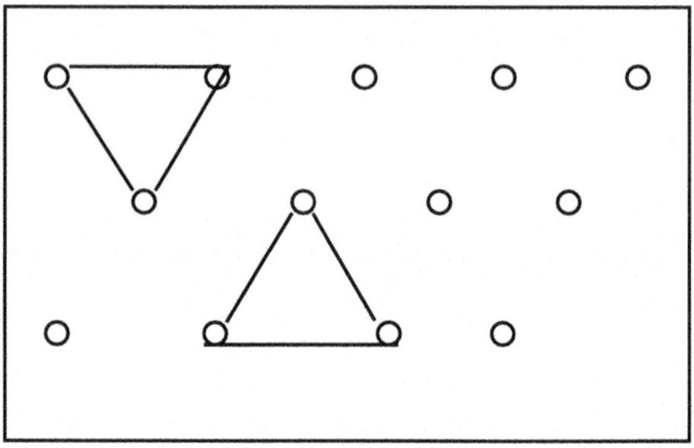

Fig. 4.25 Anchors orientation

This creates a "scrubbing" action on the object under drying, which hastens the heat transfer process through the refractory thickness. It ensures that both physically and chemically bonded moisture is rapidly and safely driven off.

The burners, with very high turndown ratio (100:1), are used enabling temperatures to be accurately controlled at all stages from 80 to 1000 °C. The burners with fully adjustable output and generating up to 2.5 million kilocalories per hour through hot gases, when connected to its combustion air fan, are used. These hot gases are emitted at an outlet velocity of some 150 meters per second [20]. This creates a turbulent atmosphere, promoting highly effective scrubbing action necessary to drive off moisture rapidly. The quick conversion of residual moisture to steam may cause explosive spalling, if the lining material is very dense and impermeable.

Besides the drying, slow preheating is a requirement to stabilize the refractory lining. Refractory lining always experiences more abuses on its hot face rather than cold face. When preheating starts, the hot face receives the heat, first. Heat is then conducted through refractory lining from the hot face to the cold face.

When preheating starts, temperature of refractory rises much faster than the steel shell. As temperature rises, refractory tries to expand, but the steel shell, behind, not yet heated up, resists the expansion of the refractory lining. Under such condition, refractory at the hot face will experience a high compressive stress. Refractories readjust themselves to mitigate the causes responsible for development of such stresses.

In a similar way, during shutdown, when the heat source is put off, the furnace is on cooling, and the refractory hot face starts losing temperature fast, by radiation and convection of heat. As the refractory hot face becoming cool, it will try to contract but the refractory face in contact with steel shell, still hotter, will not allow it to contract, and thus the hot face will experience a strong tensile stress. Stresses developed inside refractory lining, which appear as a result of rapid temperature changes, can be calculated [21–23].

$$\text{Hooke's law states that } \sigma / \varepsilon = E \tag{4.13}$$

where σ = stress, ε = strain, and E = modulus of elasticity.

$$\text{On the other hand, } \varepsilon = \Delta l / L \tag{4.14}$$

where Δl = the increase in length of refractory due to thermal expansion and L = original length of refractory.

$$\text{If } \alpha \text{ is the Thermal Expansion Coefficient of the Refractory then, } \Delta l = L.\alpha.\Delta T \tag{4.15}$$

where ΔT is increase in temperature of the refractory, which changes the length by Δl. From Eq. 6.8 and Eq. 6.9

$$\varepsilon = \alpha.\Delta T \tag{4.16}$$

The thermal stress can be calculated by putting the value of ε in Eq. 6.7

$$\sigma = \alpha.\Delta T.E \tag{4.17}$$

However, a slow temperature change does not give an effect of thermal stress growth, because of stress relaxation effect [22–24]. The stress disappears with the course of the time according to the equation

$$\sigma = \sigma_0.e^{tE/K} \tag{4.18}$$

where σ = the stress in the material, σ_0 = the stress after time of starting of relaxation process, E = modulus of elasticity, and K = a variable which characterizes the velocity of relaxation. Value of K differs from 10^{15} at room temperature to 10^{7-9} at 1550–1700 °C and down to $10^2 - 10^3$ at softening temperature [23].

4.6.1 *Preheating of Refractories*

Refractory lining when installed contains huge quantity of moisture both as physical form and in chemically bonded. The shaped refractories by virtue of its porosity absorb moisture from atmosphere, and normally the bricks are joined by mortars which contain 25–35% of moisture. The castables contain much larger quantity of moisture when installed. The reactor or furnace needs to be heated inside to remove slowly that moisture present in the refractory lining, and the process is called preheating of the furnace or reactor. The preheating is a very important part of the installation, and the successful performance of a refractory lining depends to a large extent on the quality of preheating. The larger is the size of the reactor and the larger is the thickness of the refractory lining, the more critical is the process of removal of water from the system.

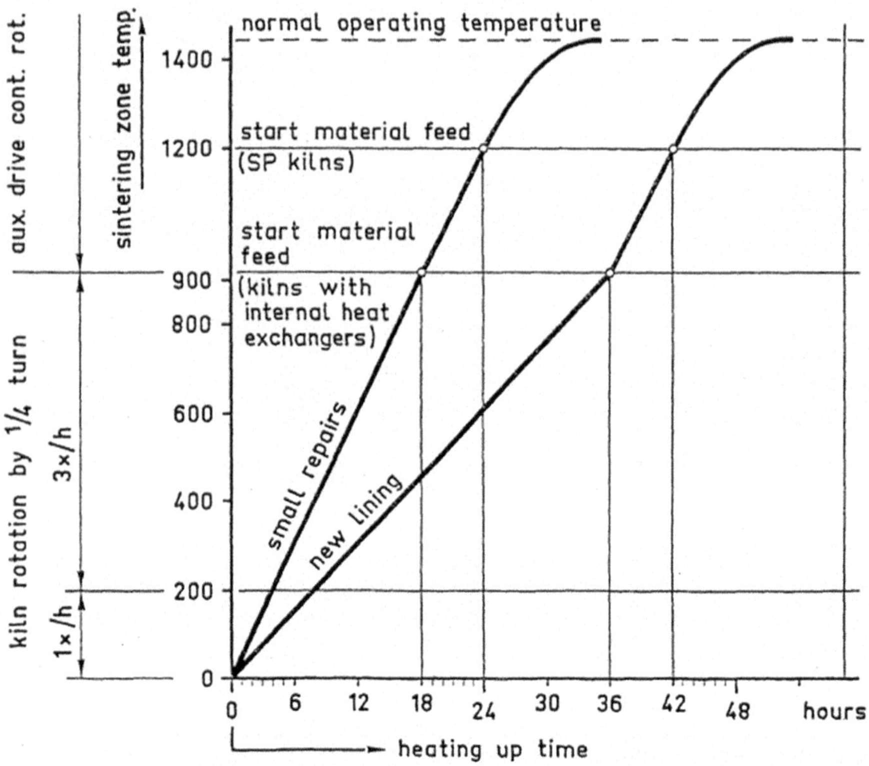

Fig. 4.26 Heat up schedule of rotary kiln refractory lining

The second important aspect of the preheating is that a mechanical stress is generated in the refractory lining when it starts getting heat on one surface of the refractory, which is the working side of the lining. Refractory materials are normally bad conductor of heat, and thermal diffusivity is also low. When the refractory lining is heated on one side, the heat propagates slowly inside, and there remains a considerable difference of temperature between hot face and cold face of the lining. As a result the thermal expansion of hot face of the lining will be more than that of the cold face. The dissimilar thermal expansion in the same body causes the generation of mechanical stresses and that can damage the refractory lining. The generation of the stresses must be slow so that the refractory lining can have movement inside to adjust it and mitigate its effect. The process is called stress relaxation. The larger is the installation, the more is the stress development due to preheating process and the more critical is the process of preheating. It must be slow enough so that the refractory lining can find time to adjust the stress. The slow preheating of the brick lining is done for stress relaxation at high temperature.

The magnitude of the stress generated depends on the rate of preheating, thermal expansion, and the thermal diffusivity of the refractory.

Bricks – For example, preheating of the cement rotary kiln is done following heating curve shown in Fig. 4.26 [25]. In case of complete new lining, the time temperature schedule will be longer than that in the case where a repairing or partial lining is done. In case of high alumina refractory lining, the time required for heat up is lesser because of lesser thermal mass. The preheating can be done faster also, because, lower thermal expansion coefficient of high alumina refractories, develops lesser thermo-mechanical stresses. Moreover the stress relaxation starts at lower temperature in case of high alumina refractories. In case of basic refractories, because of higher thermal expansion coefficient and onset of stress relaxation at higher temperature, it needs to be preheated at slower rate.

Too fast heating of the refractory lining can cause high compressive stress, on the working surface of the brick because the expansion occurs on the surface only when the brick other end on steel shell is still at much lower temperature. This can be identified by typical convex surface of the brick after cracking. Figure 4.27 shows a high alumina brick lining in upper transition zone cracks due to fast preheating [26].

The fuel used for preheating can be either gas or liquid fuel. Basic refractory lining is sensitive to reaction with water vapor, generated from the evaporation of moisture in the lining, as well as the water from the combustion product of fuel (natural gas or oil), especially in the temperature range 60–295 °C. This is a critical temperature zone, where the MgO in the basic bricks tends to react with the water vapor to form $Mg(OH)_2$. The reaction $MgO + H_2O = Mg(OH)_2$ is associated with a volume expansion and causes crack in the lining. This phenomenon can happen when the water vapor condenses on the refractories. In practical application, when natural gas and oil are used for preheating, the care is to be taken to pass the critical temperature range 60–295°*C* as quickly as possible to prevent hydration reaction of basic lining to occur [27]. The reaction tendency decreases with increasing temperature.

4.6.2 Preheating of Castable

The main purpose of preheating of the castable lining is to remove the water from the system. Castable sets and gets hard because of the formation of different hydrated compounds of calcium aluminates. These hydrates, e.g., $CaO.Al_2O_3.10H_2O$ (CAH_{10}), CAH_6, C_3AH_6, and AH_3, decompose when heated. The decomposition temperatures of these hydrates are different. The water, gets released during the thermal decomposition of the hydrates, is to be removed slowly. The fast preheating of the castable lining may lead to sudden formation of steam inside and can generate a very high steam pressure (Table 4.9) [1]. This may cause crack, even explosive bursting of the lining. This is more relevant and important for low cement castables, which, in general, have low permeability to hinder the passage of the liberated water vapor on heating through the refractory body. The water remains in castable both as physically bonded and chemically bonded water. The physical water gets removed within 200 °C, but the chemically bonded water gets dissociated at a temperature of

Fig. 4.27 Convex spalling of high alumina refractories at upper transition zone

about 550 °C. The higher is the content of cement in the castable, the more is the amount of chemically bonded water. The relationship between the heat up rate of a castable lining and the percentage of the elimination of moisture depends upon the lining thickness. The addition of organic fibers in castables helps a lot for easy removal of water vapor formed in preheating. Whatever is the thickness of the lining, it is the surface of the lining that is immediately affected by the initial heating. The castables generally have low thermal conductivity, and therefore the lining cold face is protected by the thickness of the lining.

Experiment shows that a thickness of 100 mm from the heating surface gets affected by the initial heating up. Too slow rise of temperature is safe but uneconomic, and too fast rise of temperature may cause explosion and crack in the lining to damage it. In the most desirable heat up schedule, the time vs temperature and the time vs moisture removal curve should go very close. The type of castable refractory has

Table 4.9 Temperature vs saturated steam pressure

Temperature °C	Saturated steam pressure in bar
20	0.02
50	0.12
100	1.01
150	4.76
200	15.55
250	39.77
300	85.92
350	165.37

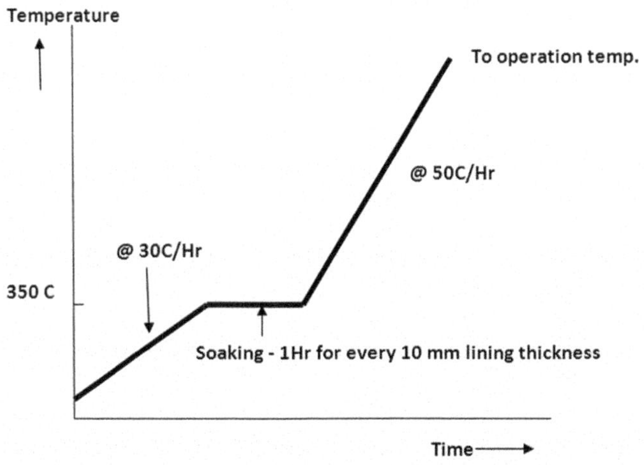

Fig. 4.28 Preheating curve of castable lining

got a bearing on the temperature rise and the moisture removal percentage. The heat up curve shown in Fig. 4.28 is safe and efficient and is commonly applicable to different types of castables [2].

The castable lining, after the shuttering is removed, should be allowed to get air-dried at ambient temperature for 24 h before the starting of preheating by hot air.

In case of reformer operating at a very high pressure of about 40 bar, if the water is not removed properly during preheating and the reformer starts operation, the moisture inside cannot come out. But in case of uncontrolled operation or by any means the pressure falls, the water will be released suddenly with big volume expansion, and there will be an explosion [28].

4.7 Pressurization and Depressurization

In case of the reactor operated at very high pressure, e.g., secondary reformer operated at about 40 bar pressure, the pressurization during start-up and depressurization during shutdown, are to be done slowly and carefully for the stabilization of the refractory lining, which is basically inelastic in nature and having lower gas permeability. Refractory lining with more than one layer will take more time in pressure equalization. The time requirement for equalization of pressure is a function of the permeability of the dense refractory layer only. In case of fast depressurization, the gas trapped in between the dense and insulation layer cannot escape that fast, and the builtup pressure in between the layers can cause substantial damage to the lining [28].

4.8 Carbonation

If the refractory lining, especially made of the conventional or insulating castables, is left without preheating for long time, it starts deteriorating because of efflorescence and carbonation reactions.

Soluble oxides, in castable, from used raw material or water, leading to diffusion of Na ions in the solution, reacts with aerial CO_2 to form sodium carbonate. This sodium carbonate reacts with calcium aluminate hydrates of the castable to form $CaCO_3$ as shown in reactions below:

$$2CO_2 + 2NaOH + CAH_{10} = Na_2CO_3 + Al_2O_3.xH_2O + CaCO_3 + (11 - x)H_2O \tag{4.19}$$

$$3CO_2 + 2NaOH + C_2AH_8 = Na_2CO_3 + Al_2O_3.xH_2O + 2CaCO_3 + (9 - x)H_2O \tag{4.20}$$

$$4CO_2 + 2NaOH + C_3AH_6 = Na_2CO_3 + Al_2O_3.xH_2O + 3CaCO_3 + (7 - x)H_2O \tag{4.21}$$

Although the equation shows sodium carbonate as anhydrous material, it remains associated with 1,7 or 10 molecule of water of crystallization which varies in terms of density (Table 4.10) [29].

Depending on the quantity and occurrence, it can destroy the castable structure because of volume expansion as evident from Table 6.6. Formation of $Na_2CO_3.H_2O$ can cause surface efflorescence (Fig. 4.29) without affecting the structure but with loss of strength of the castable.

In absence of alkali, the aerial carbon dioxide reacts with calcium aluminate hydrates to form calcium carbonate and alumina hydrates, as per the reactions below [30]:

$$CAH_{10} + CO_2 + xH_2O \rightarrow CaCO_3 + Al_2O_3.yH_2O + (10 + x - y)H_2O \tag{4.22}$$

$$C_2AH_8 + 2CO_2 + xH_2O \rightarrow 2CaCO_3 + Al_2O_3.yH_2O + (8 + x - y)H_2O \tag{4.23}$$

$$C_3AH_6 + 3CO_2 + xH_2O \rightarrow 3CaCO_3 + Al_2O_3.yH_2O + (6 + x - y)H_2O \tag{4.24}$$

Table 4.10 Specific gravity of sodium carbonates with different molecules of hydration

Reaction product	Sp. gravity	Effect
$Na_2CO_3.\ H_2O$	2.26	Surface efflorescence
$Na_2CO_3.\ 7H_2O$	1.51	Can cause cracking and bursting of castable
$Na_2CO_3.\ 10H_2O$	1.44	-do-

Fig. 4.29 Efflorescence, on casting surface

The rate and extent of the reaction depends upon the type of calcium aluminate hydrate with the order $CAH_{10} > C_2AH_8 > C_3AH_6$.

The practical observations of refractory castable reveal the following tendencies:

(i) Carbonation occurs easily in lightweight castables, but hardly at all in dense castables.
(ii) Carbonation occurs easily in not dried products and hardly at all in dried castables.
(iii) Carbonation occurs easily in an environment with high temperature and humidity.
(iv) When not dried products are exposed to rain, the carbonation rate is highly accelerated.

Therefore the lining made with conventional or insulating castables should not be allowed to remain without drying and preheating for long time, especially where the humidity and temperatures are high. The drying and preheating of the castable makes it stable, and it does not deteriorate further.

4.9 Insulation Refractory

Selection of the insulation refractory is very important to decide the economy of a process. It involves both the selection of quality of insulation refractory and its thickness. A very careful approach should be made to arrive at the right design of insulation refractory. Theoretically, the shell temperature can be maintained at very close to ambient temperature. But the cost involved is important. It is not only the thickness of the insulation material, which is important, when the insulation thickness increases, the furnace or reactor or duct's outer dimension increases. The increase in dimension increases the cost of steel; moreover when the dimension increases, the structural load increases along with that cost. In many equipment, if the steel shell temperature goes below a certain value, the volatile matter in the

Fig. 4.30 Flue gas dew point corrosion of SRU reactor steel shell

combustion gases, e.g. water, starts condensing, and it dissolves acidic gases like SO_2 and corrodes the steel shell (Fig. 4.30) [31]. The thickness of the insulation must be optimized to balance the cost through fuel saving and the increase of cost due to insulation and its depreciation cost as discussed (Fig. 4.31) [32]. To determine the optimum thickness of the insulation, the following considerations must be taken into account:

1. Cost and calorific value of the fuel.
2. Cost of the insulation refractory.
3. Surface temperature targeted.
4. The thickness and quality of the working lining.
5. Duration of operation on annual basis.
6. Ambient temperature.
7. Surrounding wind velocity.
8. Dew point of the combustion gases.

4.9.1 Installation of Ceramic Fiber Lining

Ceramic fiber lining is used widely in heaters in petrochemical industry. The ceramic fiber veneer and modules are used to line those furnaces.

Ceramic fiber modules are produced from ceramic fiber blanket that is accordion folded or stacked and precompressed to form rectangular modules shown in Fig. 4.32. The fixing arrangements of the blocks on furnace wall with various anchor system made of alloy steel or through the use of ceramic cup locks are shown in Fig. 4.33. Modules are available in a pre-weld or shoot on application form for dif-

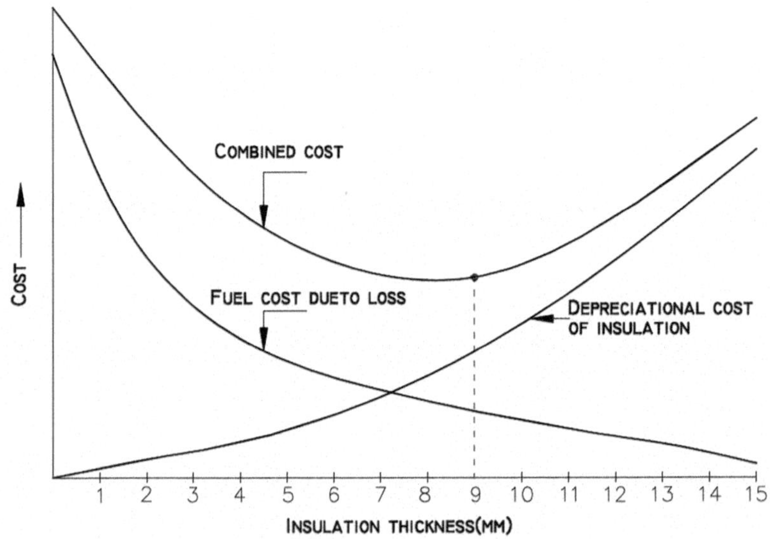

Fig. 4.31 Optimum cost of insulation lining

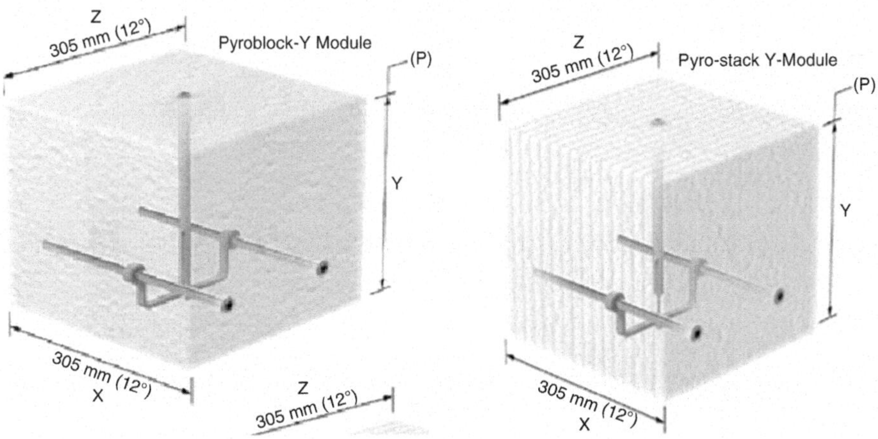

Fig. 4.32 Fixing arrangements for ceramic fiber blocks (Courtesy - Morgan advanced materials)

fering installation methods. Modules are great solutions for thick-walled high-temperature insulation needs in refining and petrochemical industry. Besides this, the ceramic blanket joint in the lining is shown in Figure 4.34.

The performance of the ceramic fiber lining is affected by:

1. Chemical corrosion in reducing atmosphere, for example, by hydrogen atmosphere, because the hydrogen attacks the SiO_2 of the lining as was mentioned in

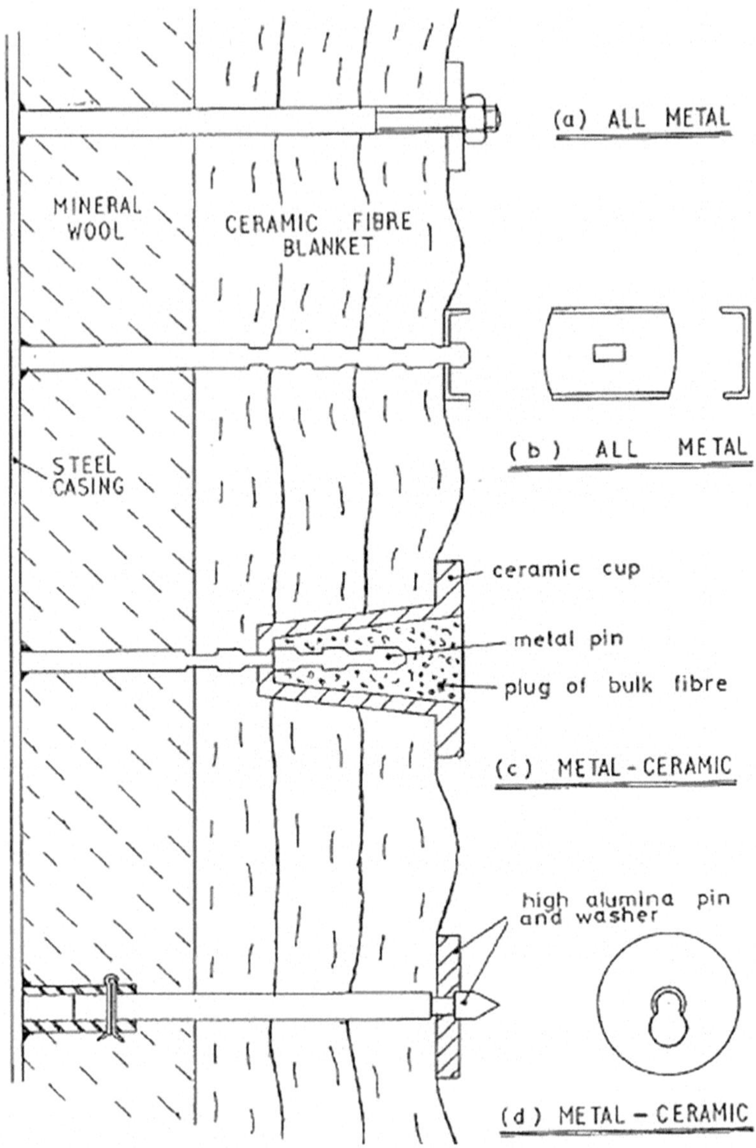

Fig. 4.33 Method of anchoring the ceramic fiber module

Chap. 1 (Sect. 1.1). Fibers with higher Al_2O_3 content are more stable in highly reducing atmosphere.

2. Glaze formation on the surface of the lining because of the attack of alkali, PbO_2, or B_2O_3 vapor.
3. Incorrect installation may lead to low life of the lining.

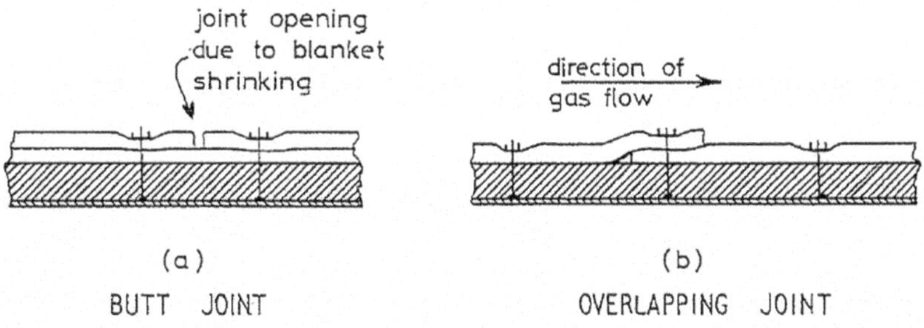

Fig. 4.34 Method of joining the ceramic fiber blankets

4. Presence of alkali in the furnace atmosphere enhances the devitrification of the fibers by which it changes slowly from glassy to crystalline structure and becomes brittle.

4.10 FEA (Finite Element Analysis)

Finite element method is a very useful tool used for designing and analysis of the refractory lining and its failure. It is a numerical method used in solving the engineering problem. The whole idea is to develop a 2D or3D mapping of the stress developed in a composite structure made up of different refractory layers on a steel shell or casing (Fig. 4.35) [33]. This tool is very successfully used in actual field application to predict the behavior of refractory in non-steady state and steady state when the structure is heated up or in thermal equilibrium.

Figure 4.35 shows the temperature profile of the lining at the top right hand side. The stress developed inside the lining is shown in the figure on bottom right hand side. This mapping of stress clearly shows how the compressive stress is concentrated in the hot face (colored blue), and the tensile stress is concentrated at the back layer (colored red). One more example on this is shown in Sect. 6.10.

The major steps in carrying out this analysis are:

1. Evaluation of physical properties by thorough testings
2. To develop a mathematical model as simplified as possible
3. To apply correct boundary conditions
4. To test input parameters on the basic mathematical model developed
5. To interpret and evaluate the results
6. To scale up the model
7. To apply the result in practice
8. To get the feedback
9. To improve further if required

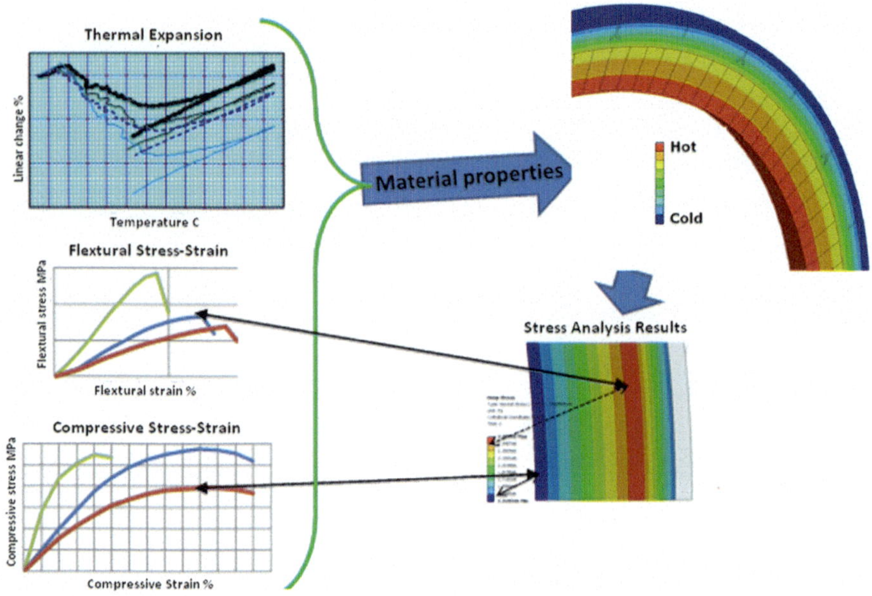

Fig. 4.35 Visual demonstration of the temperature profile vs stress profile in a refractory lining

4.11 Refractory Lining Monitoring

Most of the reactor, furnace, or kiln once put into operation runs continuously for a long time. Besides mechanical or electrical fault, the other major reason which can force a unit to take an unscheduled shutdown is the premature life of refractory lining or a refractory failure. Therefore it is very important to monitor the refractory lining thickness from outside during an equipment is in operation. Two important information required to be generated are:

1. The location of damage, if any
2. The extent of damage, e.g., area and remaining thickness of refractory

Normally it is done by different methods but most common way is to take help of both thermometry and acoustic method.

The thermometric methods are basically infrared photography or heat photography. Due to heat transfer from inside to outside, the reactor or furnace shell has a higher temperature than the ambient. If the refractory thickness is reduced in any area inside the reactor or furnace, the temperature of the shell at that location will shoot up compared to other areas where the refractory lining thickness are more. When infrared photography is done of a shell of the reactor or furnace from outside, it gives a color pattern depending upon the temperature, and it is very easy to locate the area where the temperature shoots up due to reduced refractory thickness.

Next if it is required to know the thickness of the refractory lining in the damaged area, then the help of acoustic method is to be taken. In this method an acoustic wave is allowed to propagate through the thickness of the refractory between the hot face and the cold face when multiple reflections are generated depending upon the refractory property and local conditions like voids, crack, impregnation of slag or metal, etc. The reflected waves can be analyzed to compute the refractory thickness with the flaws and anomalies [34]. The equation used to measure the thickness is

$$\mathrm{T} = \frac{\alpha \beta V_p}{2 f_p} \tag{4.25}$$

where, T = refractory thickness, f_p = frequency of the P wave reflecting between two solid interfaces, V_p = velocity of primary stress wave, α = temperature correction for refractory, and β = the vessel shape factor.

As an example, a company in the job of monitoring the FCCU refractory lining in a refinery in Linden, New Jersey, apprehended a reduced refractory thickness In the regenerator cone area, from the Thermometric photograph (Fig. 4.36) in their routine monitoring, and the shutdown was taken, and the condition of the lining found is shown in Fig. 4.37 [35].

In another example, the airway pipe from compressor to regenerator passes through a heater. During the start-up of regenerator, the catalysts are heated to a temperature to enable autoignition of the oil injected in the catalyst bed to raise the temperature prior to feed of oil. Downstream of the air heater line is refractory lined to protect it from high temperature. During routine qualitative thermographic survey, a hot spot was observed (Fig. 4.38) [36].

The shutdown was taken, and the damaged refractories were examined for probable reason of failure, and the damage was concluded as the result of spalling because of fast temperature rise of the refractory (Fig. 4.39).

One very easy and convenient method to monitor the shell outside temperature is to paint it with temperature indicating paint (TIP). This paint changes color depending upon the temperature, and the temperature distribution across the surface of the

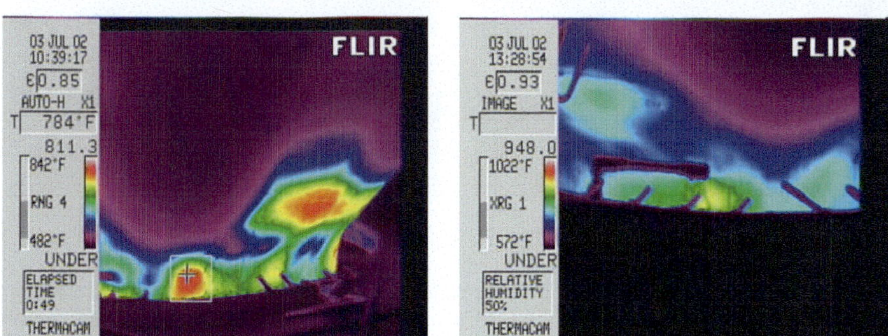

Fig. 4.36 Thermometric photos show the hot spots (yellow to red) in the regenerator cone

Fig. 4.37 Condition of the refractory lining in the hot spot area indicated by thermometry

A B

Fig. 4.38 (**a**) The air line visual image. (**b**) Air line thermographic image shows hot spot

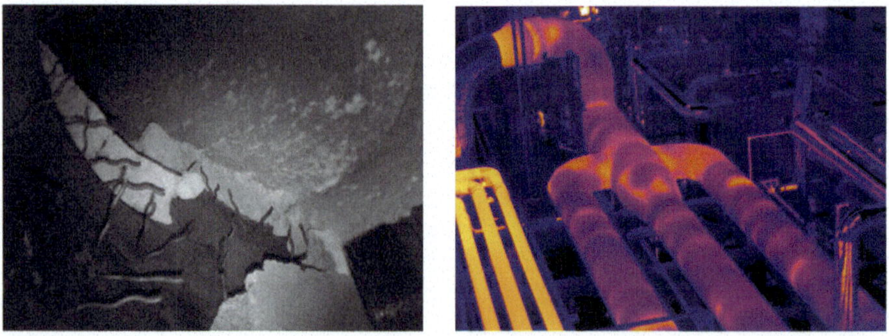

A B

Fig. 4.39 (**a**) Damaged refractory at the detected hot spot. (**b**) Thermographic image after repair

equipment can be judged looking into the color pattern on the steel surface. Different formulations are patented for this type of paint [37]. Some information is there that this paints are based on poly-diacetylene (PDA) polymers with suspended nanoparticles of ZnO (0.1–2%) alloyed with oxides of other transition metals. Monomers making up the PDA are colorless, but the color changes as the polymerization progresses. These paints are used widely in monitoring the temperatures of gas turbine and air engines.

References

1. Refractory engineering- materials, design-construction, Techbook International publisher, New Delhi
2. Technology of monolithic refractories by Plibrico Refractory Company, 1984
3. Installation guide – refractory bricks in rotary kiln- RHI Refractories
4. Hamon RJ (2008) Refractory shotcrete installation. Shotcrete, Spring
5. Allen B (2008) State of the art refractory shotcrete techniques and practices. Shotcrete, Spring
6. Hundere A et al (1998) Wet shotcreting refractory castables with varying cement content. In: XIII Conference on Refractory Castables, Nov-4-5, 1998, Prague, Czech Republic
7. Zhang M, Kannabiran S Pumping and shotcreting materials for alkali and abrasion resistance application. 60th International Refractory Colloquium
8. Sofis TW (2008) Gunite refractory installation. Shotcrete, Spring
9. Parr C et al (1998) Calcium aluminate cements for refractory gunning application. 3rd international symposium on refractories, Beijing, China
10. Wolf C, Wolf C (2005) Refractory gunning. Millenium Steel, pp 120–122
11. Zhang M, Kannabiran S Pumping and shotcreting materials for alkali and abrasion resistance application. 60th International Refractory Colloquium
12. Krischanitz R et al (2008) Monolithic lining technologies- a comparative assessment. RHI Bull 2
13. Weidhaas M Development in monolithic refractories for the Portland Cement Industry, PPT. Refratechnik North America Inc
14. Walton MC, Plater PA When a refractory failure isn't – an anchor issue. Refmet, Prahran
15. Fevbin T Well and truly anchored. IRIS, France
16. Forni G et al (2007) The influence of thermal bridges on refractory lining. La metallurgia Italiana, Giugno
17. Anchor selection for monolithic refractories- Technical bulletin, Harbison-Walker Refractories Company
18. Stainless steel technical handbook- Elga
19. Goulart CA et al (2016) A critical analysis of anchor spacings in Refractory lining design. Refractories Worldforum 8:1
20. Cobane I A 50 year history of contract refractory dryout/heat up services. Hotwork, USA Services
21. Zabolotsky AV (2011) Thermal crack growth modeling in refractory linings of metallurgical installations. Int J Math Models Methods Appl Sci 5(3)
22. Hulett C (1993) Design procedure for high temperature component using finite element method. Thesis, Dept of Mechanical Engg, University of Cape Town
23. Schacht CA Refractory linings- Thermo-mechanical design and application. Marcel Drekker Inc, New York
24. Toda Y (2011) Evaluation of stress relaxation in Basic bricks for rotary cement kiln, et al. UNITECR Proc

25. Cement H (2000) Refractory materials and coating formation. Seminar
26. Muhammed MA (2010) A case study of types of failures in refractory brick lining in Cement kilns. Kufa J Eng 3(1):103–123
27. Steencamp JD et al (2011) Magnesia refractory dry out – managing the risk of hydration. J South Afr Inst Mining Metallur 111
28. Ballon M et al (2006) Design and maintenance of Refractory linings in the reformer section of the Ammonia plants. Ammonia Technical Manual
29. Bhattacharya G et al Degradation of structure of calcium aluminate based Castable in ambient condition. Kerneos, Kolkata
30. Carbonation of Alumina cement bonded conventional Refractory castable in the fireplace – Ina Pundiene et al, Mater Sci (Medziagotyra), Vol12, No4, 2006
31. Alkasem A et al (2017) Corrosion control in Sulfur recovery units-Claus process. In: Abu Dhabi international Petroleum Exhibition and Conference, Abu Dhabi, UAE, 13-16 Nov 2017
32. Sengupta P (2020) Refractories for the Cement Industries. Springer. ISBN- 978-3-030-21339-8
33. Theron JDP (2017) FEA as a tool for design and failure analysis of the refractory structures. LUCIDEON, Digital Technologies in Ceramic Manufacturing
34. Gebski P et al (2009) Chapter 4: Non destructive testing technique for determination of Refractory deterioration in smelting furnace. In: Molten. Vessel integrities and Refractories, Santiago, Chile, pp 1358–1369
35. Weigle RK. Infrared thermography to detect the Refractory damage in a Fluid Catalytic Cracking Unit. file:///D:/technical%20papers/Chemical%20Industries/Petrochemical/FCCU/Thermography.htm
36. Whitcher A (2004) Thermographic monitoring of Refractory lined Petroleum Refinery equipments. InfraMation, Proceedings
37. Yang L, Min LZ (2015) The research of temperature indicating paints and its application in Aero-engine temperature measurements. Proc Eng 99:1152–1157

Chapter 5
Refractories for Lime Calcination

5.1 Introduction

Lime (CaO) is one of the most important industrial chemical used for various purposes, e.g., production of soda ash (by Solvay process); for making pulp and paper; for metallurgical processes, e.g., making of iron and steel; for construction purposes; for soil conditioning in agriculture; for sugar refining; for environment control to absorb oxides of sulfur; for making precipitated $CaCO_3$ as filler material; etc. Figure 5.1 shows the different use of lime (CaO) [1]. World production of lime in 2017 was 350 million ton [2].

5.2 Process

Lime is produced by calcinations of limestone ($CaCO_3$) at a temperature around 950 °C. $CaCO_3$ decomposes at high temperature according to the reaction 5.1 to produce lime (CaO).

$$CaCO_3 + \text{Heat} \rightarrow CaO + CO_2 \quad (5.1)$$

Kilns of various designs are used for calcinations, but the most important are vertical shaft kiln and rotary kiln. The flow sheet of the manufacturing process is shown in Fig. 5.2.

Different stages of manufacturing of quick lime and slaked lime are the following:

(a) Quarrying or mining limestone.
(b) Preparation of limestone by crushing and screening to get the required size (25–160) mm [1, 3].

P. Sengupta, *Refractories for the Chemical Industries*,
https://doi.org/10.1007/978-3-030-61240-5_5

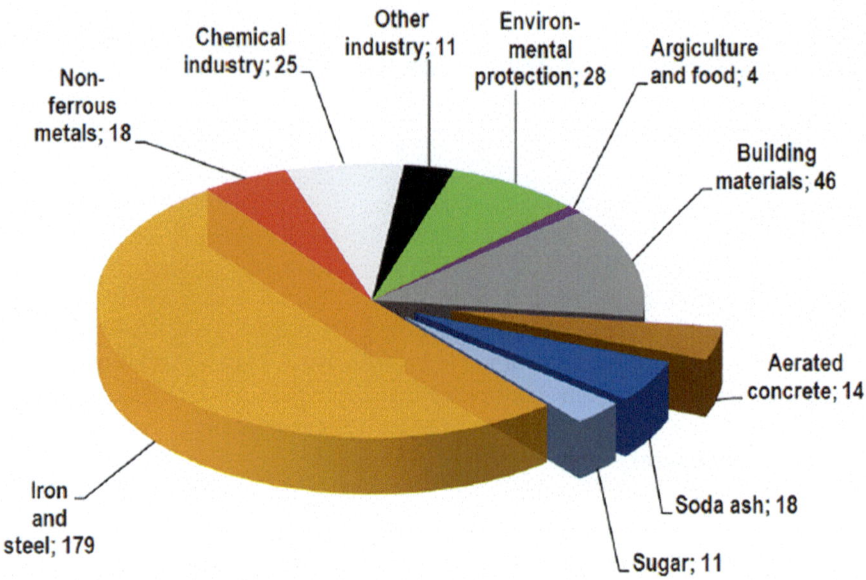

Fig. 5.1 Uses of lime in different industrial sectors

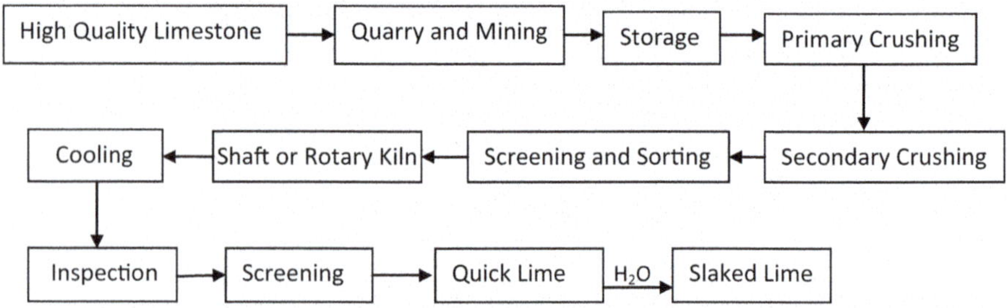

Fig. 5.2 The flowsheet of lime calcinations process

(c) Calcination of limestone at high temperature to decompose it to CaO (quick lime) and CO_2.
(d) Hydrated or slaked lime is produced by reaction of quick lime with water.

5.3 Vertical Shaft Kilns

Vertical shaft kilns are most efficient kilns found for the calcinations of limestone to produce lime. Vertical shaft kiln is a vertical hollow cylindrical furnace fitted with some burners either hang vertically from top or are placed horizontally along the circumference of the kiln in one or two plane perpendicular to kiln axis. In the shaft kilns, the limestone are charged from the top of the kiln at a certain rate and the calcined lime flows downward by gravity and taken out from the bottom opening in a control manner. This type of kilns can be used when the limestone is available as dry hard rock. Shaft kilns of various designs are available, e.g., single shaft kiln, twin shaft kilns, etc. The quality of the calcined lime and the energy efficiency of calcination depend upon the design of the kiln.

5.3.1 Single Shaft Kiln

This kiln design consists of vertical tubular shell of 2–7 m diameter and up to 28 m high fitted with small open top with automatic feeding, and air is blown through larger bottom opening. The production capacity normally varies from 100 to 500 ton quick lime per day. The firing is done through horizontal oil or gas burners fitted at almost in the middle section of the total height of the kiln. The burners are placed perpendicular to the wall of the kiln. The different zones of the kiln based on the temperature profile are shown in Fig. 5.3a. The temperature profiles of the limestone, flowing through the kiln and the gas, moving upwards, countercurrent to the flow of the limestone, are shown in Fig. 5.3b.

Operating condition prevailed in such kiln is high temperature in the burning zone which fluctuates because of difficulty in controlling the burning zone. This causes the generation of high level of thermo-mechanical stresses in the refractory lining. Chemical condition inside the kiln is also difficult to control because of changing raw material and fuel composition. Besides this, throughout the length of the kiln, the refractory lining experiences a very high abrasion because of continuous ascending of large volume of hard limestone rock.

The refractory lining in the preheating and the cooling zone can be of high abrasion-resistant brown fused alumina or high-density calcined bauxite based 70% alumina refractories (Table 5.1). In the burning zone, high-thermal shock-resistant magnesia-spinel refractory is found to be the best (Table 5.1). In the inlet and outlet zone, dense fireclay refractory with 35–40% Al_2O_3 can be used.

5.3.2 Twin Shaft Kiln

These are also called Parallel Flow Regenerative (PFR) kiln because the burden and the hot combustion gas flow in the same direction. In this process, the peak temperature of calcinations remains low enough not to allow the CaO crystallites to get fused, and the product obtained has high reactivity and suitable for certain applications like an ingredient for slag making in steel refining (Maerz). The temperature profile inside the kiln for this process is shown in Fig. 5.4a [1, 4], and the principle of operation is shown in Fig. 5.4b.

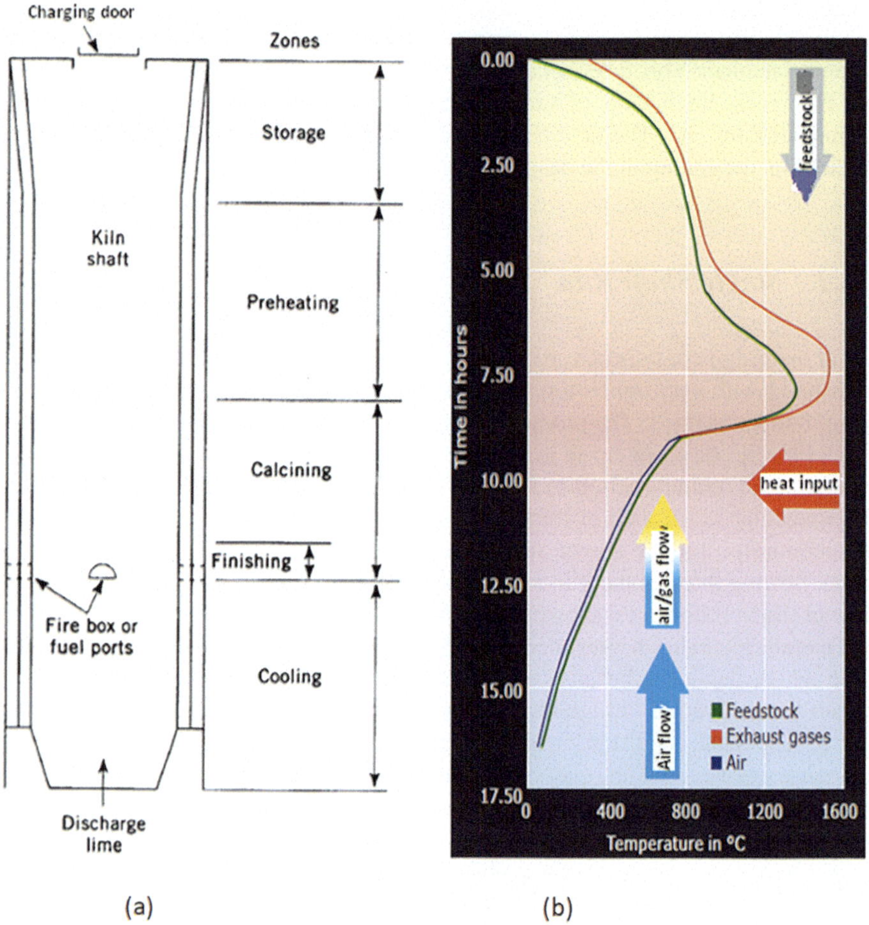

Fig. 5.3 (**a**) Different zones in the vertical shaft kiln and (**b**) the temperature profile

This type of kiln has got two vertical shafts connected by a crossover channel. Both the shafts are fed with raw limestone from the top (Fig. 5.4b). Both the shaft has got separate burners to heat the burden inside. When in one shaft, the burning process is on, in other shaft, the preheating of the burden is going on, and in next cycle the shaft under preheating will be in burning mode and the other one will be in preheating mode and this cycle continues [4].

When the material in one shaft gets burnt, the flue gas travels downwards, in parallel flow with the burden material. The hot gas then leave that shaft through the crossover channel or ring channel (according to different design) to enter the non-burning shaft and starts traveling upwards and leaves that shaft from the top (Fig. 5.4b). On its way upwards, the combustion gas transfers its heat energy to the limestone in that shaft and preheats it. The shifting between burning and non-burning

Table 5.1 Properties of bricks for lining of single shaft lime kiln

	Brick type		
Properties	Fireclay	High alumina	Basic
Chemical analysis %			
Al_2O_3	35–40	65	4–5
SiO_2			0.5–1.0
MgO	–	–	93
Fe_2O_3	1.5	2.5	0.5–1.0
Physical properties			
BD (gm/cc)	2.0–2.1	2.5	2.95
AP %	18–20	18–22	16
CCS (MPa)	40–45	55–60	75
RUL	1400	1450	>1700
TSR (cycle)	80	80	>100

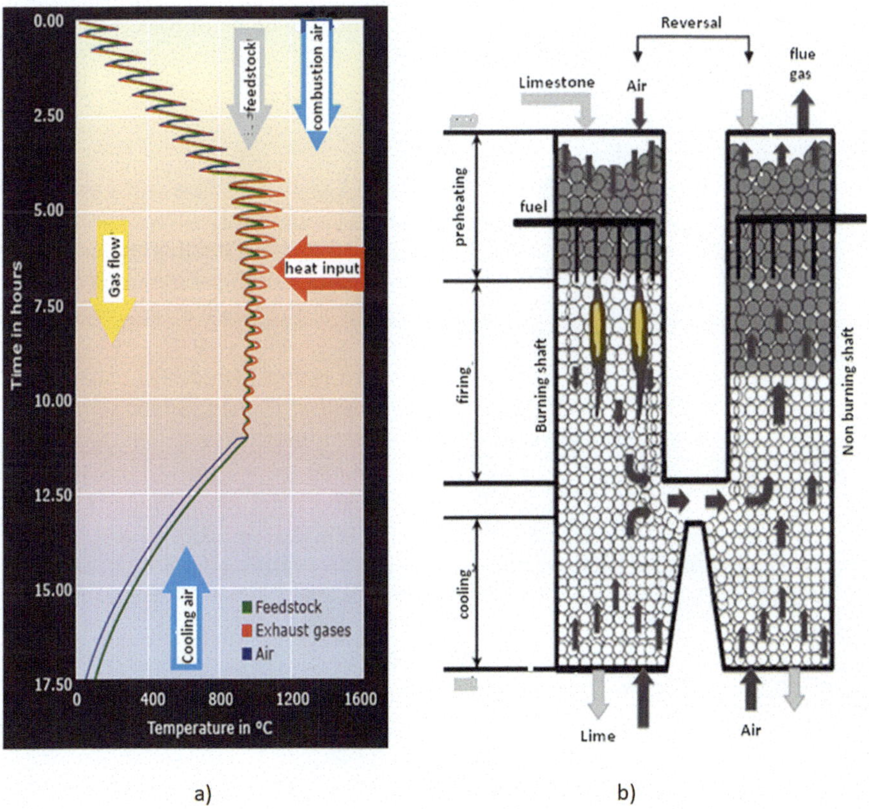

Fig. 5.4 (**a**) Temperature profile in PFR Kiln. (**b**) Principle of PFR calcining operation

mode is called inversion and takes place at a preset intervals. The shafts may have circular or rectangular cross section.

5.3.3 Refractory

The design of the twin shaft furnace is much more complicated than the single shaft kiln. The kiln works under high positive pressure, and therefore the dimensional tolerance of the refractory shapes is important to maintain the accuracy and tightness of the lining. The inner refractory lining pattern of the PFR kiln is shown in Fig. 5.5.

Different qualities of refractories are used to line the different zone of twin shaft kilns of PFR or TSR design. The charging and preheating zones are constructed with dense fireclay bricks because of heavy mechanical abrasion in that are because of charging of limestone.

Unlike in single shaft kiln, the refractory here experiences continuous temperature fluctuation because the two shafts work alternate as burning and non-burning

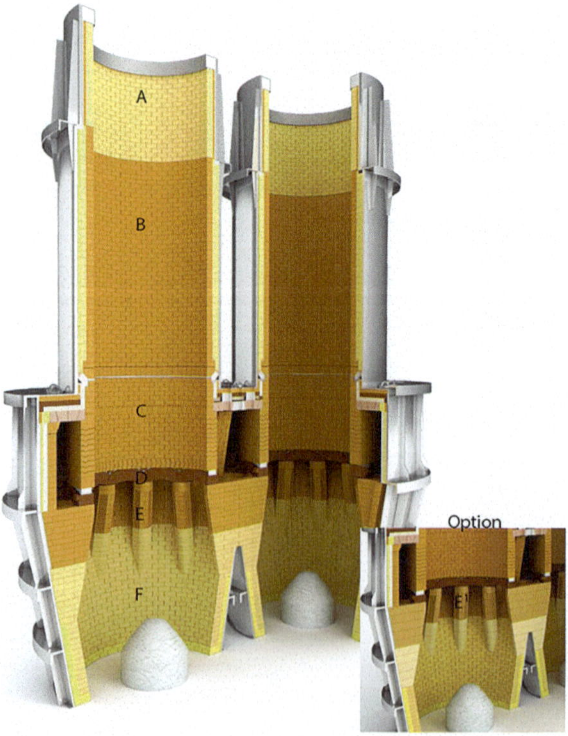

Fig. 5.5 Refractory lining of PFR lime kiln. *A* – Heating zone, *B* – Firing zone, *C* – Wall of inner cylinder, *D* – Arches, *E* – Pillars, *F* – Cooling zone

shaft. The firing zone is invariably lined with high quality magnesia-spinel or magnesia-zirconia bricks, which can withstand high flame temperature. As the process suggests, the process of burning in the shaft alters at a definite interval, and the refractory must have high thermal shock resistance to withstand that temperature fluctuation. The stability of the construction of the whole kiln depends upon the stability of the arch and pillar bricks. Therefore both the quality of both the bricks and the installation must be of topmost standard [5]. The main wear factors of the refractory in this zone are abrasion, redox condition, and thermo-mechanical and chemical stresses and thermal shock. The refractories used here are shown in Table 5.2.

The refractories in the arch (D) and pillar (E) area sometimes fail because the radius increases due to reversible or irreversible refractory expansion which specially becomes more prominent in case of discontinuous operation of the kiln [6].

5.4 Rotary Kiln

Rotary kilns are long horizontal steel tube with diameter 2.0–4.0 m and length 50–120 m and lined inside with refractory brick. The kiln rests on 2–3 rider rings and rotates slowly (3–5 rpm) in slightly inclined position (about 3–4 degree with horizontal plane). The concentric burner is fitted at the downhill end, and the limestone is fed at the uphill end of the kiln. The material proceeds slowly to the downhill end because of gravitational movement assisted by the rotational movement of the kiln [7]. The temperature inside the kiln does not remain the same across its length. Highest temperature remains at the burning zone and decreases slowly towards the feed end. $CaCO_3$ decomposes at a temperature of about 820 °C. The feed and combustion takes place at two different end of the tubular rotary kiln, and countercurrent heat exchange takes place between the feed material and the combustion gases [8].

Table 5.2 Properties of refractories used in twin shaft lime kiln

	Refractories used for different zone of the shaft kiln		
	Inlet zone	Burning zone, crossover channel and pillar	Outlet zone
BD (gm/cc)	2.1–2.2	2.95–3.0	2.15–2.2
AP %	13–15	14–16	14–16
CCS (MPa)	55–65	75–90	50–60
RUL (°C)	1300	+ 1600	1300
TSR/1000 °C (air)	>100	>80	>30
Chem analysis %			
Al_2O_3	30–35	3.0–5.0	35–40
SiO_2	–	1.0–1.5	–
MgO	–	88.0–91.0	–
Fe_2O_3	1.5 max	2.95–3.0	1.5

Table 5.3 Thermal energy used for calcinations of lime in different types of kilns

Type of calcining kiln	Kiln capacity TPD	Limestone size mm	Specific energy consumption Kcal/kg	Specific power consumption KWh/ton
Rectangular PFR kiln	100–400	30–120	810–870	Around 20
Circular PFR kiln	300–800	30–160	810–870	Around 20
Annular shaft kiln	200–600	15–200	910–980	Around 30
Single shaft kiln	50–300	10–100	980–1100	Around 12.5
Rotary kiln with preheater	300–1200	10–50	1150–1350	Around 30
Long rotary kiln without preheater	300–1100	20–50	1600–1700	Around 20

The energy consumption for calcinations in rotary kiln is much higher compared to shaft kiln (Table 5.3) [9]. This is one of the reasons that rotary kiln is not preferred for calcinations of limestone; another reason is the rotary kiln is not a static unit but is a dynamic unit, and naturally the maintenance cost is more. Rotary kiln is used nowadays mostly for calcination of lime sludge, available from paper industry, calcium carbide industry, etc. In the causticizing plant in a paper mill, CaO is used to causticize sodium carbonate in the green liquor to produce caustic soda (NaOH).

Calcium carbonate is precipitated in the process (Reaction 5.2) which is separated from the liquor, washed, and filtered to get cake of solid content above 65%. This is calcined in rotary kiln to produce CaO for reuse to generate caustic soda [10].

$$Na_2CO_3 + Ca(OH)_2 = CaCO_3 + 2NaOH \tag{5.2}$$

The composition of the lime varies from mill to mill, but typically it contains about 95% CaO and 5% impurities [11]. The typical analysis of the sludge is given in Table 5.4.

5.4.1 *Refractory*

The main operating parameters which are to be taken into consideration for selecting the Refractory for the rotary kiln are:

1. Temperature.
2. Chemical reaction of the refractory with CaO and other impurities at the operating temperature.
3. Abrasion of the lime, moving from feed end to discharge end.
4. Thermal shock, due to fluctuating temperature in the kiln, which can be due to different factors, e.g., relatively fixed flame direction and gas flow, and the cool-

Table 5.4 Analysis of the lime sludge from paper mill

Oxides	Average Wt %	Range Wt %
$CaCO_3$	95	92–97
MgO	1.04	0.4–1.6
SiO_2	0. 5	0.1–1.0
Al_2O_3	0.14	0.05–0.4
Fe_2O_3	0.05	0.01–0.4
P_2O_5	0.86	0.2–1.4
Na_2O	1.14	0.5–1.6
K_2O	0.09	0.04–0.12
SO_3	0.91	0.3–2.5

ing effect on the lining when it passes below the bed of material. The thermal oscillation lead to local stresses which risk spalling of the lining [11].

The aluminosilicate brick can be used as lining inside the kiln because pure CaO does not react significantly with alumina or silica below 1360 °C, which is much higher than the operating temperature of the kiln. But the quantity and quality of impurities in the mud are important which may react with the refractory at high temperature to damage it. In the burning zone 60–70% Al_2O_3 refractory can be used, and in other areas 40% Al_2O_3 bricks are more suitable. The sodium content of the sludge is generally below 1%, but the soda percentage gradually gets increased inside the kiln because of volatilization and condensation mechanism, and at equilibrium condition it becomes as high as 10%. The sodium gets evaporated from the kiln exit end due to higher temperature and is carried by the flue gas to the exit end where the temperature is low, and it gets condensed on the lime sludge and travels along with it to the exit end and the cycle goes on. Therefore the lining in the burning zone of the kiln must be alkali resistant.

Basic refractory is more compatible for lime calcinations and magnesia chrome or magnesia-spinel refractory are good choice for lining of the burning zone of the rotary kiln calcining lime. The properties of the refractories are given in Table 5.5.

The use of magnesite chrome is declining gradually due to environmental reason because it is not environment-friendly. The used magnesite chrome bricks react with alkalis during use to form hexavalent water-soluble chromates which are carcinogenic by nature. When the used refractories are disposed in open, the soluble chromate salts get carried away by rainwater and contaminate the groundwater and may pose serious health hazards.

In case of rotary kiln lining, unlike in cement rotary kiln, no coating or very little coating is formed inside, over the refractory lining, to protect the refractory and to reduce the heat loss from the kiln shell to ambience. It causes higher energy loss in case of lime calcinations, and secondly shell temperature may increase substantially. Increase of shell temperature leads to other kinds of problems, e.g., the ovality and deformation of the shell which has a tremendous negative influence on the refractory performance.

Table 5.5 Properties of different refractory qualities for different zone of rotary kiln

	Inlet zone	Preheat zone	Burning zone	Outlet zone
Desirable properties	High resistance to impact and abrasion	High alkali and abrasion resistance	High-temperature volume stability and low MOE	Ability to withstand high temperature and high abrasion
Type of product	Dense castable	Fireclay brick	Basic bricks	High alumina
BD (gm/cc)	2.6–2.8	2.15–2.2	2.92–2.96	2.70–2.75
AP %	–	13–15	15–16	18–20
CCS (MPa)	90–100	55–65	60–70	65–75
RUL (°C)		1300	>1700	1450
TSR/1000 °C (air)		>100	>100	>100
Chem analysis %				
Al_2O_3	56–58	30	7–10	75–80
SiO_2	–	–	0.3–0.7	–
MgO	–	–	87–90	–
Fe_2O_3	0.5–1.0	1.0–1.5	0.5–0.7	2–3

5.4.2 *Insulation*

To reduce the shell temperature and to protect it, high strength insulation refractory board of thickness 10–15 mm can be put over the shell before refractory brick lining. It has been found that the use of insulation board can bring down the shell temperature considerably. The insulation used must have high strength so that it does not get crushed when the refractory expands due to high temperature in service. The properties of such insulation material are given in Table 5.6.

A trial was done in a lime calcination kiln, in Australia, having diameter 3.35 m and length 84 m with capacity of 400 tons of calcined lime per day. The kiln was supported on three tires [12].

Phosphate-bonded magnesium silicate base insulation shapes (Table 5.6), of dimensions 1267 × 76 × 12.5 mm, were laid on the shell, in three rows (total length, 1267 × 3 = 3.8 m) throughout the circumference (Fig. 5.6). The lining of the insulation was spread equally on both the sides of tire 1. Magnesia-spinel bricks of thickness 220 mm were used as the working lining.

After a long operation, the shell temperature was found to be 360–370 °C, in case of use of insulation, compared to 420–440 °C in case of without insulation. The infrared photography recorded the effect (Fig. 5.7). In case of working lining worn to 120 mm, the shell temperature has been calculated as 352 °C, when insulation is used, in comparison to 453 °C, in case of without insulation. Therefore benefit is still more in case of worn-out lining.

It was calculated that if 46 m length of the kiln (55%) is lined with the insulation mentioned, the saving is 0.2 GJ/ton of calcined lime, which ultimately saves large quantity of fuel.

Table 5.6 Properties of magnesium silicate insulation used in rotary kiln for calcinations of lime

Properties	Phosphate-bonded magnesium silicate insulation
Bulk density (gm/cc)	1.7
Apparent porosity %	53
CCS (MPa)	15.0
Thermal conductivity (W/mK) at	
200 °C	0.23
600 °C	0.31
PLC % at	
500 °C	0.07
900 °C	0.64

Fig. 5.6 Lining of the insulation shapes on the rotary kiln shell over which the 220 mm working lining bricks are laid

When the temperature profile inside the working lining in case of, with and without insulation (Fig. 5.8) was calculated, it has obviously shown that the temperature profile is much flatter inside the working lining, when insulation is used. The flatter temperature profile across the thickness of the refractory means the temperature is more uniform across the refractory which results in less thermo-mechanical stress generation in the refractory body due to differences in thermal expansion and ensures better performance.

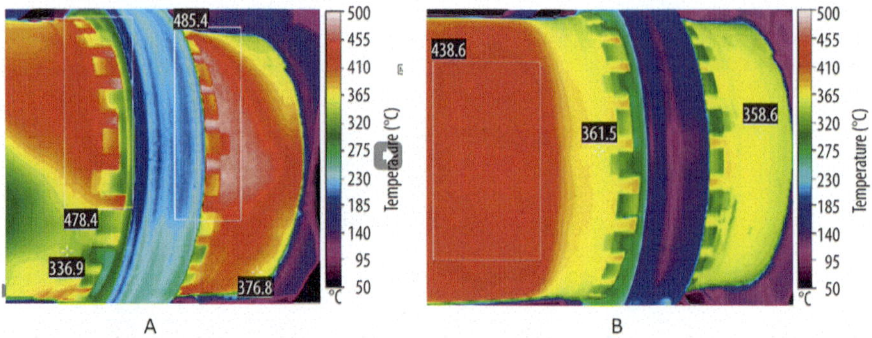

Fig. 5.7 Infrared thermography shows the comparison of shell temperature in case of (**a**) without insulation and (**b**) in case of with insulation

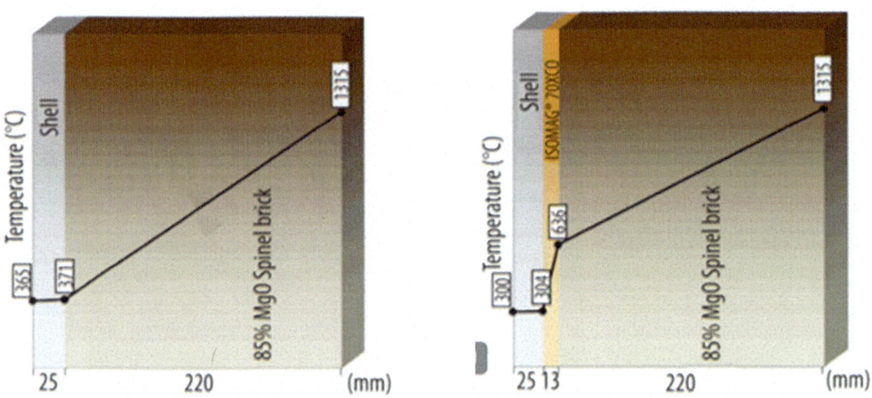

Fig. 5.8 Temperature profile inside the working lining (**a**) without insulation (**b**) with insulation

References

1. Lime Shaft Kilns – Hans Piringer, Energy Procedia, 120 (2017) 75–95
2. World Lime production – International lime Association
3. Lime Production – Industry Profile, Final Report, Eric L Crump, US Environmental Protection agency, Sept, 2000
4. We are on fire for the most efficient Lime kiln – Maerz
5. Optimized lining concept for the pillar zone of the Parallel Flow Regenerative Line kiln – Axel Belau Klemt and Ulrich Marschall, RHI Bulletin, 1, 2006
6. Twin shaft regenerative Lime kiln – Cimprogetti

7. Refractories for Cement Industries – Prasunjit Sengupta. Springer Book, 2019
8. Modeling of rotary kiln and application to lime calcination – Uwe Kussel et.al. Proceedings, 7th Modelica conference, Como, Italy, Sept 20–22, 2009
9. Comparison of Lime kiln types – Blitzco GmbH Stuttgart. www.blitzco.de.info@blitzco.de
10. Lime Kiln – Principle and operation – Terry N Adams, Technical consultant, 900 Leonara Street Unit, 200, Seattle, WA 99121, USA
11. Lime kiln Chemistry and its effect on kilns operations – Honghi Tran. https://www.researchgate.net/publication/290320968. Conference paper, 2007
12. Improving the performance of a lime kiln burning zone – Mario Taddeo and Michael Dubovokich, Global Cement Review, Nov, 2016

Chapter 6
Refractories for Syngas Manufacturing

6.1 Introduction

Syngas is the abbreviation of synthetic gas which is a mixture of hydrogen, carbon dioxide, carbon monoxide, and traces of other impurities. The name is derived from its process of synthesis from a carbon or hydrocarbon feedstock. Syngas is used as an intermediate for production of many useful chemicals, e.g., ammonia, methanol, synthetic gasoline, and many other chemicals. This process was developed during World War II in Germany to synthesize fuel oil. Since then, the process is gaining importance day by day because it is one of the cleanest source of energy and can use different types of feedstocks which, otherwise, cannot be used as a source of energy or for producing any chemicals. The different chemicals that can be produced from syngas are shown in Fig. 6.1. Global syngas capacity share, based on its use, is 74.5% for producing chemicals, 10.6% for producing gaseous fuel, 11.6% for producing liquid fuel, and 3.3% for power generation [1].

The share of different chemicals produced globally from syngas is shown in Fig. 6.2.

It has been projected that till 2024, its production will remain lower, but after then its CAGR for chemical, liquid fuel, gaseous fuel, and power will be 5%, 20%, 12%, and 15% [1].

The major feedstock for syngas production is coal and natural gas, whose shares are 48.3% and 46.5%. The rest 4.6% are from refinery residue like petcoke, and 0.9% is from biowaste [1].

P. Sengupta, *Refractories for the Chemical Industries*,
https://doi.org/10.1007/978-3-030-61240-5_6

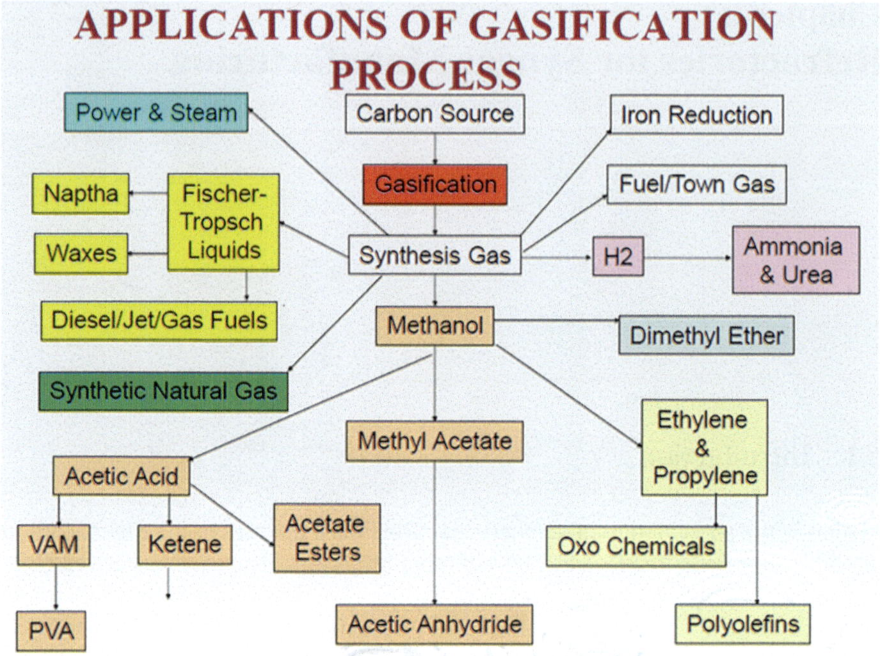

Fig. 6.1 The chemicals produced from syngas

6.2 Chemical Process

There are a number of technologies available to produce syngas. These technologies are summarized in Fig. 6.3. Of the different technologies shown in Fig. 6.3, steam methane reforming (SMR) is the most common. In this process light hydrocarbon feedstock and steam are converted in an endothermic reaction over a nickel catalyst. Heat to the reaction is provided in a radiant furnace. Due to the sulfur content in heavier hydrocarbons, their use in the SMR process can cause problems with the catalysts, and also there is a risk of tar and coke deposition. The second most common technology is partial oxidation (POX) which proceeds exothermically. Non-catalytic oxidation reaction allows a wider range of feedstock; however, an oxygen source is needed in most applications. Carbon dioxide reforming is a less common technology, which is primarily used for the production of syngas with a low H_2/CO ratio. The combined and advanced processes, also detailed in Fig. 6.3, are either a combination or an enhancement of the three basic processes. In the following sections, firstly the three basic processes are further described; shorter descriptions of the combined and advanced processes follow later in the section [2, 3].

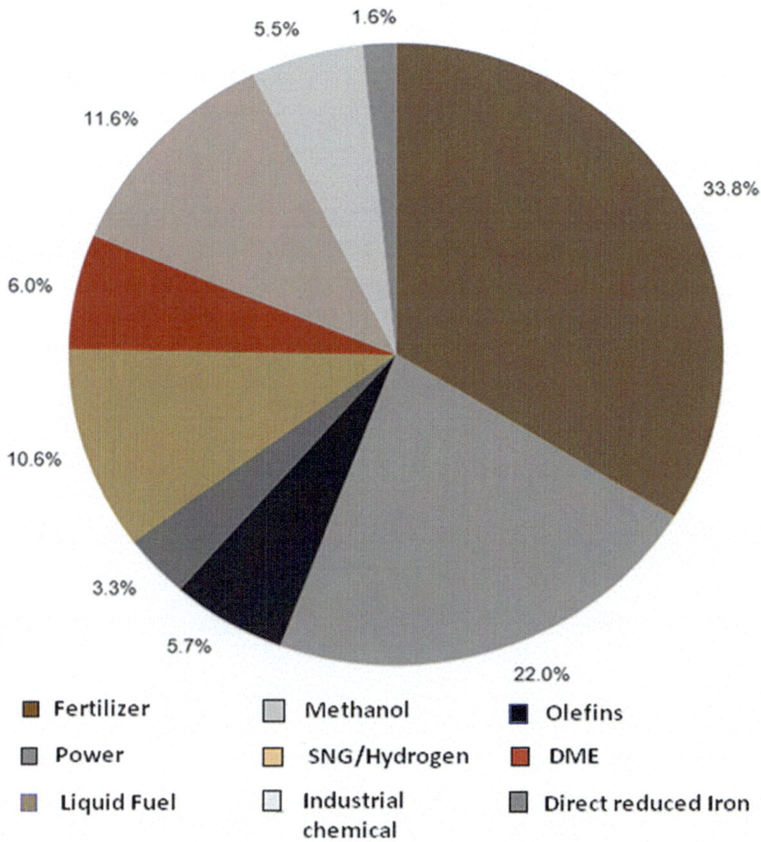

Fig. 6.2 Global share of different chemicals produced from syngas

6.2.1 Steam Methane Reforming (SMR)

The steam methane reforming (SMR) process can be described by two main reactions:

$$CH_4 + H_2O = CO + 3H_2, \Delta H = +198\ \text{kJ / mol}\ (\text{Endothermic}) \tag{6.1}$$

$$CO + H_2O = CO_2 + H_2, \Delta H = -41\ \text{kJ / mol}\ (\text{Exothermic}) \tag{6.2}$$

The first reaction is reforming itself, while the second is the water-gas shift reaction. Since the overall reaction is endothermic, some heat input is required. This is accomplished by combustion of natural gas or other fuel in a direct-fired furnace. Reaction (6.1) favors high temperature and low pressure and proceeds usually in the presence of a nickel-based catalyst. The first patents on steam methane reforming were awarded to BASF in 1926, and the first reforming plants were built in the

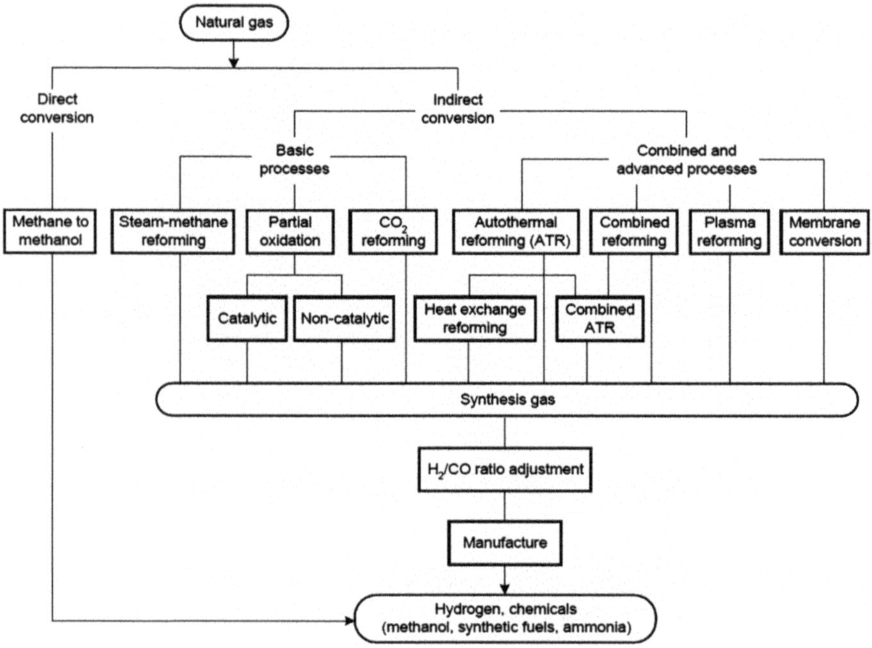

Fig. 6.3 General overview of the routes to convert natural gas to chemicals

1930s. Large-scale production had begun only in the beginning of the 1960s following the discovery of large gas fields in Europe with subsequent changeover from use of coal to natural gas as a feedstock. In the early days, reforming was done at atmospheric pressure; later, the process conditions were changed to increase to pressures up to 30 bar and temperatures of up to 1000 °C. The increased pressure saves compression energy in the downstream synthesis stage; however, the high temperature necessitates an extensive heat recovery system.

6.2.1.1 Process Description

In a direct-fired furnace, a preheated mixture of natural gas and steam is passed through catalyst-filled tubes, where it is converted to hydrogen, carbon monoxide, and carbon dioxide (Fig. 6.4a, b). It is of a great importance to control the maximum tube temperature and heat flux in the reformer to maintain a reliable and prolonged performance. To obtain this, several burner arrangements are employed: top-fired, bottom-fired, side-fired, terrace-walled, and cylindrical type shown in Fig. 6.5 [4].

35–50% of total energy input in SMR process is absorbed due to endothermic nature of the reforming process half of which is required for temperature rise and

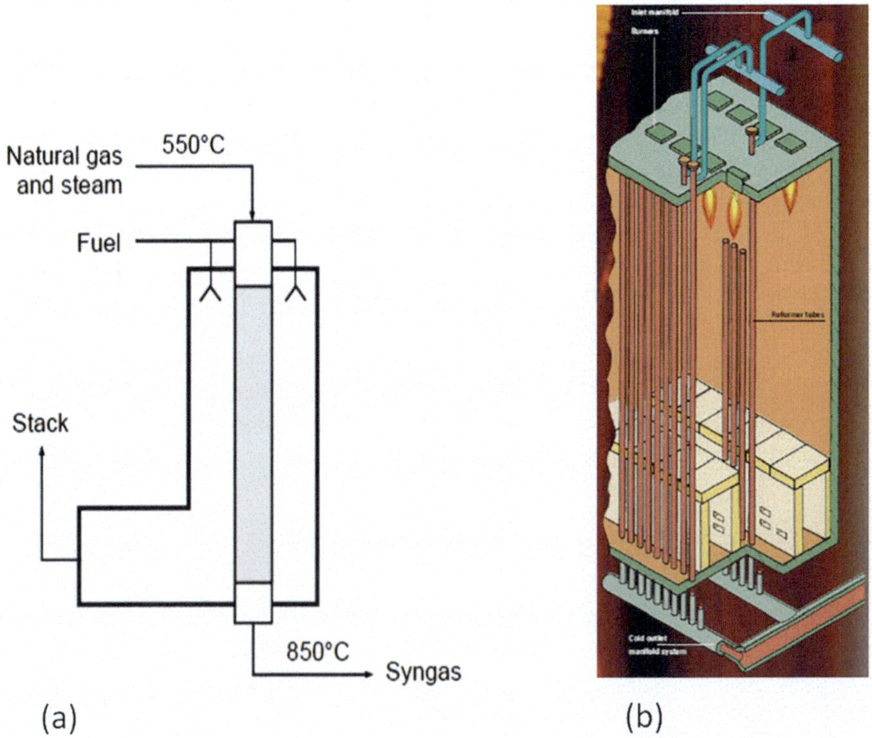

Fig. 6.4 (**a**) Schematic diagram of reformer, (**b**) arrangements of burners and tubes inside a reformer

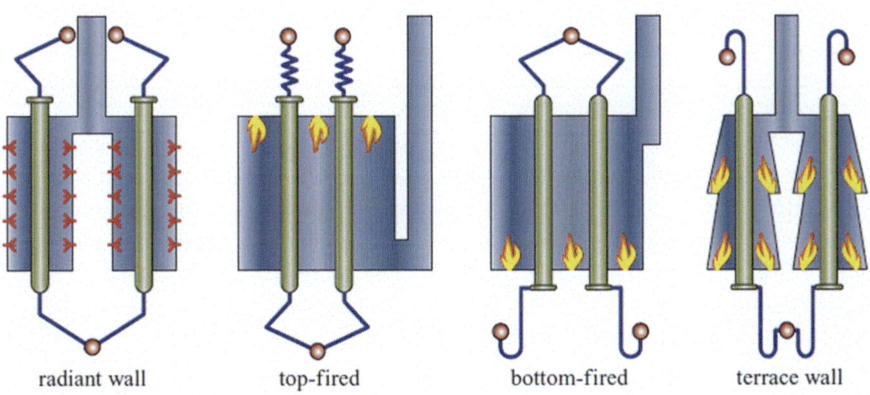

Fig. 6.5 Different arrangements of burner in the reactors

the other half for the reaction itself. The produced syngas leaves the reformer at a temperature of 800–900 °C. The heat of the flue gases is usually utilized in the convective part of the reformer by generating steam and preheating the feedstock, thus bringing the overall thermal efficiency to over 85%. Only a portion of the steam generated in the boiler is required for the reforming process, while most of the steam (about 60%) is consumed elsewhere. There are also arrangements to recover the heat of the syngas such as the regenerative burners [5].

To avoid catalyst poisoning, a de-sulfurization stage is usually required. In addition to the usual nickel-based catalysts, cobalt and noble metals are often used in SMR processes. Non-metallic catalysts have not proved their feasibility due to their low activity (Rostrup-Nielsen et al., 1993).

Another catalyst related problem is carbon deposition, which is especially present when processing higher hydrocarbons. In this case, ruthenium, which can effectively resist carbon formation in steam reforming, can be used.

Typical operating parameters of the SMR process are:

Pressure: 20–26 bar
Temperature: 850–950 °C
H_2/CO ratio: 2.9–6.5

Complete conversion cannot be obtained in the SMR process, typically 65% of methane is converted, at best it is about 98% (Appl, 1992); therefore a secondary reforming must be used if a higher conversion rate is desired (see Section on combined reforming).

In view of the high H_2/CO ratio, steam reforming is the most effective means for hydrogen production. The carbon monoxide then is oxidized to CO_2 in the shift reaction (6.2), producing even more hydrogen, which is subsequently purified in a pressure-swing adsorption (PSA) unit.

Summarizing the advantages of the SMR process, it should be noted that this is the most proven technology with a great deal of industrial experience; it requires no oxygen and produces syngas with a high H_2/CO ratio. It also has relatively low operating temperatures and pressures in comparison to other technologies. Nevertheless, expensive catalyst tubing and a large heat recovery section make an SMR plant a costly investment that can only be justified for very large-scale production. The large size of the reformer and the potential risk of local overheating leads to a complicated heat management system with a slow response; furthermore, external heating results in relatively high atmospheric emissions from the combustion process. The presence of catalyst imposes other problems. Care should be taken to avoid sulfurization and carbon deposition. The catalyst also should be regularly refilled due to its deterioration. All these problems prevent the use of heavy hydrocarbons as a feedstock in the SMR process.

6.2.2 Partial Oxidation (POX) and Catalytic Partial Oxidation (CPO)

In the partial oxidation reaction, which proceeds exothermically according to the reaction 6.3.

$$CH_4 + \frac{1}{2}O_2 = CO + 2H_2, \Delta H = -44\ \text{kJ/mol}, \tag{6.3}$$

complete conversion of methane is obtained above 750 °C, resulting in a mixture with H_2/CO ratio of 2. Since the reaction is exothermic; no fuel is required. When applied to heavy hydrocarbons, coal, or vacuum residue, this process is commonly referred to as gasification.

Temperatures of the reaction varied between 750 and 900 °C, and the pressure was slightly above atmospheric. Nowadays, pressures up to 75 bar and temperatures up to 1400 °C are employed. In addition to the non-catalytic oxidation, a lower-temperature catalytic process has been developed. The catalytic partial oxidation (CPO) has a higher flexibility and is less susceptible to soot formation.

6.2.2.1 Process Description

A refractory-lined pressure vessel is fed with natural gas and oxygen at a typical pressure of 40 bar (Fig. 6.6). Both natural gas and oxygen are preheated before entering the vessel and mixed in a burner. Partial oxidation reaction occurs immediately in a combustion zone below the burner. To avoid carbon deposition, the reactants should be thoroughly mixed, and the reaction temperature should not be lower than 1200 °C. Sometimes steam is added to the mixture to suppress carbon formation. In the case of catalytic partial oxidation, steam is not required, and the temperature can be below 1000 °C.

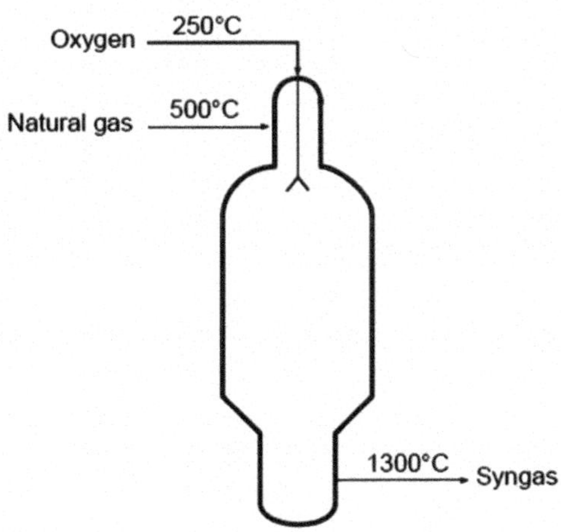

Fig. 6.6 Partial oxidation reactor

The syngas produced leaves the reactor at temperatures of 1300–1500 °C. Since the natural gas is usually supplied from a network at high pressure and oxygen is delivered in the liquid form, the costs of upstream compression are almost negligible. In this manner, compression work is saved in the downstream synthesis process. In practice, syngas from the POX process has a H_2/CO ratio between 1.6 and 1.8, so a shift converter or steam injection should be employed to increase this ratio, for instance, for methanol synthesis.

The non-catalytic process allows the use of a broad range of hydrocarbon fuels from natural gas to coal and oil residue and remains the only viable technology for heavy hydrocarbons. Almost 100% conversion is obtained in the POX reaction, and this is the reason why a POX reactor is used in combined reforming to complete conversion of methane coming from a steam methane reformer. At high temperatures carbon particles are burned completely, so that no steam is required; this simplifies the process operation. However, the need for oxygen results in high operating costs, and also provisions should be made to minimize the risk of explosion.

The catalytic process has a reduced size and consumes less oxygen, but runs the risk of catalyst destruction by local thermal stress.

6.2.3 *Autothermal Reforming (ATR)*

This process combines partial oxidation and steam methane reforming in one vessel, where the hydrocarbon conversion is driven by heat released in the POX reaction. The process is used for methanol and ammonia production. Both light and heavy hydrocarbon feedstocks can be converted. In the latter case, an adiabatic pre-reformer is required.

6.2.3.1 Process Description

A preheated mixture of natural gas, steam, and oxygen is fed through the top of the reactor. In the upper zone, partial oxidation proceeds at a temperature of around 1200 °C. After that, the mixture is passed through a catalyst bed, where final reforming reaction takes place (Fig. 6.7). The catalyst destroys any carbon formed at the top of the reactor. The outlet temperature of the catalyst bed is between 850 and 1050 °C.

The main advantages of ATR are a favorable H_2/CO ratio (1.6–2.6), reduction of emissions due to internal heat supply, a high methane conversion, and the possibility to adjust the syngas composition by changing the temperature of the reaction. However, it requires an oxygen source.

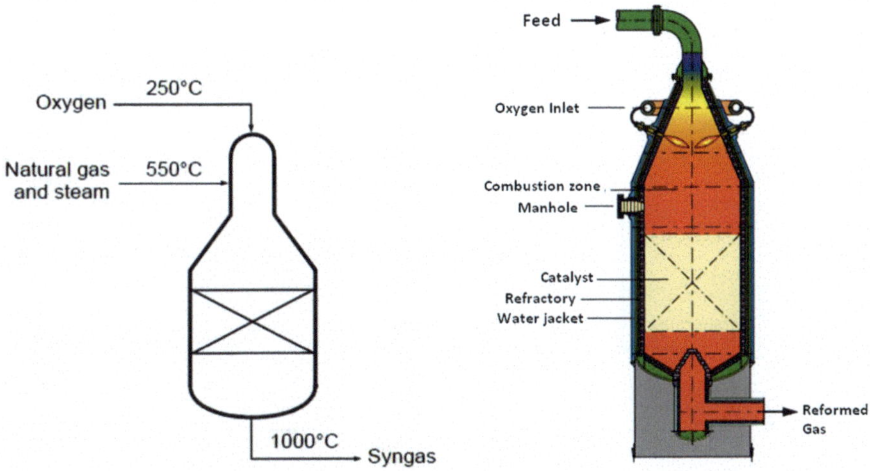

Fig. 6.7 Autothermal process

6.2.4 *Combined Reforming*

Since total methane is not converted in the SMR reaction, a secondary reformer behind the SMR unit can be installed to provide complete conversion of methane and to adjust H_2/CO ratio. In addition, the size of the costly SMR plant can be reduced by shifting part of its load to the secondary, oxygen-fired reactor. This is why this process is also called two-step reforming, or oxygen-enhanced reforming.

6.2.4.1 Process Description

As the schematic of the process shows (Fig. 6.8), it consists of a primary SMR unit and a secondary ATR or POX reformer with oxygen supply.

A pre-reformer is often installed before the main reformer (Fig. 6.9). Pre-reforming, which proceeds at low temperatures, is usually adiabatic. The pre-reformer widens the range of hydrocarbons suitable for reforming. It also takes over some duty from the primary reformer, in order that it can operate under less severe conditions. This, in turn, allows for less costly materials and a smaller heat transfer surface, which results in reduced costs of the equipment. However, the need for an oxygen plant might overweigh this advantage.

Table 6.1 gives the parameters of the different types of reforming processes being used actually in industries, and the data shows that CPO and the combined processes are most attractive.

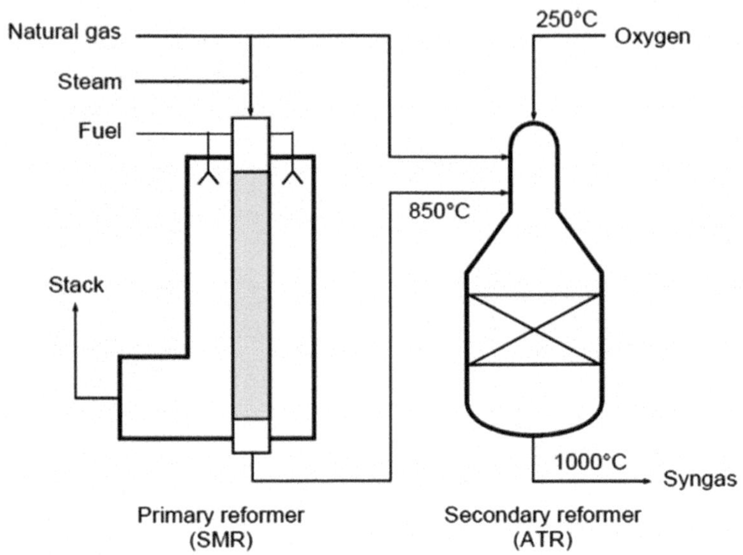

Fig. 6.8 Combined reforming

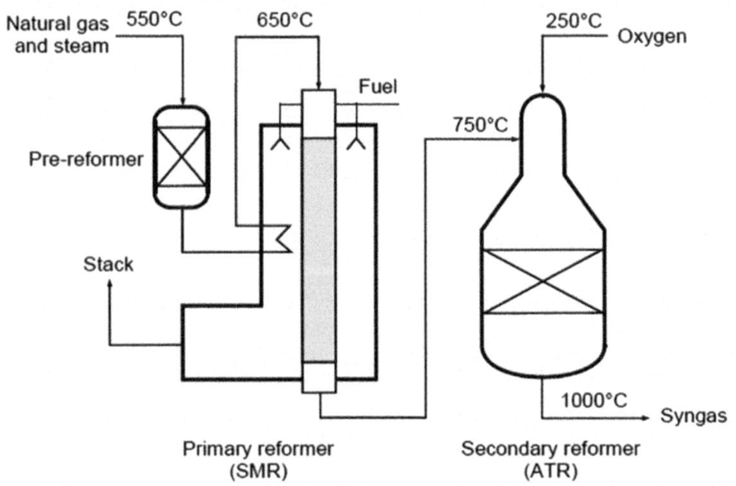

Fig. 6.9 Combined reforming with pre-forming

6.3 Refractories

In the last 10 years, the syngas manufacturing capacity, worldwide, has been increased by 400%. Refractory lining plays a major role in the successful operation of the gasifier. The production loss due to the shutdown because of outage of refractory lining or its repair can be as high as 1.5 million USD per day [6].

Table 6.1 Overview of the reforming processes

	SMR	POX	CPO	CO_2	ATR	Combined
Temperature (°C)	800–900	1000–1450	800–1000	900–1000	850–1300	P: 800 S:1000–1200
Pressure(bar)	20–30	30–85	15–40	10	20–70	20–30
H_2/CO ratio	3–6	1.6–2	1.6–2.0	1	1.6–2.0	2.5–4.0
CH_4 conversion %	65–95	95–100	95–100	High	95–100	95–100
Oxygen	None	High	High	None	High	Low
Steam consumption	High	Optional	Optional	Optional	Low	Medium
Capital cost %	100	80–110	55–80	–	65–80	75–115
Emission	High	Low	Low	Low	Low	Medium
Scale	Large	Small-large	Small-large	Medium	Large	Large

6.3.1 Primary Reformers

The arrangement of primary and secondary reformer with the different important parts that require refractory lining is shown in Fig. 6.10. Alumina refractories as dense lining material and bubble alumina refractories as the primary insulating materials are used to line the reactor carrying out partial oxidation process. Ceramic fiber boards and blankets are used as the secondary insulating material. In primary reformers, radiant wall is used to heat the tubes. Thermal shock is experienced by the refractories in the areas where the flame impinges upon the refractory wall. In such areas, the use of mullite bricks with low impurity level improves the performance [7].

The lining made with ceramic fiber board has lesser thickness and lesser weight. The lining of the entire vessel with the ceramic fiber board and blanket can be considered which will drastically reduce the structural load with subsequent reduction in the cost of construction.

The hydrogen transfer line also is an important area in the reformer, and these are normally lined by bubble alumina castables as shown in Fig. 6.11 [8].

One of the most problem-prone areas in the primary reformer is refractory lining in tunnel area [9]. The process gas and steam are fed downwards over a catalyst bed and are fired by the side burners. The flue gases are collected in tunnels along the radiant section of the floor to improve the uniformity of its distribution (Fig. 6.12a). These tunnels range from 1.2 to 3 m tall, 0.6 to 0.9 m wide, and 12.2 to 30.5 m long.

The tunnels sidewalls are constructed directly on the floor with insulating bricks at bottom followed by dense firebricks. Both flat face and tongue and groove design are prevalent. The sidewalls are constructed with bricks having ports for distribution of the gas, and the top is covered by the large refractory slabs called lids or coffins (Fig. 6.12b).

The expansion gap between the slabs is very critical feature to accommodate the thermal expansion of the slabs in operation. One of the major modes of failure of tunnel is found to be the collapsing of refractory slabs. FEA technique was applied in solving the problem.

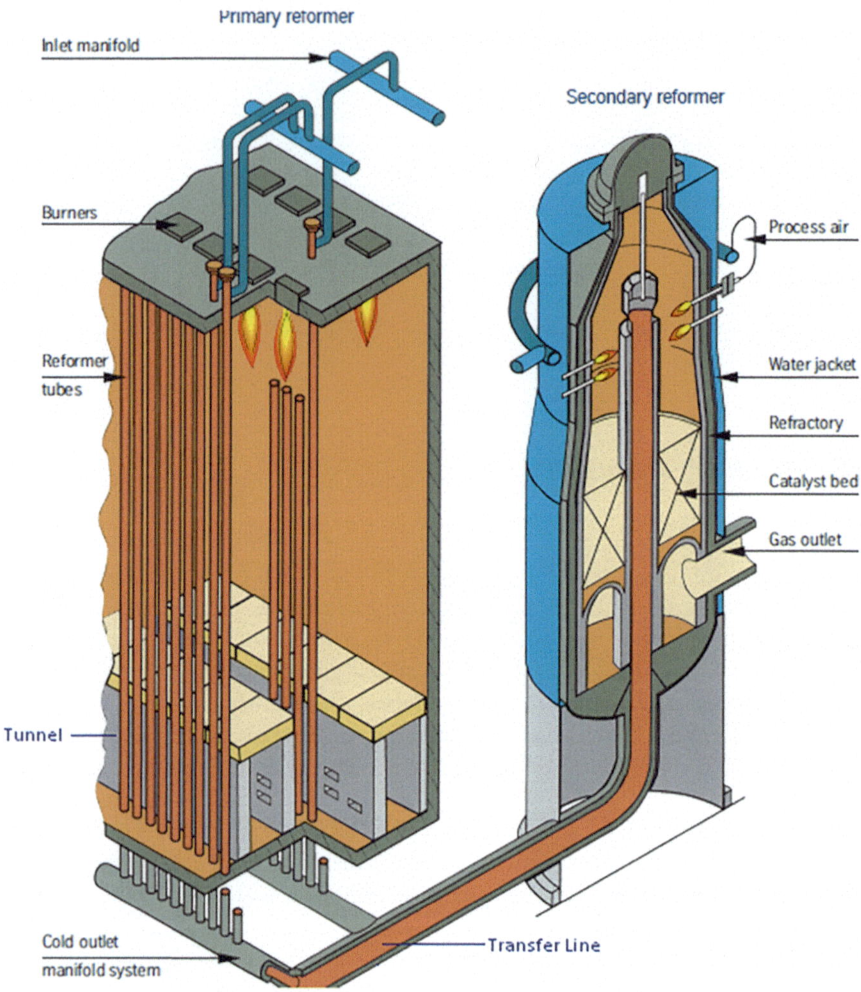

Fig. 6.10 Arrangement of primary and secondary reformers (Courtesy-Thyssen Krupp)

The lids can be considered as the refractory beams, and calculation of their static load is found to be very less which is unlikely to cause the failure. The FEA done on the typical refractory slabs on the tunnel at 1038 °C (1900 °*F*) all both at top and bottom of the slab shows the maximum thermal stress developed is 0.71 Kg/cm^2 (10 psi) (Fig. 6.13) [9].

The FEA technique was applied again on the slab cover with a consideration that there is a temperature difference of 1.5 °C (10 °F). The maximum stress on the slab cover mounted (Fig. 6.14) and was found to be 79 kg/cm^2(1100 psi), which is a very high value of HMOR at a temperature of 1038 °C to be withstood by ordinary quality of refractory. Decrease of the slab thickness lowers the thermal mass and

Fig. 6.11 Castable lined hydrogen transfer line

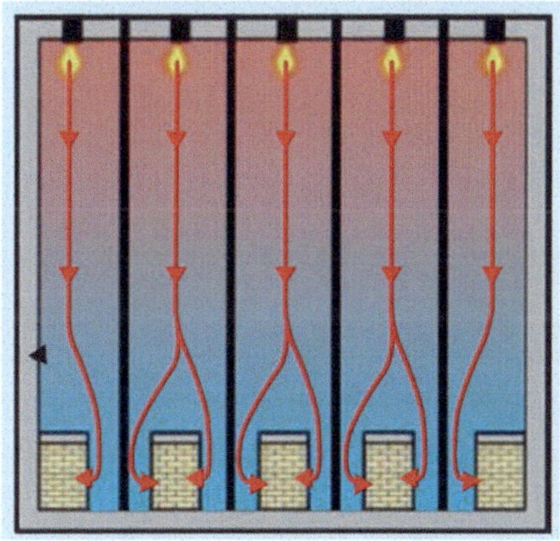

a) Distribution system of flu in primary reformer

b) Construction of the tunnel

Fig. 6.12 (**a**) Distribution system of flu in primary reformer (**b**) Construction of the tunnel

increases the bulk conductivity of the slab. This reduces the temperature gradient across the thickness and lowers the thermal stress.

Another important aspect of the performance of the lids is selection of the right material. High HMOR and low creep value at 1200 °C is an important property of the material which ensures high performance. Creep of lid causes sagging of the middle portion of the lid and alters the interaction of the forces between the lid and the wall which keeps the structure stable.

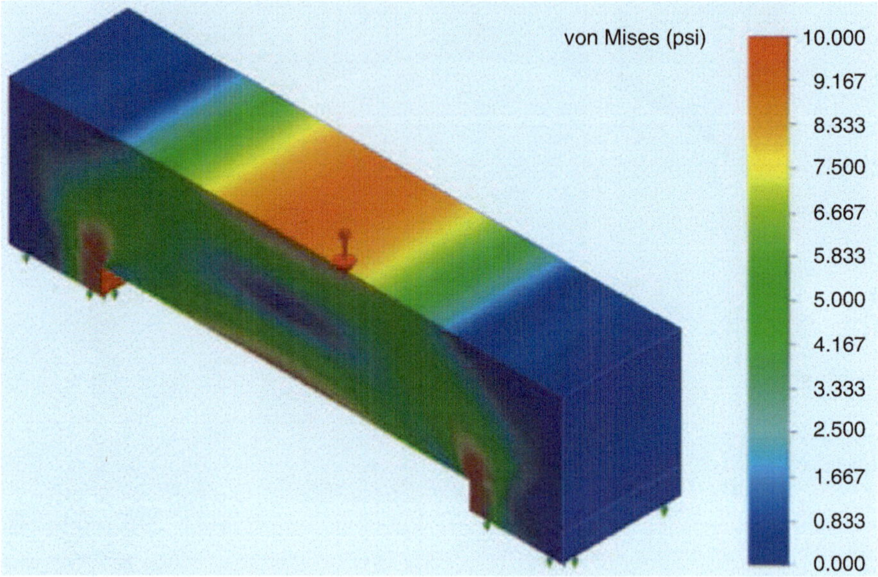

Fig. 6.13 Stress distribution in the refractory slab at an uniform temperature of 1038 C

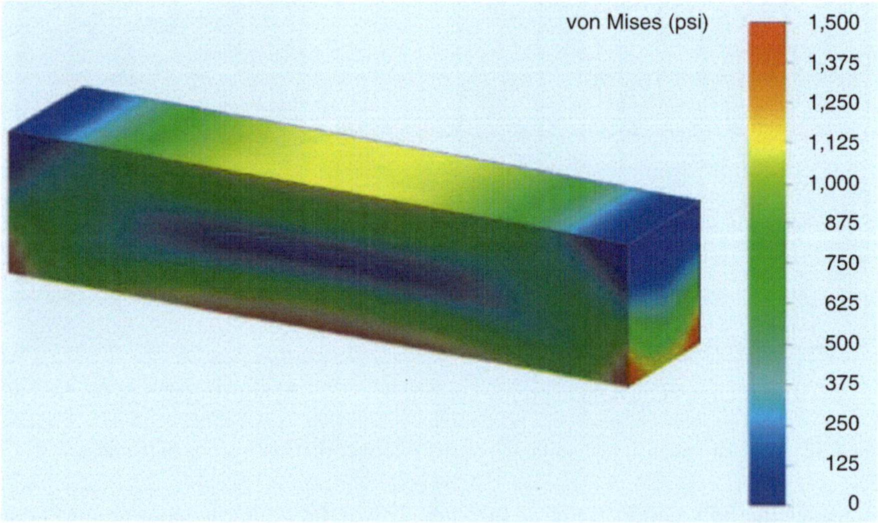

Fig. 6.14 Stress profile when the difference of top and bottom temperature of the slab is 1.5 C

It is suggested that the base, properties wall and the lid of the tunnel must be made of the same quality material with similar so that their thermal expansion character is the same. This ensures that the tunnel expands and contracts equally in both top and bottom of the wall. Cross beam supports can be provided at predetermined

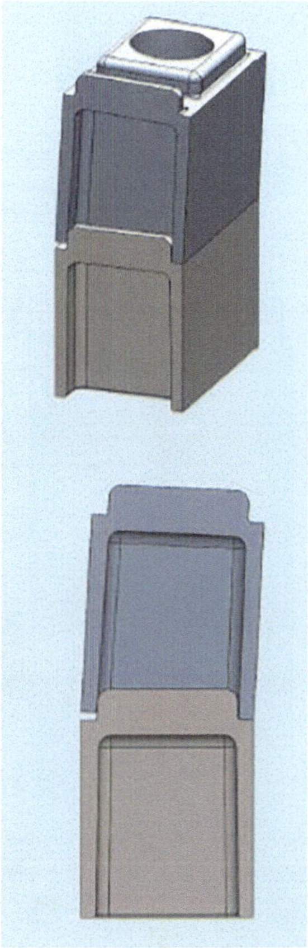

Fig. 6.15 Buckling of the refractory blocks

locations to arrest buckling of the wall (Fig. 6.12b). Buckling can also be arrested by the use of very tight tolerance of the wall element blocks (Fig. 6.15) as it arrests the rotation of the top block with respect to the bottom block.

6.3.2 Refractory for Secondary Reformer

The secondary reformer consists of:

(a) A cylindrical carbon-molybdenum steel vessel
(b) One hydrogen burner inside
(c) Conical portion lined with refractory
(d) Bed with tiles, alumina lump, and catalysts
(e) Refractory dome that supports the catalyst bed
(f) Water jacket (in some designs)

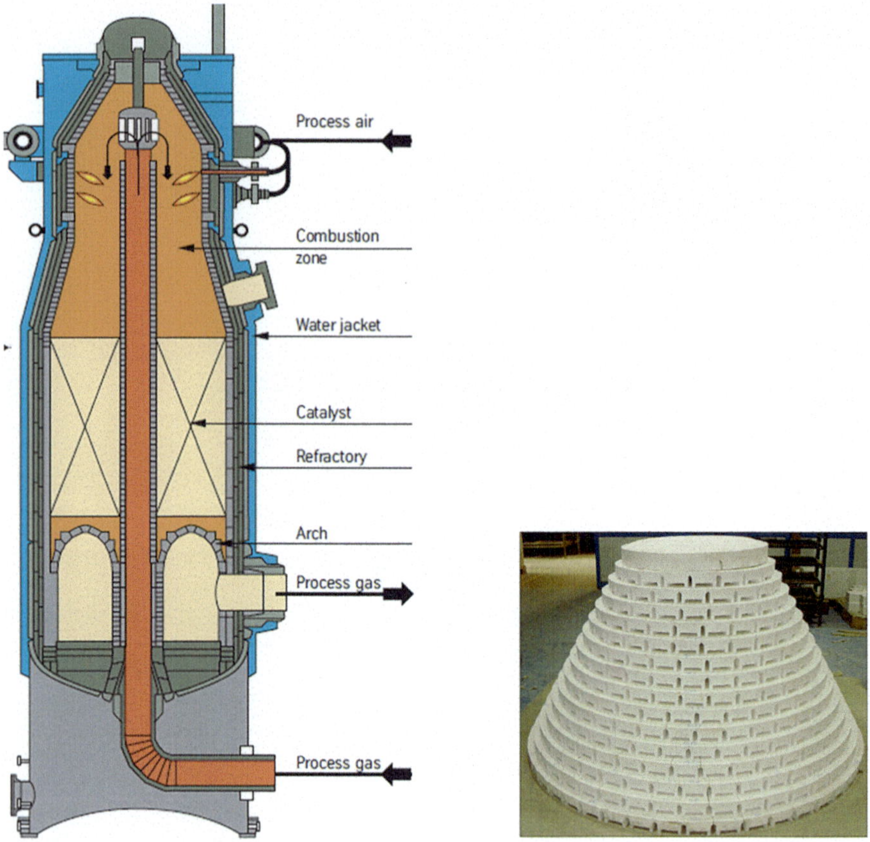

A B

Fig. 6.16 (**a**) Schematic drawing of the secondary reformer. (Courtesy-Thyssen Krupp), (**b**) Catalyst support

Process – The process gas leaves the primary reformer and is fed into the secondary reformer (Fig. 6.10). The feeding temperature of the process gas is in the range of 800–850 °C. The gas temperature increases at the secondary reformer combustion section to approximately 1300 °C(Fig. 6.16). The gas on passing through the catalyst bed temperature drops to approximately 1000 °C due to the endothermic reaction. The operating pressure of modern units is around 40 bars. Due to the catalyst in between gas inlet and outlet of the secondary reformer, a pressure drop occurs. It is extremely important that the refractory design does not enable process gas bypassing through the lining. For this reason, the hot face refractory layer has to have low gas permeability, and the expansion joints are to be closed during operation of the unit at higher temperatures.

The refractory lining of the secondary reformer is shown in Fig. 6.16 [9]. The primary function of the refractory lining system in the gasifier is to protect the shell from thermal and corrosive attack and then to insulate walls of hydrogen pressure vessels and the transfer line. The cylindrical vessel may have 5–6 layers of different refractory lining.

The thermal conductivity of hydrogen is 7 times that of air [6]. The viscosity of hydrogen gas is almost half of that of air. Under high pressure inside the reformer, hydrogen gas permeates through the pores and fissures in the refractory lining, and the overall thermal conductivity of the lining gets increased. In general the thermal conductivity of an insulating castable (with 50% porosity) can be 1.5–2 times more, in a gas, containing 50% hydrogen.

The operating condition inside the reformer is very arduous. The temperature at the top is 1200–1400 °C [5] and the mixture of gases, e.g., hydrogen, carbon monoxide, carbon dioxide, steam, and other gases, exists at a very high pressure of about 40 Bar. A typical analysis of the gas mixture inside the reformer is given in Table 6.2 [10]. The reforming reaction of methane lowers the temperature during passage of the gas mixture through catalyst bed and at the exit of the reformer temperature remains at about 1000 °C [11].

Under such condition, alumina is found to be most suitable material. The presence of SiO_2 in the refractory body is most unwanted as under the highly reducing atmosphere, SiO_2 is reduced to SiO (reaction −1. 20 and 1.21) and migrated SiO is found to get condensed in the gas transmission pipes. This phenomenon makes the refractory lining porous and weak.

Properties of different refractories used in the different areas of secondary reformer is given in Tables 6.3 and 6.5. The catalyst support in the secondary reformer is made of 99% alumina bricks as shown in Fig. 6.16.

It is important to know that the vertical and horizontal expansion joints are determined in accordance with the design temperature, i.e., during cold shutdown, the joints must be opened and during operation closed again. Operating temperatures more than the designed temperature would cause generation of additional forces due to constrained expansion. During the shutdowns, the joints are to be cleaned; grains or dust should be removed; the condition of the bricks around the joints are to be checked with regard to the edges, quantity of cracks, and spalling. The vertical and horizontal expansion joints should be checked and maintained during each shutdown period (Table 6.4).

PG Cooler- The synthesis gas of the secondary reformer will be introduced to the inlet chamber of the PG Cooler (Process Gas Cooler). The gas temperature is in the range of 1000 °C, and the gas pressure is around 40 bar. The gas will be cooled down by passing through the heat exchanger tubes to the outlet chamber. The PG Cooler inlet area is typically lined with a dense material for the hot face layer and an insulating backup layer. Due to the process conditions and the turbulent gas flow, the integrity of the refractory lining is very important. Special attention has to be

Table 6.2 Typical analysis of the gas mixture in secondary reformer

H_2	N_2	CH_4	CO	CO_2
69.0%	0.35%	9.67%	8.98%	9.67%

Table 6.3 Refractory qualities used in different areas of secondary reformer

Product	Burner	Dome	Cylinder	Bottom	Outlet	Support	Trnsf. line
Dense bricks with							
60% alumina						Y	
99% alumina		Y	Y	Y	Y		
90% alumina		Y	Y	Y	Y		
Dense castable with							
45% alumina				Y			
95% alumina				Y			
Insulating castable							
70% Al_2O_3,low Fe_2O_3	Y	Y					
Bubble alumina base			Y				Y
Insulating bricks							
45% alumina		Y	Y	Y	Y		
55% alumina		Y					
65% alumina		Y					
98% alumina		Y					
Pre cast shape							
80% alumina	Y				Y		
99% alumina	Y	Y			Y		
Mortars							
90% Al_2O_3 heat set			Y	Y	Y		
40% Al_2O_3 heat set	Y	Y	Y	Y	Y		

paid to the transition area of the tube sheet lining and cylindrical brick lining as well as the lining of the conical section. Maintenance activities are limited to checking of the expansion joints during each shutdown period. The outlet section of PG Cooler because of lower temperature of the gas is lined with medium heat duty refractory (Fig. 6.17) [12].

6.3.3 Precautions in Refractory Lining

As the secondary reformer operates under very high pressure, some special precautions to be taken for refractory lining as given below

(a) Dry-out and initial heating-up of the refractory lining is an extremely important process which needs special attention. The water will be there in both as physically bonded water and chemically bonded water. A standard reformer for its total refractory lining consumes about 20,000 Liter of water which remains inside the lining. The physically bonded water will be removed within 200C, but it needs special care to remove all chemically boded water. If the water is not removed completely and the reformer is taken in operation under high pressure, nothing will happen, but if by any means the pressure inside drops, the

Table 6.4 Properties of different refractory qualities used in the secondary reformer

Products	Properties							
	Al_2O_3%	Fe_2O_3%	SiO_2%	BD gm/cc	AP %	CCS (MPa)	Th.Cond W/mK at 400 °C	Max. Service Temp. °C
Dense bricks								
60% Al_2O_3 brick	60	<1.0	–	2.55	18	>60	1.6	1550
90% Al_2O_3 brick	>90	<0.2	>9	3.0	<17	100	2.80	1650
99% Al_2O_3 brick	>99	<0.1	–	3.30	18	95	3.80	1700
Dense castables								
45% Al_2O_3	>45	<2.0	–	–	–	30	–	1400
97% Al_2O_3	>97	<0.05	<1.5	3.2	–	80	3.20	1700
Insulating castable								
70% alumina low Fe_2O_3	–	–	–	–	–	–	–	–
Bubble alumina base	95–96	<0.04	<3.0	1.42	–	20	1.05	1700
Insulating brick								
50% alumina	>50	<1.5	–	0.5	–	10	<0.47	–
55% alumina	>55	2.5	–	–	–	10	0.52	1450
65% alumina	>65	<1.0	–	1.0	–	4	0.59	1650
98% alumina	98	0.2	<0.3	1.50	–	8–30	1.10	1700
Precast shape								
80% alumina	>78	0.5	18	2.80	–	90	–	1600
99% alumina	>99	<0.1	<1.0	3.30	–	100	3.30	1700

water may suddenly get vaporized with high volume and will cause an explosion. During heating up in actual practice, the thermocouple is to be fitted at different temperatures, and the time temperature curve is monitored to ensure proper drying (Fig. 6.18) [13].

(b) Pressurization and depressurization always need to be done very carefully because of the nature of refractory products with the inelastic material structures and the less gas permeability of the hot face layer material typically installed in secondary reformers and PG Cooler inlet chambers. For refractory linings with more than one layer, pressure equalization is a function of time in case of non-open gaps and joints. The time required for pressure equalization is mostly a function of the gas permeability of the dense refractory layer, only. In case of a too fast depressurization, the gas trapped in the backup layers cannot escape fast enough due to the dense hot face layer. This can cause considerable damages to the lining system. The quantity of gas passing through the

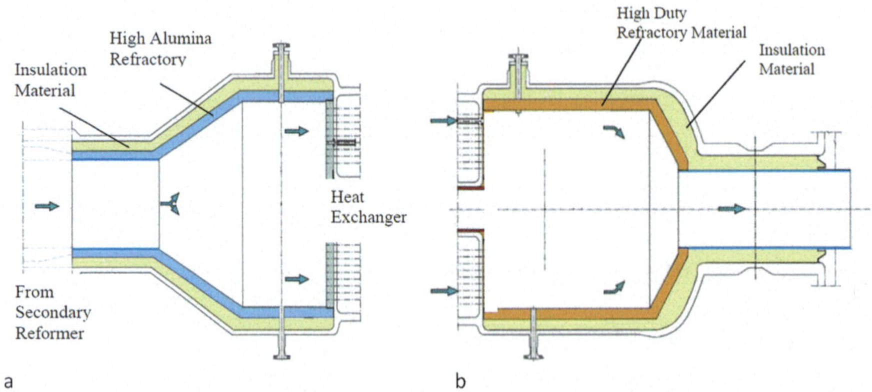

Fig. 6.17 Refractory lining scheme of (**a**) PG Cooler inlet and (**b**) PG Cooler outlet

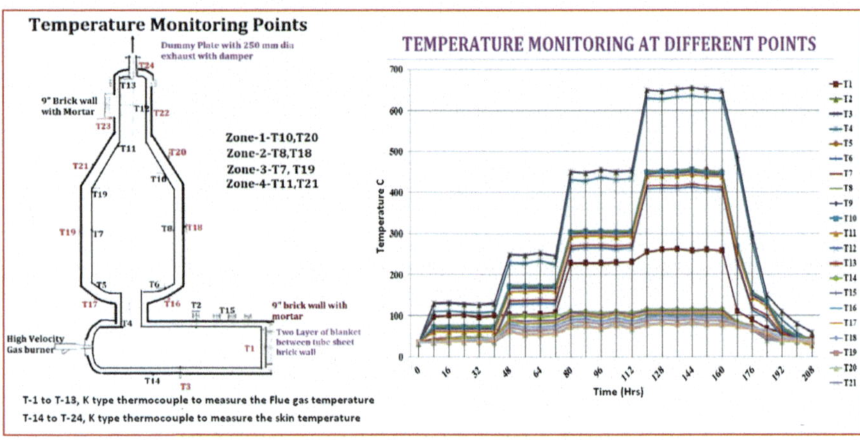

Fig. 6.18 Scheme of preheating of the secondary reformer after installation of refractory

refractory system depends on the gas permeability of the dense layer. In the normal practice, the pressure is raised at the rate of 0.5 bar/min to 2.5 bar. Then it is kept constant for 10 min, and again the pressure is increased in the same rate. In case of depressurization also the pressure is reduced in the same fashion as 0.5 bar/min and kept constant for 10 min, and the pressure again is reduced again in the same rate [13].

As it is mentioned in Chap.4, the design and installation of refractory is as important as the quality of the refractory. If the design of the lining and installation is not proper, the best quality of refractory also does not work. A case history of a refractory failure in dome area is given here as an example.

Table 6.5 Technical details of the reformer

Internal diameter	2883 mm
Shell MOC	SA516 Gr70 carbon steel
Refractory lining of shell	184 mm insulation +127 mm dense alumina brick
Water jacket	From bottom dome up the shell boiler fed water at temp 100 °C
Outlet temperature	950–980 °C
Operating pressure	3400 KPa
Pressure drop	Normally 117 KPa but during failure 900 KPa
Reformer outlet gas composition	11% CO, 6% CO_2, 0.6% CH_4, 30% N_2, 50% H_2
Dew point	177 °C
Support dome material	98% alumina bricks and tiles with 120 pcs skew brick. Single layer of tapered brick in the dome of radius 1981 mm and thickness 229 mm
Catalyst bed	About 10 ton of catalyst and alumina balls

6.3.4 Case Study

There was an instance of collapse of dome refractory construction in a secondary reformer. Its dimension details are given in Fig. 6.19 [14]

The details of its operational parameters are shown in Table 6.5.

The temperature rate of heat up during starting and rate of cooling during shutdown is shown in Table 6.6.

The reformer was running steadily for 5 days after the commissioning of the reformer when a shutdown was taken because of higher temperature in discharge of natural gas compressor. After fixing the problem, it was started, and after 8 hrs, the pressure drop shooted up to 825 KPa from 104 KPa in 5 min and went further to 909 KPa in another 3 min. The outlet gas temperature also went up to 1080 °C about 100 °C above normal and the plant tripped. The furnace was cooled down. The manway below the dome was opened, and it was found the dome was failed in number of locations. The investigations were made, and the following was found as the reason of failure.

When the reformer operation started, the steel shell was heated from room temperature up to 120 C (the temperature of the water cooled jacket). For a diameter of 2883 mm, it expanded by 3.5 mm considering its coefficient of expansion as 12.5×10^{-6} mm/mm/°C. At the same time, the alumina brick lining temperature went up to 950C, and its expansion was 22 mm considering its expansion coefficient as 8×10^{-6} mm/mm/°C. To keep the structure stable, it is imperative that the excess expansion of 22–3.5 = 18.5 mm expansion is to be accommodated. This could also be accommodated by the rise of height of the dome, but the load of 10 ton of catalyst and alumina ball did not allow it to do that, and as a result, the lining collapsed.

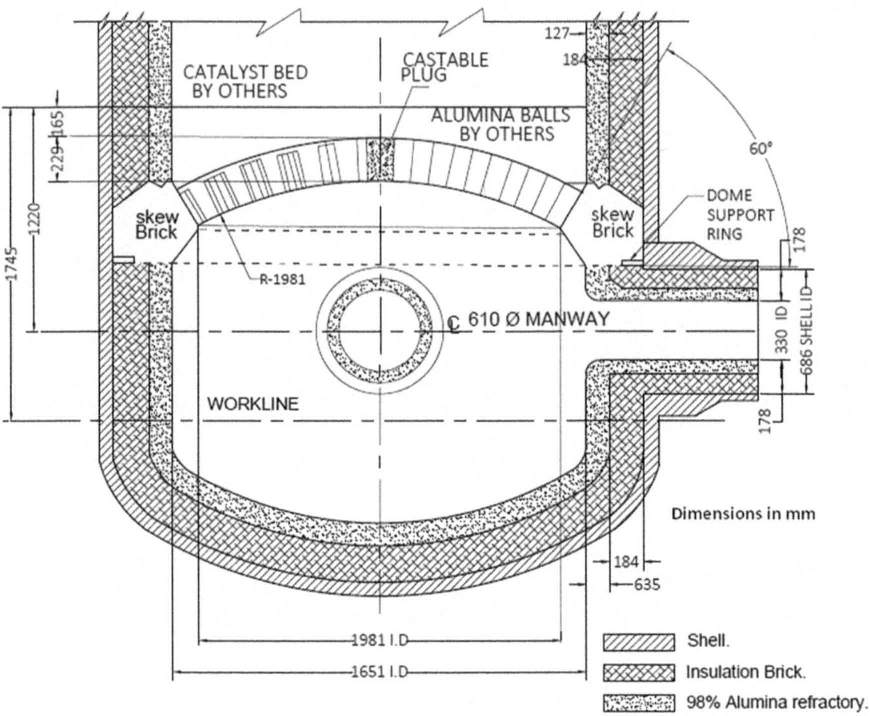

Fig. 6.19 Bottom portion of the secondary reformer

Table 6.6 Heating up and cooling down rate of the reformer

Heatup			Cool down		
450 °C	670 °C	@ 100 °C/h	980 °C	760 °C	@ 315 °C/h
670 °C	925 °C	@ 230 °C/h	750 °C	550 °C	@ 25 °C/h

References

1. Global Syngas Overview (2018) PPT by STRATAS ADVISORS
2. Korobitsyn MA, Van Berkel FPF, Christie GM (2000) Review of synthesis gas processes
3. Encyclopedia of hydrocarbons, p 473
4. Jhonson RC, Crowley MS (2005) State of the art refractory lining for hydrogen reformer vessel. UNITECR
5. Dial RE High Alumina Refractory material for gas reforming. Carborundum Co, USA
6. Primary reformer, PPT – Hampa Energy Engineering and design company (HEEDCO)
7. Dey R et al (2005) Characterisation of thermo mechanical behavior of High alumina Refractories in secondary reformer application. UNITECR
8. Bansal IN, Shukla RL (2008) Sudden failure of the transfer line and in situ repair of Refractory and replacement of SS liner, Ammonia technical manual
9. Peirson M, Millsape D (2014) Reforming furnace reliability: engineered refractory solution, Ammonia technical manual

10. Quintilliani J, Russel W Improving flue gas tunnel reliability. Blasch Precision Ceramics
11. Ahmed SKS, Haque ME (2012) Operation and maintenance of secondary reformer – lesson from the plant experience. J Chem Eng. IEB 27(2)
12. Ballon M et al (2006) Design and maintenance of refractory linings in the reformer section of the Ammonia plants, Ammonia technical manual
13. Baboo P (2019) Secondary reformer refractory dry out operation. Int J Eng Res Technol 8(05)
14. Alibadi N et al (2016) Failure of secondary reformer dome and its repair, Ammonia technical manual

Chapter 7
Refractories for Coal Gasification

7.1 Introduction

Coal gasification produces H_2 and CO the same as syngas, and the basic principle and chemistry involved are the same. Coal gasification units partially burn coal with pure oxygen and steam to generate synthetic gas or "syngas." Coal gasification units offer unique flexibility in that the syngas can also be chemically converted to liquid fuels and other chemicals like ammonia, methanol, etc.; alternatively the hydrogen can be refined from the syngas to produce hydrogen fuel as shown in Fig. 7.1.

The key advantage of coal gasification process is the reduction in environment pollution. The problems inherent in coal combustion are well-known from the particulates to the acid rain caused by high-sulfur coal, to mercury pollution, not to mention the CO_2. One of the major environmental advantages of coal gasification is the opportunity to remove impurities such as sulfur and mercury (by 95%) and soot *before* burning the fuel, using readily available chemical engineering processes. The sulfur can be collected in elemental form for further use. 90% of the CO_2 emission from the gasifier can also be captured and processed for utilization and storage. The nitrogen in the coal is liberated as N_2 with small amount NH_3 and HCN which can be removed easily.

Modern gasification process can convert any carbon containing feedstock to syngas which in turn can be converted to valuable chemicals like ammonia, methanol, Fischer-Tropsch liquid, synthetic natural gas (SNG), etc. or can be utilized to produce electricity (Fig. 7.2) [1]. Presently the proportion of different products made out of gasification process is shown in Fig. 7.3.

In addition, the ash produced is in a vitreous or glasslike state which can be removed in a much easier way than the ash usually formed in normal combustion process and can be recycled as concrete aggregates. The other important advantage of this process is that the low-grade coal can be used for generation of power through syngas.

P. Sengupta, *Refractories for the Chemical Industries*,
https://doi.org/10.1007/978-3-030-61240-5_7

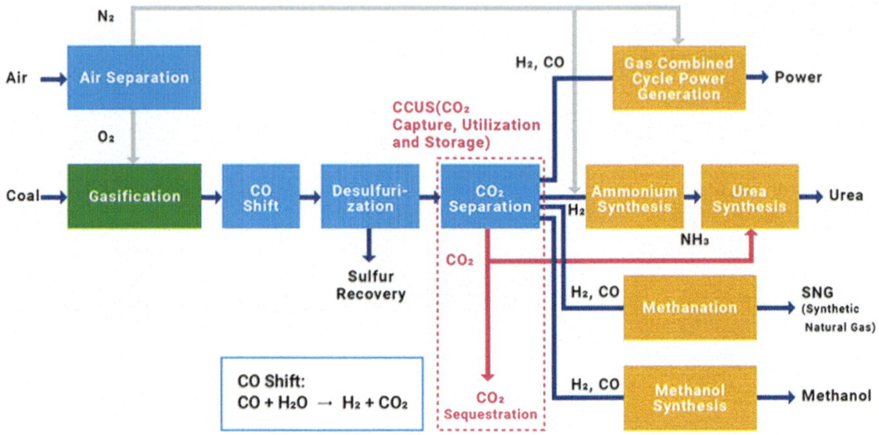

Fig. 7.1 The flow sheet of gasification process to produce chemicals and power

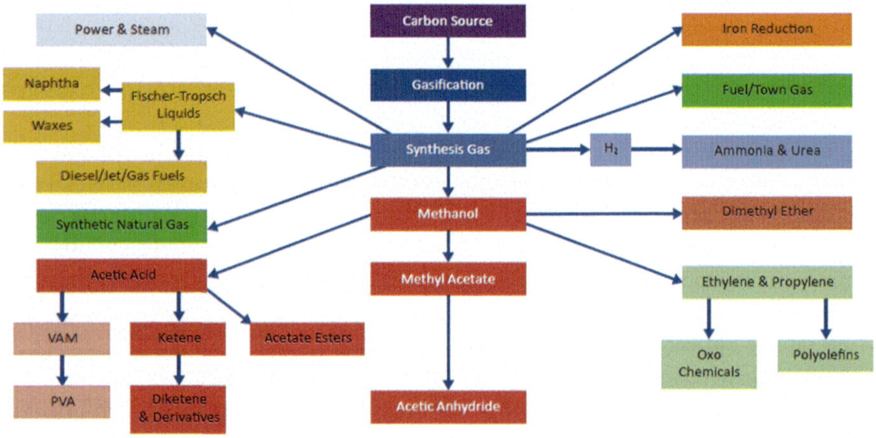

Fig. 7.2 Products that can be obtained from gasification process

The difference between the manufacturing of syngas by reforming of natural gas and by coal gasification is that unlike the reforming of natural gas which uses very expensive catalysts and alloy steel tubes, coal gasification process does not require those.

Gasification plants are becoming larger in capacity with the passing days as shown in Table 7.1 [2].

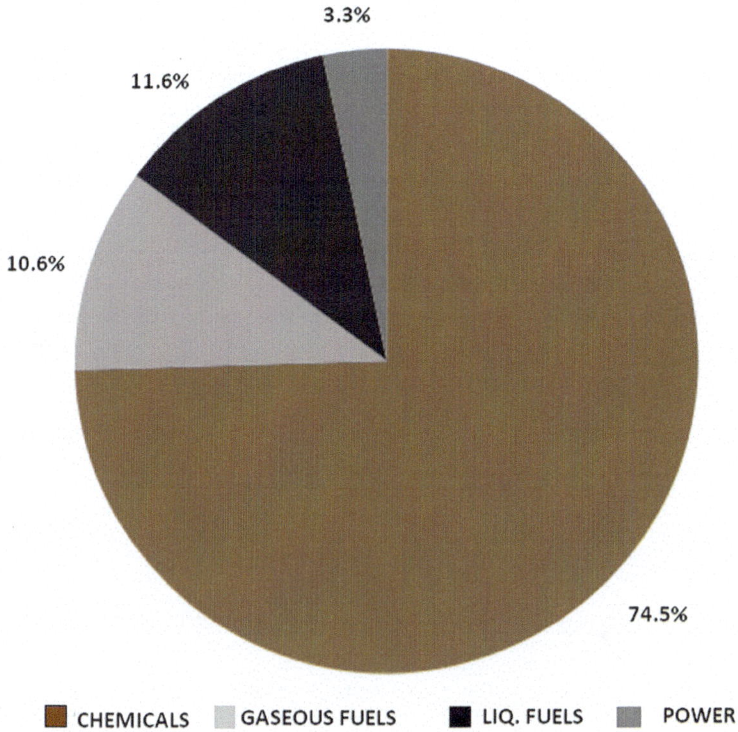

Fig. 7.3 Proportion of the syngas used for manufacturing different products

7.2 Process

Although the modern gasification process can use any carbon-bearing feedstock, the majority of the feedstock used are coal and natural gas as shown in Fig. 7.4 [3]. Refinery residuals like petcoke can be used as well as feedstock.

These materials are ground in ball mill or any other mill to fine powder of about −250 mesh and fed into the reactor chamber of gasifier, at high temperature and high pressure under reducing condition. Within the gasifier, a series of complex chemical reactions take place producing heat and syngas, as per the chemical reaction:

$$C + H_2O + O_2\,(\text{insufficient}) \rightarrow CO + H_2 + CO_2 \tag{7.1}$$

By-products from the gasification reaction are H_2S, CH_4, HCN, NH_3, N_2 [4, 5], and unreacted carbon, mineral, and organo-metallic impurities. The impurities itself and the products formed due to their interaction can be in solid, liquid, and gaseous forms. The liquid, when formed, adheres on the wall and flows down the gasifier wall and corrodes the refractory. Gasifier operates at high temperature and pressure

Table 7.1 Top 27 gasification projects across the globe

Gasification plant	Location	Technology	No. of gasifier	MWtHr SG output	Start-up year	Feed/product
Pearl GTL	Qatar	Shell	18	10,936	2011	NG/FTL
Yinchuan CTL plant	China	Siemens	22 + 2	9300	2016	Coal/FTL
Datang Ningxia SNG plant	China	SEDIN	45 + 3	7125	2015	Lignite/SNG
Sasol Synfuels West	S. Africa	Lurgi FBDB	40 + 0	7048	1977	Coal/FTL
Sasol Synfuels East	S. Africa	Lurgi FBDB	40 + 0	7048	1982	Coal/FTL
CHNG Xinjiang SNG Plant	China	TPRI	7 + 1	6450	2014	Coal/SNG
Reliance	India	E-gas	6 + 2	5000	2015	Petcoke/ Electricity
Jazam IGCC	Saudi Arabia	Shell	16	4465	2016	Refinery residue/ Electricity
Yankuan Yulin CTL	China	OMB	8	3733	2015	Coal/FTL
Yulin Methanol Plant	China	GE	10 + 4	3383	2015	Coal/Methanol
Inner Mongolia Chemical Plant	China	Shell	3 + 0	3373	2015	Lignite/ Methanol
Yitai CTL Plant	China	OMB	4 + 1	2800	2014	Coal/CTL
Yil SNG Plant	China	Siemens	7 + 1	2500	2015	Coal/SNG
Senhua Ningxia	China	SEDIN	14 + 2	2500	2015	Coal/Methanol
Xinnan Yli SNG Plant	China	SEDIN	20 + 2	2500	2014	Coal/SNG
Senhua Ningxia	China	Siemens	5	1912	2011	Coal/Methanol
Great Plain Synfuel plant	USA	Lurgi	12 + 2	1900	1984	Lignite/ Methanol
Qinghua Yili SNG plant	China	SEDIN	15 + 1	1700	2013	Coal/SNG
Xinjiang Guanghui Hami	China	SEDIN	12 + 2	1750	2012	Sub-bituminous coal/Methanol
Shenhua Baotou Coal to Olefin Plant	China	GE	5 + 2	1750	2011	Coal/Methanol
Yunnan MPG Plant	China	BGL	6 + 2	1680	2013	Coal/Methanol
China Coal Fertilizer Plant	China	BGL	6 + 1	1680	2014	Coal/ammonia
Hexigten SNG Plant	China	SEDIN	12 + 2	1670	2012	Coal/SNG
Gwanyang SNG	S.Korea	E-gas	3 + 1	1565	2014	Coal/SNG

(continued)

Table 7.1 (continued)

Gasification plant	Location	Technology	No. of gasifier	MWtHr SG output	Start-up year	Feed/product
Neimenggu Methanol Plant	China	SEDIN	11 + 1	1558	2015	Coal/Methanol
Rongxing inner Mongolia Methanol Plant	China	OMB	2 + 1	1300	2014	Coal/Methanol
Wehe Pucheng Methanol Plant	China	GE	4 + 2	1400	2014	Coal/methanol

FTL Fischer-Tropsch liquid

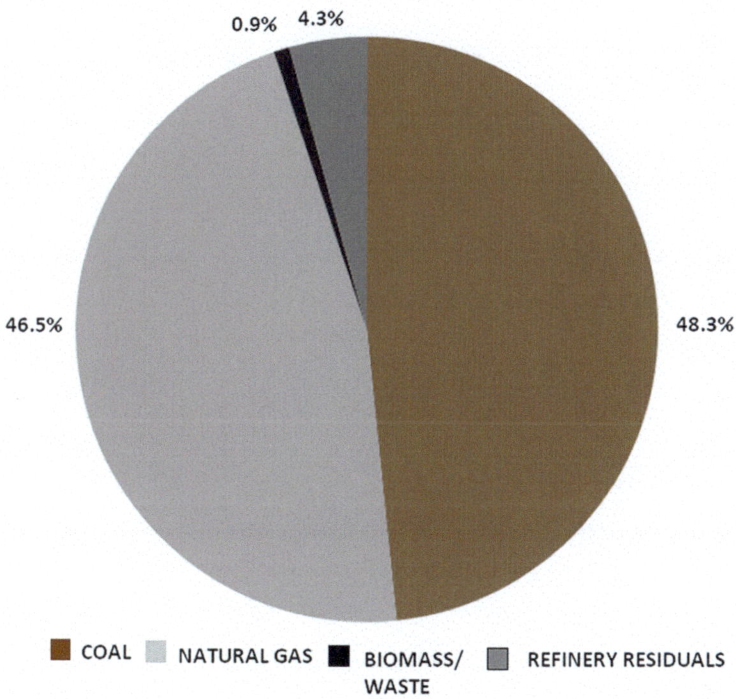

Fig. 7.4 Percentage of different feedstock used for the production of syngas

in a reducing environment (1325–1575 °C, 2–6 MPa and O_2 partial pressure between 10^{-6} and 10^{-9}). Depending upon the gasifier size, the refractory can be exposed to 200 MT of slag per day. The slag quality and quantity depends upon the quality of carbon feedstock used in gasifier.

Different designs of gasifier are commercially available patented by different companies. The major types of gasifying processes are categorized, based on the reactor bed where coal is gasified, as fixed bed gasifier, fluidized bed gasifier, and entrained flow gasifier. The schematic drawing of the different processes with the temperature profile of the coal and the gas are shown in Fig. 7.5 [6, 7]. Fluidized bed

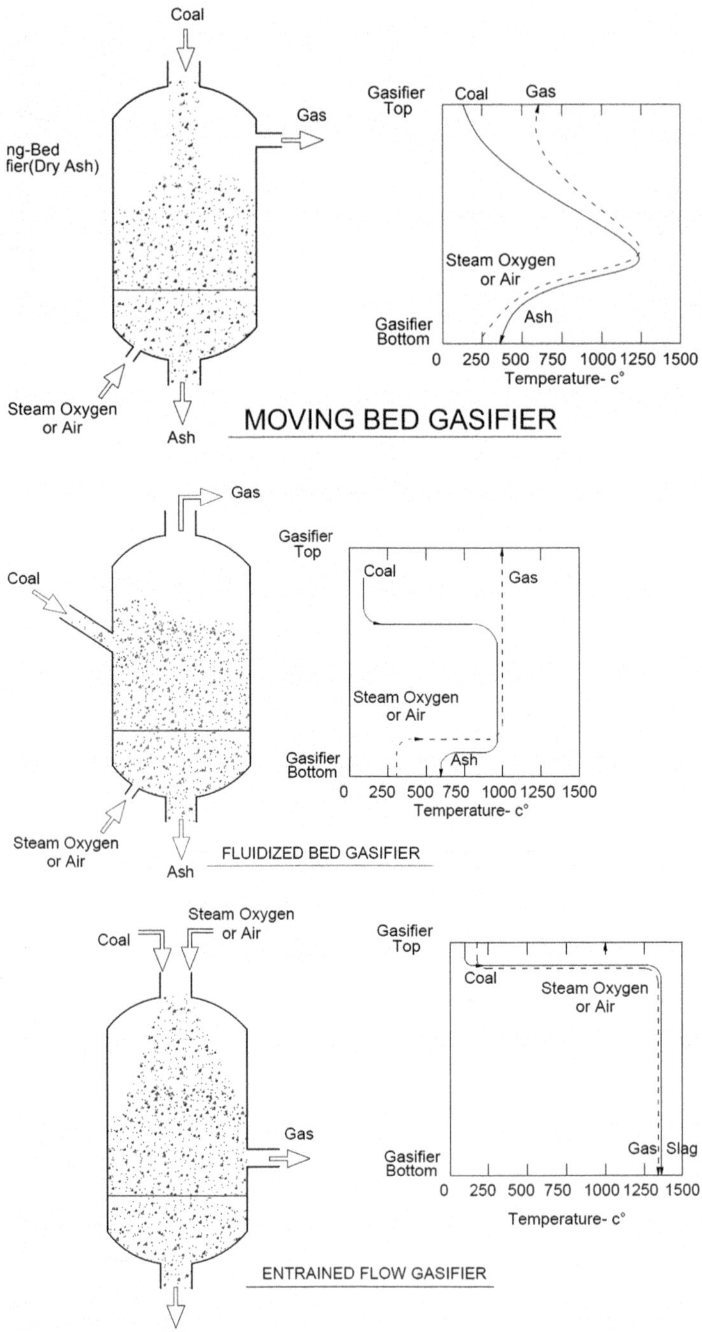

Fig. 7.5 Different processes of gasification

Table 7.2 Temperature profile of different gasification processes

Parameters	Fixed bed	Fluidized bed	Entrained bed
Residence time	1–3 h	20–150 min.	0.4–2 s
Feed (coal) size	6–50 mm	6–10 mm	< 100 microns
O_2/coal	0.14–0.81	0.25–0.97	0.28–1.17
Steam/coal	0.28–3.09	0.11–1.93	0.10–1.20
Coal type	Most types	Non-coking coal	All types
Temp. range °C	420–650	900–1050	1250–1600
Pressure range (atm)	1–10	10–30	30–80
Product gases Mol (%)			
$CO + H_2$	40–70	60–80	65–95
CH_4	5–15	3–5	0.1–2
HHV (Btu/SCF)	250–320	300–400	300–550
Efficiency	Low	Medium	High

gasifiers operate below ash fusibility temperature (AFT) and produce dry ash. Entrained flow gasifiers operate at very high temperatures and, therefore, produce vitreous slag, while fixed bed gasifiers generate dry ash or slag depending upon the design temperature. The details of the operational parameters of different processes mentioned are given in Table 7.2 [8].

The entrained flow gasifiers are called slagging gasifier where the interaction of slag with refractory is the most important. Refractories for entrained flow gasifier will be discussed here in details. The key characteristics of entrained flow gasifiers are their high uniform temperatures (more than 1000 °C) and very short residence time of the fuel within the gasifier. For this reason, feedstock to the gasifier must be very finely divided and homogeneous, which indicates that, feedstocks, such as biomass or wastes, which cannot be readily pulverized are not suitable for entrained-type gasifier.

7.3 Gasifier Design

The refractory lining used in gasifiers varies with design. In general, the refractory linings can be classified as air-cooled or water-cooled type. The water-cooled refractory lining is also called as a membrane-wall design with stainless steel tubes to carry water/steam, covered with a thin layer of Al_2O_3-SiC refractory (Fig. 7.6a) [8, 9]. When water or steam is used as cooling medium in the furnace shell, a steep temperature gradient is formed between furnace shell and thin-walled refractory layer. This steep gradient causes the molten slag generated during the gasifier operation to solidify, thereby protecting the membrane wall from corrosion and erosion.

On the other hand, air-cooled refractory lined design is traditionally a thick multi-layered refractory lining to reduce heat loss (Fig. 7.6b) [10]. This kind of design is relatively inexpensive compared to water-cooled design. The downside of

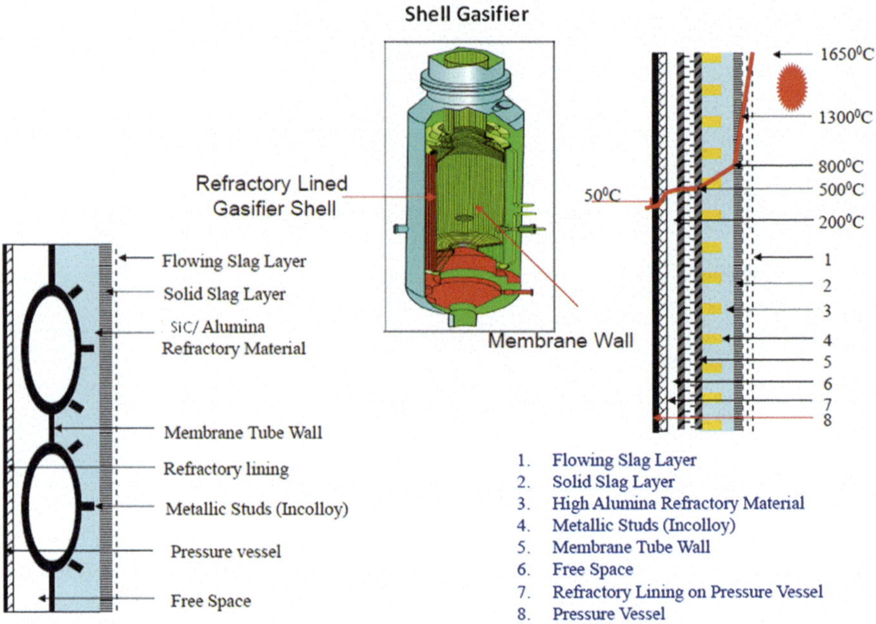

Fig. 7.6a Water-cooled refractory lining

the design is that the campaign life is sometimes extremely short, ranging from anywhere between 4 and 24 months. Studies have identified that the refractory service time as one of the factors limiting the availability of the air cooled refractory-walled gasifiers.

Refractory lining wear out and degradation in a gasifier is mainly because of its interaction with molten slag formed during the gasification process. Therefore the characterization and prediction of slag quality is absolutely necessary for the selection, control, and the improvement of refractory lining in a gasifier.

7.3.1 Characterization of Slag

To characterize a slag, the knowledge of the geological formation of the coal and the associated minerals occur with the coal deposit is very important. Coal ash originates from different impurities which occurred in coal, like, clay, quartz and its polymorphs, pyrophyllite, pyrites, mica, feldspar, etc. During burning of the coal, the minerals are converted to oxides (ash), and these oxides interact among themselves at high temperature to form slag. The impurities in the coal vary widely depending upon its origin. Therefore, the composition and character of the slag obtained from coal of different origins varies widely. The composition of ash from

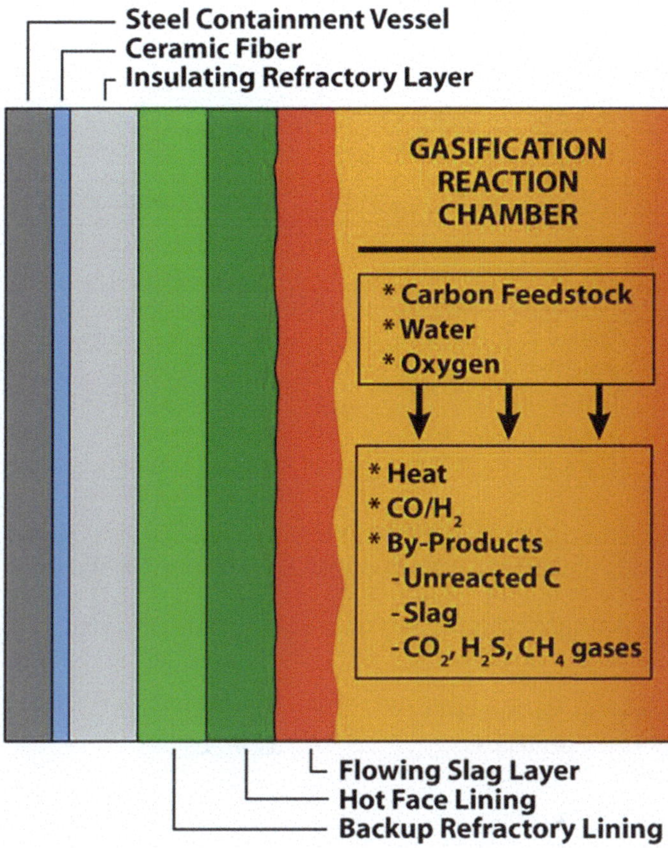

Fig. 7.6b Refractory lining – air-cooled type

petroleum coke is also quite different from that of the coal ash. The composition of ashes from different source of coal is furnished in Table 7.3 [11].

Mechanisms [12] by which the slag can degrade the refractory are:

1. Refractory dissolution
2. Refractory particles debonding
3. Slag refractory elemental replacement
4. Slag refractory reaction producing new phases
5. Slag penetration in the refractory pores
6. Temperature of critical viscosity
7. Slag-wear-abrasion
8. Stress and thermal cycle loading resulting in fragmentation and spalling

Table 7.3 Chemical composition of coal ash from different origin

	Range (mass%)				
Component (%)	Europe	US	China	India	Australia
SiO_2	28.5–59.7	37.8–58.5	35.6–57.2	50.2–59.7	48.8–66.0
Al_2O_3	12.5–35.6	19.1–28.6	18.8–55.0	14.0–32.4	17.0–27.8
Fe_2O_3	2.6–212	6.8–25.5	2.3–19.3	2.7–14.4	1.1–13.9
CaO	0.5–28.9	1.4–22.4	1.1–7.0	0.6–2.6	2.9–5.3
MgO	0.6–3.8	0.7–4.8	0.7–4.8	0.1–2.1	0.3–2.0
Na_2O	0.1–1.9	0.3–1.8	0.6–1.3	0.5–12	0.2–1.3
K_2O	0.4–4	0.9–2.6	0.8–0.9	0.8–4.7	1.1–2.9
P_2O_5	0.1–1.7	0.1–0.3	1.1–1.5	0.1–0.6	02–3.9
TiO_2	0.5–2.6	1.1–1.6	02–0.7	1.0–2.7	1.3–3.7
MnO	0.03–0.2	nd	nd	0.5–1.4	nd
SO_3	0.1–12.7	0.1–2.1	1.0–2.9	nd	0.1–0.6
LOI	0.8–32.8	0.2–11.0	nd	0.5–5.0	nd

The characterization of the slag formed in the gasifier and its interaction with the refractory is very important for the proper design and selection of the refractory for the gasifier.

The characterization of the ash is done by studying of the following properties [13]:

1. Ash fusion temperature (AFT)
2. Viscosity vs temperature
3. Flow property
4. Basicity

The conventional method of determination of ash fusion temperature (AFT) involves heating of a pyramid of ash at 5–10 °C/min from 1000 to 1600 °C. The time taken for the test is about 1–2 h. The results of an AFT analysis consist of four temperatures; these are the initial deformation temperature (DT), softening temperature (ST), hemispherical temperature (HT), and flow temperatures (FT) [14].

(i) Deformation temperature, when ash just begins to flow as shown by the first sign of deformation of pyramid.
(ii) Softening temperature is recorded when the height of the ash becomes equal to the width of the base.
(iii) Hemispherical temperature is noted when the height of the fused ash becomes equal to half of its width.
(iv) Flow temperatures is noted when the height becomes a 16th of the width.

Basicity is another important index to characterize the ash which ultimately forms slag on fusion. Theoretically acidic refractories should be used where the slag is acidic and the basic refractory where the slag is basic. But in practice, neutral refractories are used when the slag is acidic.

The ratio R = $SiO_2/(SiO_2 + CaO + MgO + Fe_2O_3)$ should be more than 0.86 for an ash having good slagging behavior.

The basicity vs acidity ratio, B/A = $(Fe_2O_3 + CaO + MgO + Na_2O + K_2O)/(Al_2O_3 + SiO_2 + TiO_2)$, is an important indicator which describes the slag behavior at high temperature. The value above 0.11 indicates higher tendency to slag formation [14].

Slag behavior at high temperature is very important for smooth operation of gasifier. The viscosity of the slag is very important from the view point of its interaction with refractory, and it should be low enough to ensure a good flow, and it should be high enough to resist the penetration to the refractory pores ensuring lesser corrosion. Viscosity between 15 and 25 Pa.s ensures consistent flow of the slag necessary for smooth operation of gasifier [12, 13, 15]. The behavior is affected by the properties of both gaseous and solid phases, gasifier wall structure, fluid dynamics, and operating conditions. Under reducing environments (gasification), the viscosity at a given temperature is generally lower than that under oxidizing atmospheres (conventional combustion). Slag behaves as Newtonian fluid at high temperature, and its viscosity gets reduced logarithmically with increasing temperature. Crystallization or separation of a phase, at lower temperature, increases the viscosity of a liquid slag dramatically. Slag viscosity depends strongly on chemical composition of the slag and temperature, and the relevant data can be used to design the gasifier and its operational parameters.

Oxygen potential in gasifier affects the:

(a) Valence of the transition metal in the oxides
(b) Oxide basicity
(c) Basicity of the slag
(d) Melting point of the slag

Therefore, the oxygen potential in the gasifier influences the slag refractory interaction and experience indicates that oxygen potential varies in gasifier from 10^{-7} to 10^{-9} [10]. Figure 7.7 shows the schematic drawing of a gasifier [16].

The slag residence time in the entrained type gasifier is short and therefore complete equilibrium may hardly be attained. The slag residence time depends upon the dimension of the gasifier, and the slag viscosity depends upon the slag chemistry and temperature of operation. The average velocity of a cross section of a slag film over the wall can be calculated using equation below [17]:

$$v = \frac{\rho g \delta^2 \cos \beta}{3\mu} \tag{7.2}$$

v the average velocity of the slag m/s
ρ the density of the slag Kg/m^3 considered as 2500 Kg/m^3
g acceleration due to gravity 9.81 m/s^2
β inclination angle relative to vertical (for vertical wall, it is zero)
μ viscosity of the slag Pa.s

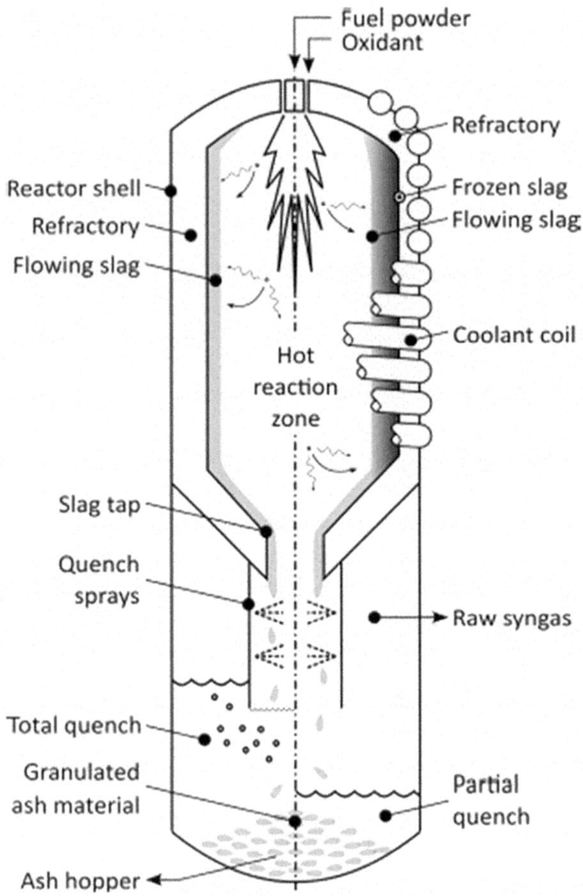

Fig. 7.7 The entrained type gasifier – a schematic diagram

δ the thickness of slag film(m) which can be calculated using an equation proposed by Bird et al. as follows:

$$\delta = \left(\frac{3\mu\omega}{\rho^2 g W \cos\beta} \right)^{\frac{1}{3}} \tag{7.3}$$

W = circumference of inner wall for vertical sections of reactor or average circumference of inner wall for inclined sections (m)

ω = slag mass flow (kg/S)

The depth of corrosion of the slag in the Al_2O_3-Cr_2O_3 ceramics depends upon the CaO content of the slag because the viscosity of the slag decreases with increasing CaO content as shown in Fig. 7.8 [18].

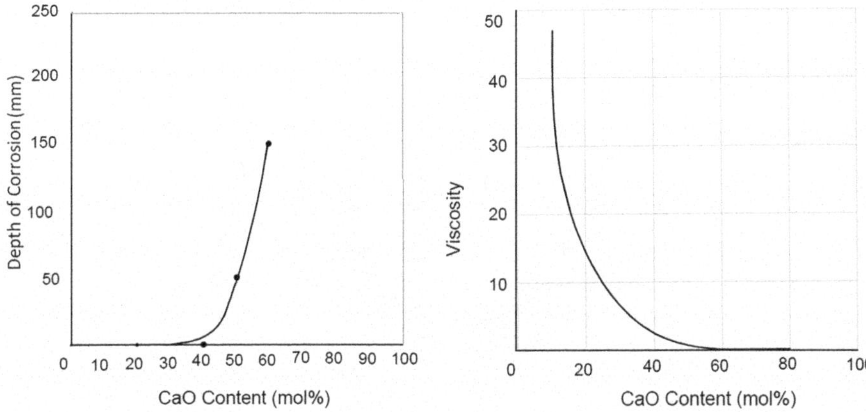

Fig. 7.8 Effect of CaO content in the slag on (**a**) corrosion and (**b**) viscosity

Table 7.4 The wall type in the entrained flow gasifier of different design

Process design by	Stages	Feed	Flow	Wall type
Shell	1	Dry	Up	Membrane
GE	1	Slurry	Down	Air cooled
Siemens	1	Dry	Down	Membrane
E-gas	2	Slurry	Up	Air cooled
Eagle	2	Dry	Up	Membrane
Prenflo	1	Dry	Up	Membrane
MHI	2	Dry	Up	Membrane

7.4 Refractories Used for Lining

The refractory selection for the gasifier wall depends primarily upon the design of the gasifier, whether it has water-cooled (membrane wall) or air-cooled wall. Table 7.4 shows the refractory design in reactor wall of entrained flow gasifier of different design [19].

7.4.1 Refractories for Membrane Wall

The conventional approach to contain the aggressive molten slag in metallurgical furnaces is to use high-quality dense refractory material that is in direct contact with molten slag. Under this condition the refractory progressively is degraded and must be replaced periodically to run the furnace continuously. The refractory life depends majorly on the rate of reaction of the refractory and the slag at operational temperature although many other factors play their role.

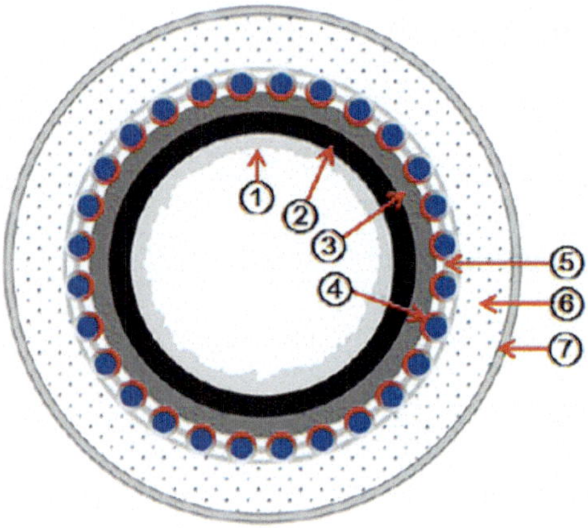

1. Liquid slag, 2. Solidified slag 3. Refractory 4. Steel tubes 5. Tube fins 6. Pressurized gas 7. Steel shell of the reactor

Fig. 7.9 Horizontal section of the reactor vessel

An alternative solution is the freeze lining in which the solid layer of the bath material is deliberately formed on the surface of the refractory hot face to provide protection from corrosive liquid slags. The technology involves the extraction of heat through the lining by cooling the face of the refractory opposite to the hot face.

In case of membrane wall, the wall is constructed basically by stainless steel tubes through which water is circulated and water actually serves the function as refractory (Figs. 7.9 and 7.6a). To protect the stainless tube directly from the temperature and abrasion of moving slag particles, a thin layer of refractories are applied.

This is called "freeze lining," and the advantages are thin refractory lining and thereby to increase the furnace volume and productivity keeping the exterior diameter same. The lining life is more and maintenance is less. But it loses the energy through water cooling [20, 21].

7.4.2 *Concept of Freeze Lining*

When the furnace wall is cooled below the liquidus temperature, the slag in immediate contact with the surface will start getting solidified, and the solid layer will get thicker gradually until the equilibrium is reached. The process of solidification will proceed until the thermal equilibrium is reached between the molten slag bath to the freeze lining (Q_{in}) and the heat transfer (Q_{out}) through the freeze lining and refractory wall to the flowing water through the tubes. Mathematically, at steady state

$$Q_{in} = Q_{out}\,[22,23] \tag{7.4}$$

Let us consider a plane wall (Fig. 7.8), where

T_{Liq} = liquid slag temperature
(degree C) = Furnace temperature
T_{freeze} = slag freezing temperature (°C)

$$T_{freeze} = \left(T_{liquidus} + T_{Solidus}\right)/2 \tag{7.5}$$

h_{liq} = heat transfer coefficient-liquid slag to solid slag (W/m^{2}K)
Q_{in} = heat flux (KW/m^2)

Then,

$$Q_{in} = h_{liq}\left(T_{Liq} - T_{freeze}\right) \tag{7.6}$$

Equation 7.5 indicates that the lower the slag freezing temperature relative to the furnace temperature, the higher is the heat transfer load.

Heat flux out from the furnace wall is

$$\begin{aligned} Q_{out} &= \left(T_{freezing} - T_{cooling}\right)/\left(R_{freeze} + R_{cooling}\right) \\ &= \left(T_{freezing} - T_{cooling}\right)/\left(x_{freeze}/k_{freeze} + 1/h_{fc} + 1/h_{lcs} + 1/h_c\right) \end{aligned} \tag{7.7}$$

where

$T_{cooling}$ – temperature of cooling medium
x_{freeze} – thickness of freeze lining
k_{freeze} – thermal conductivity of freeze lining
h_c – cooling medium heat transfer coefficient
h_{fc} – heat transfer coefficient between freeze lining and cooling system
h_{lcs} – effective lining cooling system heat transfer coefficient

Combining Eqs. 7.4, 7.6, and 7.7, the freeze lining thickness can be decided as

$$x_{freeze} = k_{freeze}\left[\frac{T_{freezing} - T_{cooling}}{T_{bath} - T_{freezing}}\left(\frac{1}{h_{bath}}\right) - \left(\frac{1}{h_{fc}} + \frac{1}{h_{lcs}} + \frac{1}{h_c}\right)\right] \tag{7.8}$$

Although the exact equation for lining thickness for the furnace of cylindrical geometry differs from Eq. 7.8, the behavior will be similar.

The factors upon which the thickness of the freeze lining depends can be tabulated as in Table 7.5.

The actual situation is more complex, and lot of factors influence the process of heat transfer which ultimately decides the equilibrium condition and effectiveness

Table 7.5 Factors upon which the heat transfer and the thickness of freeze layer depends

	Thermal factors	Physicochemical factors
Furnace environment	**Bulk heat transfer**	**Homogeneous nucleation rate**
	$N_u = f(R_a\ R_e\ P_r\ G_r)$	Solid composition
	T_{Liq} -T_{freeze}	Undercooling
	Natural convection	Interfacial energy
	Forced convection	**Heterogeneous nucleation rate**
	h_{liq}	Pre-existing solid
	C_p liquid slag	% solid
	K liquid slag	Crystal structure
	ρ of liquid slag	Crystallization rate
	Geometry	Undercooling
		Mass diffusivity
		Natural convection
		Forced convection
		Crystallization path
		Shape of liquidus
		Shape of solidus
		Max % of solids above solidus
		Solid solution
		Glass transition temperature
Freeze lining	k_{Liq}(Thermal conductivity of freeze lining)	Phase transformation reactions
	Porosity	Phase transition temperature
	Phases present	Change in molar volume
	Phase proportion	Chemical reactions
	Microstructure	
Refractory	Thermal conductivity	
	Geometry	
Coolant	Temperature	
	$h_{coolant}$	
	C_p	
	ρ	

of the design. Table 7.5 has tabulated the list of the factors that decides the heat transfer process.

The refractory must have high thermal conductivity to enable the water cooling more effective and must have high abrasion resistance to withstand the abrasion of high velocity impingement of slag particles.

As shown above the most important aspect of the freeze lining is to provide a configuration to achieve the desired thermal equilibrium, and its properties and

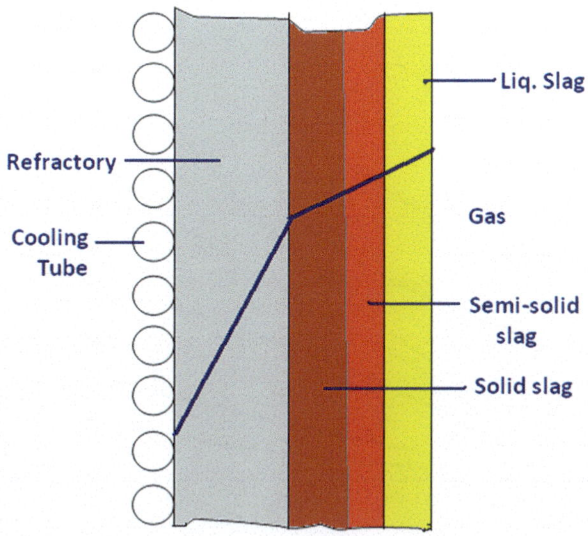

Fig. 7.10 Temperature profile across the sidewall lining of the gasifier

integrity should not deteriorate with time. These requirements are not only fulfilled by material but also contributed by concept, geometry, configuration, and the execution.

Proper freeze lining requires a refractory material with high thermal conductivity and thermal diffusivity that maximizes the thermal gradient (Fig. 7.10). A very high temperature gradient exists between the hot face and the cold face in contact with the cooling tubes. As a result, the hot face will try to expand much faster than the cold face, and a high thermo-mechanical stress will set up because of difference in thermal expansion of hot face and cold face which ultimately causes the formation of crack. Therefore, the other requirement is that the refractory should have low thermal expansion coefficient. The refractory must be resistant to the coal ash slag, and it should have low reactivity with and low dissolution rate in the liquid slag. Figure 7.11 shows the solubility rate of different oxides in coal ash slag at high temperature [24].

From Fig. 7.11 it is evident that Cr_2O_3 and Al_2O_3 have the minimum solubility in coal ash slag. Alumina chrome or alumina SiC phosphate-bonded plastic material is found to be good candidates for this application [25, 26]. The material can be troweled over the tubes fitted with small studs on its surface (Fig. 7.6a) to anchor the refractory. Castables can also be applied by gunning method. A steep temperature gradient is formed between the water-cooled tube wall and the refractory layer. This steep gradient will not allow the slag to penetrate deep in the refractory and will form a solidified layer permanently on the refractory wall which will prevent the refractory from further attack by molten slag. The typical properties of the said refractory products are shown in Table 7.6.

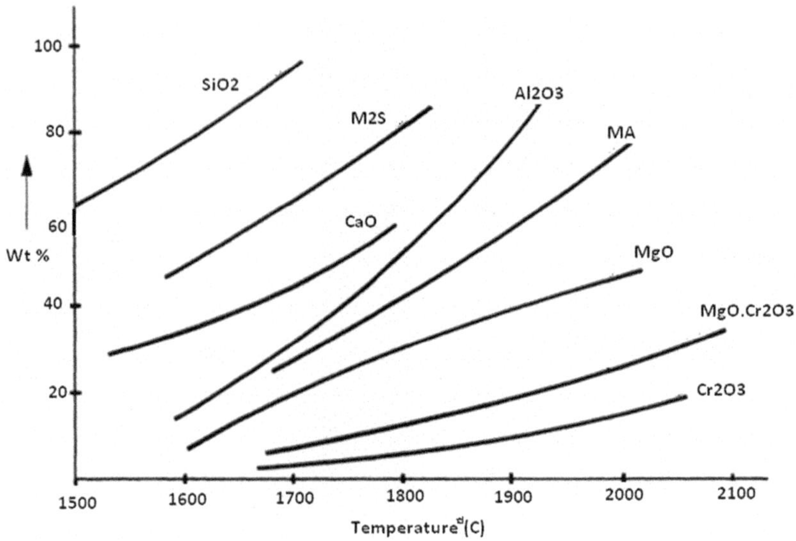

Fig. 7.11 Solubility of different oxides in coal ash slag

Table 7.6 Typical properties of refractories for membrane cooling

	Product	
Properties	Alumina chrome plastic	Alumina silicon carbide plastic
Al_2O_3 %	84–86	85–86
Cr_2O_3 %	9–10	–
SiC %	–	10–11
SiO_2 %	0.5	1.0
Fe_2O_3 %	0.3	0.3
CaO %	0.9	0.7
P_2O_5 %	1.5	1.5
Max grain size (mm)	5	5
Bulk density (gm/cc)	3.1	2.7
Permeability	0.1	
CCS (kg/cm^2)/110C	500	600
/1000 C	700	800
Abrasion loss (cc) as per ASTM 704/110 C	6	5
Reversible thermal expansion at 1000 C	8.6×10^{-6}	–
Permanent linear change /1500C	0.6%	0.5%
Th. conductivity /1000C (W/mK)	2.3	3.0

7.4.3 Refractories for Air-Cooled Wall

In this design, conventional multilayer refractory lining is used with dense refractory at the working layer and insulating refractory on the shell and medium duty refractory layer in between. The failure of refractory in a gasifier is very expensive and may cost one million US dollar depending on the extent of relining and the size of the gasifier and production loss. The process of relining involves the cooling of the gasifier (up to 4 days) removal of the old existing lining (1–3 days) and partial repair (1–3 days) or full replacement of lining (7–10 days) [27]. Chemical corrosion and thermal and structural spalling have been found to be the most prominent cause of wear of refractories in the gasifier.

Wear of refractory by spalling can be of different types:

1. Pinch spalling originated from the compressive hoof stress due to restraining of expansion of the refractory lining by the steel shell.
2. Spalling due to rapid temperature change.
3. Spalling due to complex combination of slag infiltration, temperature fluctuation, and stress due to restraining the expansion of refractories by the steel shell.

Examples of spalling of different kind are shown in Fig. 7.12 [28]. The temperature designed for a slagging gasifier is very critical and selected within a narrow band of temperature which will allow the slag to get liquefied and flows out from the reactor but not to make the slag fluid enough to enhance the corrosion of refractory lining by the molten slag. Ash chemistry and quantity can vary a lot depending upon the feedstock and the understanding of the slag chemistry and its interaction with refractory is very important to design the refractory lining for the gasifier. Figure 7.13 [29] shows the factors that influence the refractory lining life in a gasifier using solid carbon feedstock.

The coal slags are normally acidic in character and acidic or neutral refractories should work best. Aluminosilicate or alumina refractories are not suitable for lining of gasifier where feedstock used is coal or petcoke because of low corrosion resistance of this kind of refractories towards molten slag formed in the gasifier. Vanadium present in petcoke or in coal is very detrimental for aluminosilicate refractories. V_2O_5 forms low melting liquid at 640 °C reacting with Al_2O_3.

Experience shows that the refractories of chrome oxide containing systems, such as Cr_2O_3- Al_2O_3, Cr_2O_3-Al_2O_3-ZrO_2, and Cr_2O_3-MgO, work best as the working hot face lining of the gasifier.

7.4.4 Corrosion of Refractory

During the corrosion reaction of the refractory by slag, the refractory material gets dissolved in the slag. The difference between the concentrations of the refractory material, in the refractory saturated slag and in the original slag, is assumed to be

Fig. 7.12 (**a**) Pinch spalling along the refractory joint. (**b**) Thermal spalling. (**c**) Structural spalling

the driving force of corrosion by diffusion. In this study Fick's first law of diffusion in combination with Stokes-Einstein equation such that

$$\text{J is proportional to F} = T\Delta C / \eta$$

where J is the diffusion coefficient of the refractory in the slag, F = diffusion factor, T = absolute temp, and η = viscosity of the slag and ΔC = concentration driving force. A greater diffusion flux indicates a higher degree of corrosion [25].

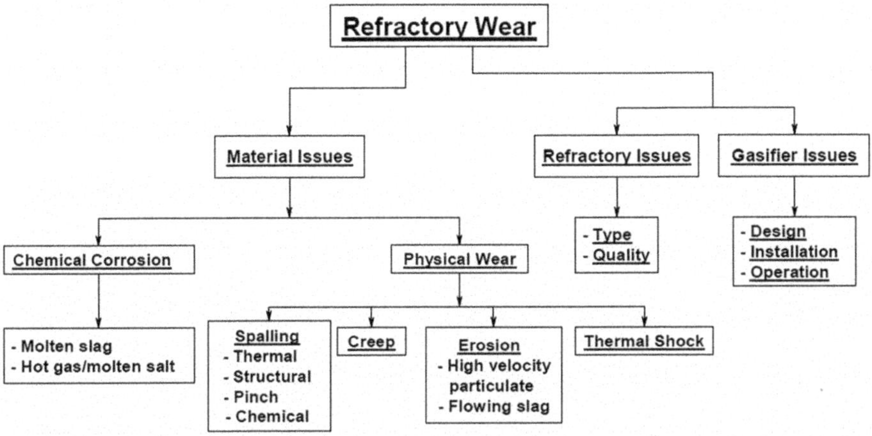

Fig. 7.13 Factors responsible for refractory performance in a gasifier

The slag becomes saturated with refractory when large amount of refractory is dissolved in the slag. The phases present in the slag and their concentration can be predicted by using Factsage software. Table 7.7 provides an example of various relevant data like slag viscosity, refractory equilibrium phase, other solid phases present, concentration driving force, and diffusion coefficient [25]. The best refractory for a gasifier can be developed based on theoretical modeling and experiments focused on the different properties of the refractory, e.g., corrosion, etc., as shown in Fig. 7.14 [30]. The properties of the typical refractories for the gasifiers are given in Table 7.8.

Table 7.7 An example of usage of Factsage to generate different data to predict slag attack

Fuel Slag	Refractory material	Temperature (degree C)	Viscosity (Pa.s)	Refractory equilibrium solid phase	Other solid present at saturation	Concentration driving force (g/g)	Diffusion factor ($K.Pa^{-1}.s^{-1}$)	Metastable concentration driving force (g/g)	Metastable diffusion factor ($K.Pa^{-1}.s^{-1}$)
F1	Alumina	1250	29.1	Corundum	Spinel, Anorthite Ca2Mg2Al28O46, CaMg2Al16O27	0.29	15	0.103	5
F1	Alumina	1500	3.2	Corundum	none	0.248	137	0.2	137
F3L1	Alumina	1250	22	Corundum	Hibonite,Spinel, Anorthite, Leucite, $Ca_2Mg_2Al_{28}O_{46}$	0.407	28	0.1	7
F3L1	Alumina	1500	2.2	-Do-	None	0.231	187	0.2	187
F4L1	Alumina	1250	47	Do	Spinel, Hibonite, Anorthite $Ca_2Mg_2Al_{28}O_{46}$,	0.289	9	0.1	3
F4L1	Alumina	1500	3.4	do	None	0.209	109	0.2	109
F1	SiC	1250	29	Slicon Carbide	FeSi, Graphite	0.065	3	0	1
F1	SiC	1500	3.2	-do-	FeSi, Graphite	0.065	36	0	11
F1	Al_2O_3-Cr_2O_3	1250	29	Corundum	Spinel, Anorthite	0.248	13	0	1
F1	Al_2O_3-Cr_2O_3	1500	3.2	Corundum	Spinel	0.242	134	0.1	36

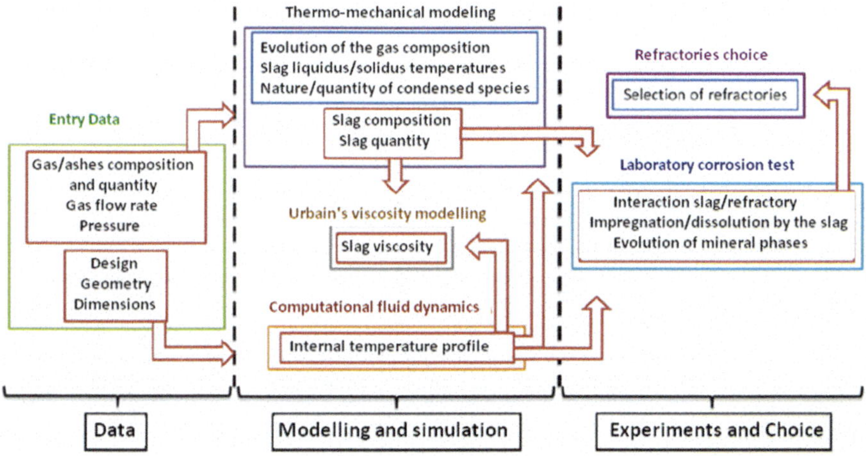

Fig. 7.14 The cycle of development of refractories for gasifier

Table 7.8 Properties of refractories for the gasifier application

	Alumina-chrome	Alumina-chrome-zirconia	Chrome-alumina-zirconia	Alumina
Al_2O_3%	87	41–42	17	99.5
Cr_2O_3%	12	30	61–62	–
ZrO_2%	–	17–18	12	–
SiO_2%	–	10	9	–
Others %	–			0.5
BD (gm/cc)	3.35	3.45	3.84	3.32
AP %	17	14	10	16
Th. Conductivity (W/mK)	4.4	2.9	2.9	3.8
Maximum service temperature	1750	1750	1800	1800

References

1. Kerester A (2014) Gasification can help. Cornerstone 2(3)
2. Higman C (2013) State of the gasification industry- the updated worldwide gasification database. 2013 International Pittsburgh Coal Conference, Sept 16–19, Beijing, China
3. (2018) Global Syngas Overview. PPT by STRATAS ADVISORS
4. Phillips J Different types of gasifiers and their integration with gas turbines. EPRI/Advanced coal generation, Charlotte, NC 28221
5. Othmer H Coal gasification- Encyclopedia of Chemical Technology, Kirk. Wiley
6. Maurstad O (2005) An overview of coal based integrated gasification combined cycle (IGCC) technology. Massachusetts Institute of Technology Publication – LFEE 2005-002- WP
7. Holt NAP (2001) Integrated gasification of combined cycle power plants. EPRI, Encyclopedia of Physical Science and Technology, Academy Press
8. Mathew T Coal/ Petcoke gasification and Reliance's Jamnagar gasification project. PPT

9. Kern H et al (2012) Repair action of refractory linings in high temperature reactors. Refractory World Forum 4:2
10. Benett JA et al (2013) Mechanism of wear reduction in high chrome oxide containing phosphate additions exposed to coal slag. UNITECR
11. Stotch A (2015) Fly ash from Coal combustion- characterization. Dissertation for M.Sc, Dept. of Energy Engineering and Management, IST Instituto Superior Tecnico, Lisbon
12. Duchesne M (2011) Application of the FACTSAGE for the study of slagging in the entrained flow gasifier. University of Ottawa, CALPHAD XL
13. Wang P, Massoudi M (2013) Slag behaviour in gasifiers. Part I – influence of coal properties and gasification conditions. Energies 6:2
14. Akar G et al (2009) Relationship between ash fusion temperature and coal mineral matter of some Turkish coal ashes. Asian J Chem 21(3)
15. Kondratiev A et al (2002) Predicting slag viscosities in the metallurgical system. JOM
16. Ma C (2017) Aspects of ash transformation in pressurized entrained flow gasification of woody biomass. Doctoral Thesis – Lulea University of Technology, Dept. of Engineering and Mathematics
17. Ilyusechkin AY et al (2012) Interaction of the Vanadium rich slag with the crucible materials during viscosity measurement. J Mater Sci
18. Hirata T et al (2002) Corrosion resistance of alumina –Chromia materials against molten slag. Mater Transac 43(10)
19. Doshi S Overview of Saint-Gobain and introduction to Refractory solutions for gasification. St. Gobain Refractory Company
20. Wang Z et al (2013) Use of a reactor network model in the design and operation of a new type of membrane wall entrained flow gasifier. Energy Fuel 27
21. Hearn AM et al Freeze lining concepts for improving the submerged arc furnace lining life and performance. Proc INFACON 8:401–426
22. Crivits T Fundamental studies on chemical aspects of Freeze lining. M.Sc thesis, School of Engineering, Queensland, Australia
23. Joubert H, Dougall IM (2019) Designing furnace lining/cooling system to operate with competent freeze lining. Researchgate net publication 331393751. The Minerals, Metals and Materials Society
24. Ceramics in advanced energy technologies, Proceedings of European Colloquium, Petten, Netherland, Sept 20–22, 1982
25. Duchesne M (2012) Slagging in entrained flow gasifier. Thesis for the Ph.D in Chemical Engg, Ottawa University
26. Colombet L, Poirier J (2014) Slag formation and corrosion of SiC Refractories in Biomass gasification reactors. Refractories Worldforum 6(2)
27. Benett JP, Kwong KS (2011) Failure mechanism of high chrome oxide gasifier refractories. Metallurg Mater Transac 42A
28. Benett J et al (2005) An analysis of the causes of failure in high chrome oxide refractory materials from slagging gasifier. Proceedings UNITECR
29. Benett JP, Kwong KS Low chrome/chrome free refractories for slagging gasifier. DOE/NETL-IR-2006-119
30. Thevenin G et al (2014) Modelling and design of a refractory lining for a bio-mass gasification reactor fed by a Plasma torch. In: Waste and biomass valorization. Springer., ISSN 1877-2641, Published on 31st January

Chapter 8
Refractories for Carbon Black Manufacturing

8.1 Carbon Black

Carbon black is an industrial chemical used in large quantity in different industries. Some of the major user industries of carbon black are in automobile tire as filler, plastic, paint, and ink industries. Global production of C-black is about 19 million tons with a value of about 20 billion USD. C-black is used as filler in rubber and other polymers to impart and improve some of the desired physical properties like abrasion resistance, thermal conductivity, tensile strength, etc.

8.2 Process Principle

Carbon black is produced by pyrolysis of feedstock containing unsaturated hydrocarbons, like gas or naptha. Carbon black is produced by thermal cracking of high molecular weight aromatic hydrocarbons by controlled incomplete combustion at a temperature of 1350–1540 °C. The chemical reaction taken place during different stages of the process can be shown by the chemical reactions given below:

$$\text{Combustion reaction,}\quad C_nH_{2n+x} + (6n+x)/4O_2 = nCO_2 + (2n+x)/2H_2O + \Delta H \tag{8.1}$$

$$\text{Cracking reaction,}\quad C_nH_{2n+x} + \Delta H = C_nH_y + (2n+x-y)/2H_2 \tag{8.2}$$

$$C_nH_y \rightarrow nC + y/2H_2 \tag{8.3}$$

P. Sengupta, *Refractories for the Chemical Industries*,
https://doi.org/10.1007/978-3-030-61240-5_8

8.3 Manufacturing Process

Although many different manufacturing processes like lampblack process, furnace black process, channel process, Degussa gas black process, acetylene black process, etc. are available, but the furnace black method is most commonly used in large-scale manufacturing of C-black.

The furnace black method is a continuous process in which the liquid hydrocarbon is used as feedstock and the gaseous hydrocarbon as fuel. In the combustion chamber, heat is generated by burning the hydrocarbon gas with preheated air in a swirl burner, and the hot gas is allowed in the next chamber where liquid oil is sprayed when thermal cracking takes place and the carbon black is produced. As soon as the carbon black is produced, the gas-solid mixture is quenched with water to prevent any oxidation or secondary reaction of carbon black (Fig. 8.1).

8.4 Reactor Design

Two different types of reactor design are normally used:

(i) Tread reactor – Manufactures hard carbon black and the temperature in this reactor may go beyond 1900 °C.
(ii) Carcass reactor – Manufactures soft C-black and this reactor operates at 1550–1700 °C.

Different zones in these reactors are given below with the operating conditions; the refractories are to withstand, and the different zones of the reactor are shown in Fig. 8.2.

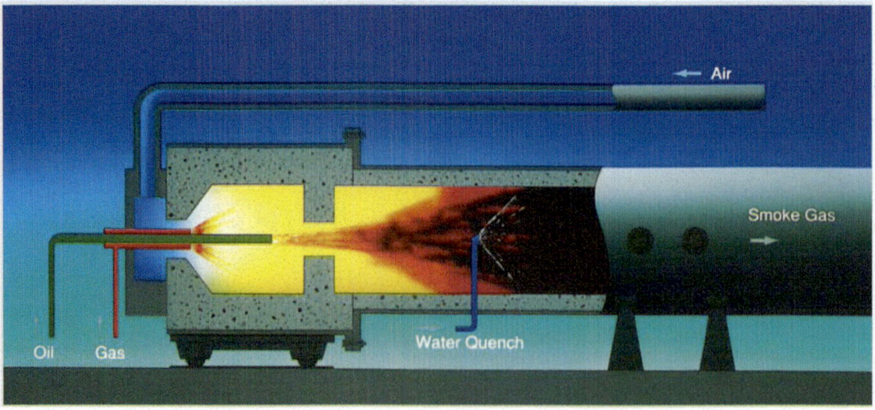

Fig. 8.1 The carbon black reactor

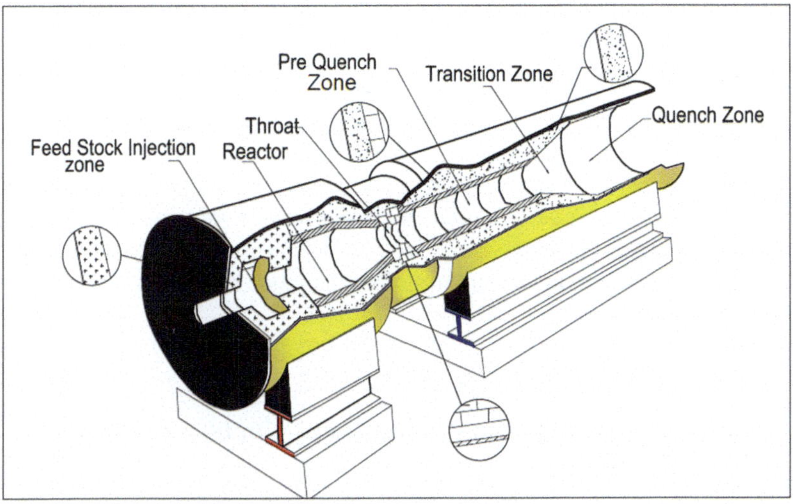

Fig. 8.2 Different zones in a carbon black reactor

1. Combustion zone – This is the zone where the combustion of the fuel gas takes place in the presence of preheated air fed to this chamber. Temperature goes very high (beyond 1900 °C) in this zone. The refractory lining here must have high volume stability at high temperature. It must also need to have high thermal shock resistance. Therefore alumina refractory with mullite bond in the matrix or with zirconia will be most suitable (Table 2.1).
2. Restriction zone – Restriction zone is a narrow chamber that separates the combustion zone and the reaction zone. Because the combustion gas with a high volume is allowed to pass through the narrow cross section, the velocity of gas goes very high to the supersonic level (400–450 m/s). The refractory must have to withstand very high temperature and very high abrasion along with high thermal shocks.
3. Reaction chamber – Here the feedstock is pyrolyzed at high temperature. The hardness and morphology of the C-black is controlled here by control of temperature, gas velocity, and the seeding material (potassium salt). The refractory lining here must have to withstand high temperature at reducing environment (presence of hydrogen) and the attack of different impurities inside the feedstocks and additives added like alkali, sulfur, vanadium, etc.
4. Quench zone – Here the water is sprayed to bring down the temperature and to arrest any oxidation of C-black or other secondary reactions of the C-black. The reaction must have to withstand the thermal shocks, attack of steam, hydrogen, and other impurities mentioned before.

5. Product gases laden with C-black are cooled to 240–260 °C in a venturi cooler, and the C-black is separated from the gas mixture containing CO, CO_2, CH_4, C_2H_2, air, and water vapor.

Carbon black having fine primary particles is produced at higher temperature than carbon black with coarser particles. Therefore for the production of finer sizes of product, the reaction temperature is higher when the reaction is faster, and the residence time in the reactor is shorter. In case of designing the reactor for coarser particle, the temperature is lower and reaction speed is low, and the reactor residence time is longer. Energy density in a modern carbon black reactor is allowed up to 100 MJ/m³ [1].

8.5 Temperature at Different Stages of Carbon Black Manufacturing

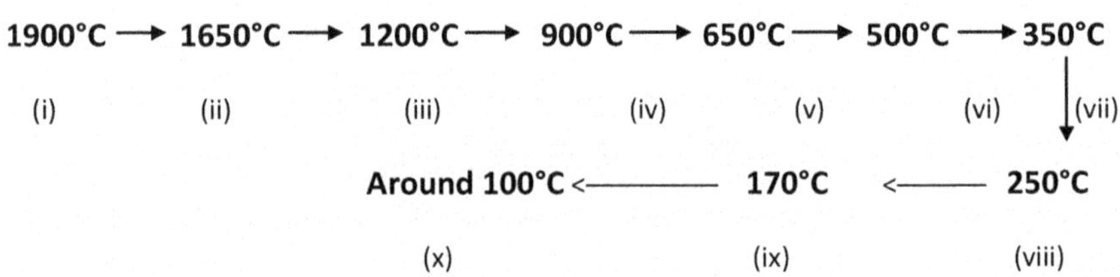

(i) Temperature reached during combustion.
(ii) Temperature after cracking of oil has occurred.
(iii) Temperature after quenching with water.
(iv) Air is heated by passing it through the APH and is sent to the reactor header at 800 °C.
(v) This temperature in the waste heat boiler (WHB) is used to generate steam. This steam is used for atomizing purposes, flashing in the APH to prevent carbon black from sticking to its sides.
(vi) The OPH (oil preheater) at this temperature preheats the Make Oil before sending it to the choke section of the reactor.
(vii) Smoke + flue gases reach the quench tower for further cooling.
(viii) It is passed through the PBC modules from where fluffy carbon black is obtained.
(ix) Temperature inside the dryer.
(x) Product temperature.

8.6 Selection of Refractories

In general, the refractory lining, in this kind of reactors, is preferred to be done in three different layers of three different quality of refractories. The refractory lining scheme is shown in Fig. 8.3. The thicknesses of different layers are decided by the reactor diameter, temperature inside the reactor, the shell temperature as per design of the reactor, and the properties of the refractories used.

The refractory to be used in the hot face, which is to withstand all the abuses of reactor, should have very low porosity, high volume stability at high temperature, high corrosion resistance, high thermal shock resistance, and high stability in reducing atmosphere. Refractories with corundum as major mineral phase with mullite or zirconia in the matrix are the most ideal. The magnesite-chrome brick has been found get damaged very severely in the lower part of the combustion and reaction zone when the feed materials are natural gas and green oil [2]. The reason of failure is mainly due to volume change of refractories under redox condition.

The next to the hot face refractory is a refractory with higher porosity but with good refractoriness, good corrosion resistance, and thermal shock resistance. It is used as a buffer layer between the hot face refractory and the insulation refractory at the outermost layer.

The refractories used in the hot face have high thermal conductivity because of their low porosity. If, instead of three layers of refractories, two layers are used, the junction temperature of the hot face refractory and the insulation refractory would be very high, and normal insulation refractory cannot withstand that temperature. In that case special insulation refractory, e.g., bubble alumina castable, is to be used to withstand that condition, and its price will be very high. For this reason, a second layer between hot face and insulation layer is required which lowers the junction temperature of the insulation refractory and the middle layer to make it tolerable for the insulation refractory. The properties of the refractory used in different layers are shown in Tables 8.1 and 8.2.

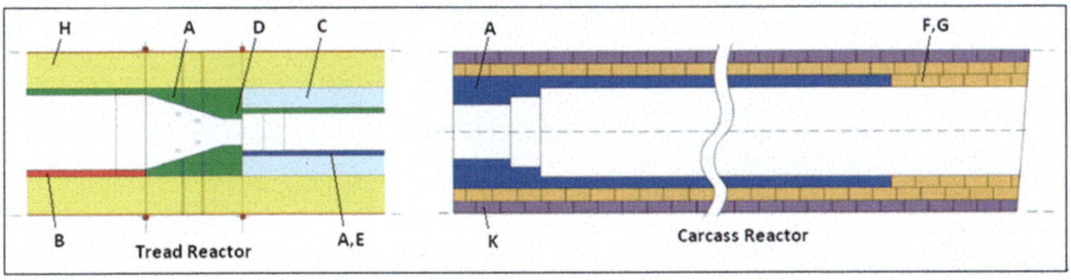

Fig. 8.3 The refractory lining scheme in C-black reactor, the quality details are shown in Tables 8.1 and 8.2. (Courtesy of St Gobain Refractories)

Table 8.1 Refractories for the C-black reactors

Refractory properties	Quality of refractory in working face						
	Combustion zone and restricted zone				Reaction zone	Quenching zone	
	Bricks		Castable	Precast		Brick	
	A	B	C	D	E	F	G
Chemistry							
Al_2O_3 %	99.5	86–88	97.5	98	95–97	90–91	77–78
SiO_2 %	0.1–0.2	–	0.1		0.2	9–10	21–22
ZrO_2 %	–	–	–	–	2–3	–	–
Cr_2O_3 %		12	–	–	–	–	–
Physical properties							
BD (gm/cc)	3.25–3.35	3.4	3.10	3.25	3.15–3.2	3.0	2.65
AP %	16–18	15	–	16–17	16–18	16	17–18
CCS (MPa)	80	100	70	120	75	120	100
RUL (°C)	1700	1700	–	1700	1650	+1650	1650
Th. expansion % (1400 °C)	1.2	1.1	–	1.2	–	1.05	0.8
Th. cond. (W/mk)/1000 °C	3.5	2.80	2.20	3.5	2.3	2.2	1.5

Table 8.2 Refractories for the safety zone of the reactor lining

Refractory properties	Quality of refractory in safety lining	
	Second layer of refractory lining	Outermost layer of insulating lining
	Castable-H	Castable-K
Base material	Bubble alumina	–
Al_2O_3 %	91–92	35
SiO_2 %	0.1	37
CaO %	7–8	20
Physical properties		
BD (gm/cc)	1.4	0.9
CCS (MPa)	10	7
PLC %	−1.2/1500 °C	−1.0/1200 °C
Th. cond. (W/mk)/900 °C	0.9	0.29

In the quench zone, temperature is low, but the refractories are to withstand high thermal shocks and high abrasion besides the attack of alkali, carbon black, and other impurities.

References

1. Kuhner G, Voll M. Manufacture of carbon black. In: Donnet JB, Bansal RC, Wang MJ (eds) Carbon black – science and technology, 2nd edn, Marcel Dekker Inc., New York
2. Zukhova ZD et al. Phase changes in the lining of reactors used for Carbon black production. Ogneupory, No 9, September, 1965

Chapter 9
Refractories for Petroleum Refinery

9.1 Introduction

Modern civilization cannot think of a world without petroleum. Besides gasoline and other fuels, different synthetic polymers and thousands of organic chemicals are derived from petroleum. The global refining capacity of crude was 120 million barrel or 19.1 billion liters per day in 2018. The crude petroleum available in nature is a mixture of many different hydrocarbons, and the hydrocarbon compounds are separated and refined from the crude in the refinery to make it usable for different purposes. The total refinery process is shown in Fig. 9.1 and is a very complex one with downstream sector.

It shows hundreds of different units are there in a refinery and some of the units operate at elevated temperatures, and the refractory lining to the reactor vessel and associated equipment is very essential to ascertain the smooth operation of the refinery.

The main units in the refinery which operate at higher temperatures are:

1. Process heaters
2. Fluidized catalytic cracking unit (FCCU)
3. Sulfur recovery unit

The brief operational parameters of the said units and the refractory requirements for these different equipment are discussed subsequently.

P. Sengupta, *Refractories for the Chemical Industries*,
https://doi.org/10.1007/978-3-030-61240-5_9

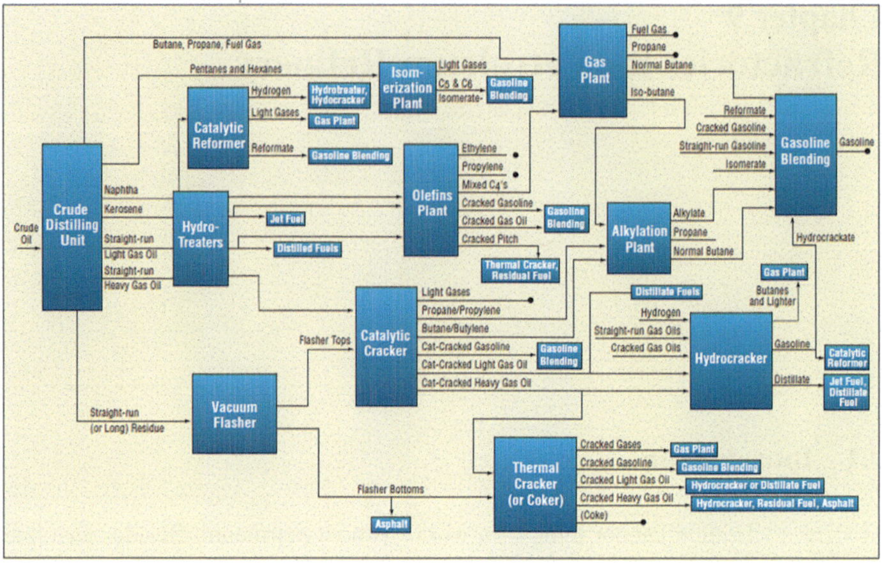

Fig. 9.1 Schematic diagram of the refining process in an integrated petroleum refinery

9.2 Process Heaters

In petrochemical industries, process heaters or furnaces are used extensively to heat petroleum feedstock for fractionation, thermal cracking, and high-temperature processing. It is used to raise the temperature of the feed material to reaction or distillation level. The heaters are designed to raise the temperature of the feed material up to a maximum of 510°C. An actual process heater is shown in Fig. 9.2a. The fuel used to burn is refinery gas, natural gas, residual fuel oil, etc. The process heater is basically a vertical, cylindrical, or rectangular furnace with carbon steel casing lined inside with refractory material, in which the burners are mounted on the floor or wall or roof according to the design of the furnace.

Steel tubes are passed though the furnace vertically or horizontally, through which the feedstock to be heated flows to attain the desired temperature (Fig. 9.2b).

The tubular furnaces are built with two distinct heating sections, namely, radiant section or combustion chamber or firebox and a convection section followed by stack. The hot flue gas generated in the radiant or combustion section is then allowed to flow to the convection section at high velocity to circulate the tube bundles before leaving to the stack. A third section known as shield or shock section separates the radiant and convection section. The shield section normally contains two to three rows of bare tubes directly exposed to hot gases to shield the tubes in the convection section from direct radiation [1]. The temperature in the radiant section is normally 750–850 °C and 450–500 °C in the convection section [2]. The typical heat transfer rate for the vertical tubes in the radiant section is about 50KW/m^2.

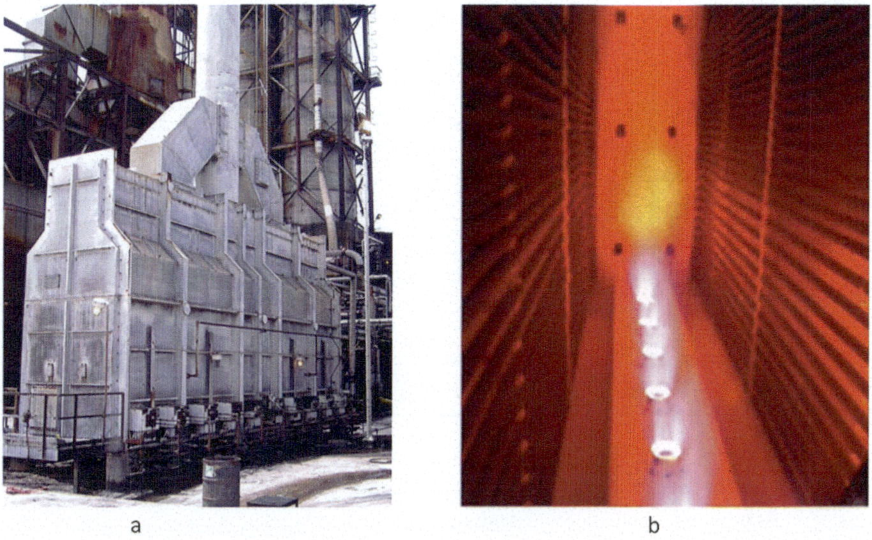

Fig. 9.2 (a) A process heater. (b) Floor-mounted burners inside, with horizontal radiant tubes

Heaters can be of various shapes in order to best fit the required radiant and convective tube arrangements. Figure 9.3 shows some of the most common design of the heaters [3]. The components of a heater are shown in Fig. 9.4 [4] and cut section in Fig. 9.5.

9.2.1 Refractory for Process Heaters

The quality of the refractory material and the installation is very important for process heaters. The refractory should be designed to get a very long life because there is no scope of maintenance or repair of the lining once the heater is in service, due to inaccessibility of the lining because of packed tubes and convection coils inside the furnace.

9.2.2 Refractory Selection

The primary objectives of refractory lining in the fired heater are to protect the steel casing from overheating and corrosion, to conserve the thermal energy and to minimize the heat flow out of the furnace. The selection of the refractory is primarily based on two most important operational parameters:

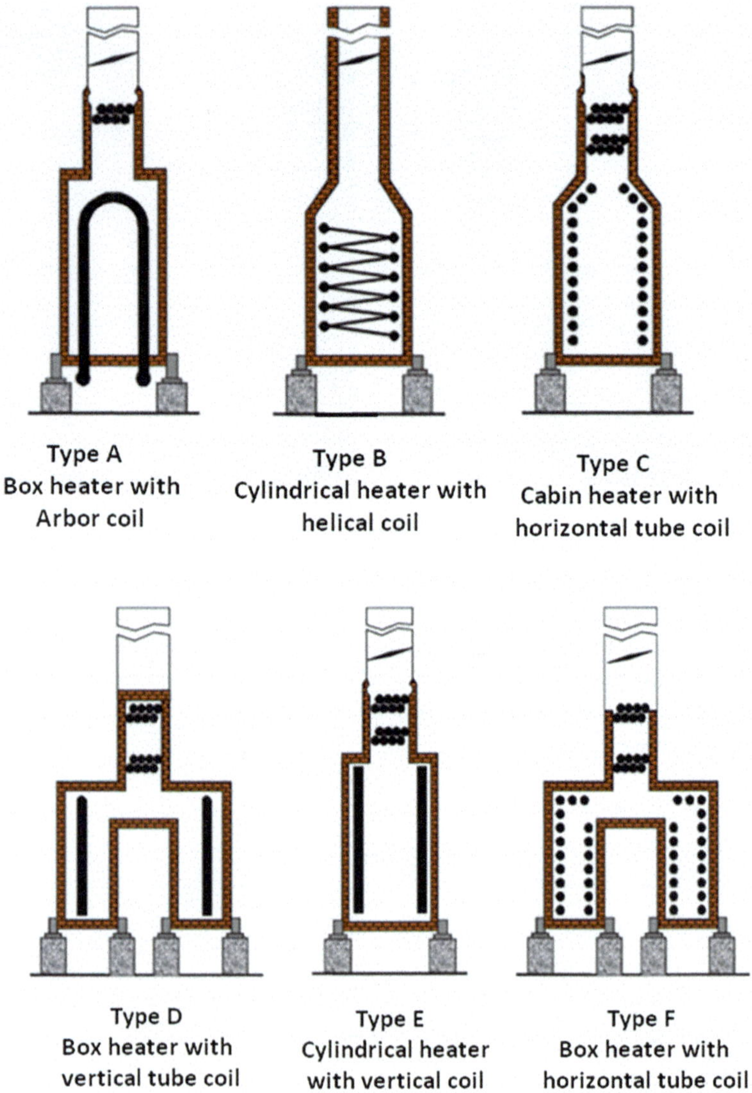

Fig. 9.3 Various designs of the heaters commonly being used

1. The type and quality of the fuel used to heat the furnace
2. The velocity of the combustion gas inside the furnace.

Furnaces fired with clean fuel like natural gas or with fuel having very low sulfur or heavy metal content should be lined with ceramic fiber modules. The ceramic fibers' lining, because of its very light weight and low thermal conductivity, is highly thermal efficient. When the fuel used has high sulfur content or has high

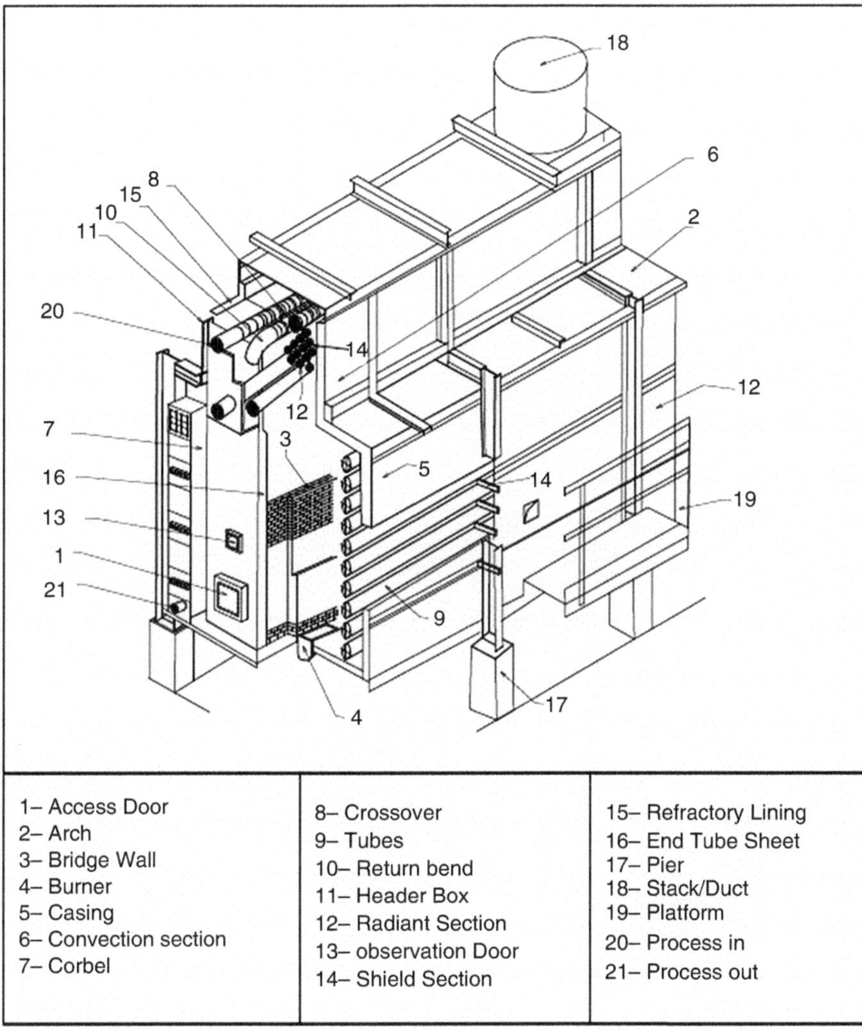

1– Access Door	8– Crossover	15– Refractory Lining
2– Arch	9– Tubes	16– End Tube Sheet
3– Bridge Wall	10– Return bend	17– Pier
4– Burner	11– Header Box	18– Stack/Duct
5– Casing	12– Radiant Section	19– Platform
6– Convection section	13– observation Door	20– Process in
7– Corbel	14– Shield Section	21– Process out

Fig. 9.4 Components of a heater

heavy metal content, bricks with high alumina content with suitable insulating backup lining are used.

In case of high-velocity flue gas, e.g., in the convection zone and above, ceramic fiber lining is not used, because of its poor abrasion resistance, and instead insulating castable lining is used.

As per the API 560 guideline, ceramic fiber system of lining is not used when the heavy metal content in the fuel is more than 100 ppm or the soot blowers or steam lancing or water wash facilities are provided, because those will damage the ceramic fiber lining.

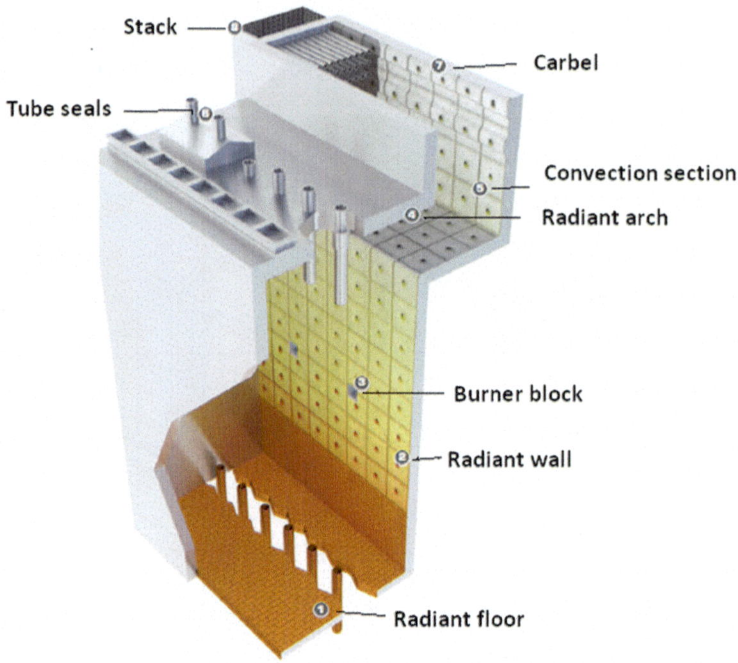

Fig. 9.5 The schematic diagram showing different parts of the heater

Ceramic fiber lining is recommended up to a flue gas velocity of 12.2 m/sec, and beyond it up to a velocity 24.2 m/sec, ceramic fiber board or module can be used, and if the flue gas velocity exceeds 24.2 m/sec, insulating bricks or castable should be used.

For fuels containing sulfur of more than 10 ppm, as per API guideline, the furnace wall should be protected inside by a coating that prevents corrosion up to 177.8°C. When the sulfur content in the fuel exceeds 500 ppm, SS 304 foil is to be used as the vapor barrier to protect the furnace shell, and the furnace wall should be maintained at 56 °C above the calculated acid dew point to prevent the condensation of the acid and the associated corrosion of steel shell.

The suitable refractories for the different areas in the heater are shown in Table 9.1.

Ceramic fiber should not be used when the hot face temperature exceeds 700 °C and combined sodium and vanadium content of the fuel exceeds 100 ppm. When the ceramic fiber is used as the backup insulation and the fuel used has a sulfur content of more than 10 ppm, then a protective coating should be applied on the steel casing. The protective coating should have the capacity to work up to a temperature of 175 °C. The lining is always designed, especially when the fuel contains high sulfur, to get a surface temperature above 175 °C. Because the fiber lining is porous, water vapor in flue gas migrates easily to the colder casing, where if it is allowed to condense on the shell dissolving the sulfur oxides, it corrodes the shell severely because of acid formation. The welding joints, e.g., the anchor weldings on the shell, to hold

Table 9.1 Different types of refractories suitable for different parts of the heater

Area of application	Type of refractory used
Radiant floor	Medium-density fire bricks or castable of 65–75 mm thickness
Target wall	It will have flame impingement from both sides. The construction should be made of fireclay brick able to withstand service temperature of 1550°C. It is to be laid with mortars and expansion joint is to be provided by ceramic fiber strip
Radiant wall	Pyroblock module-A, fiber layer blanket-A, ceramic board, insulating firebrick, superwool block, insulating castable, microporous refractory blocks
Burner block and peepside	Ceramic fiber vacuum-formed boards, precast shapes, insulating firebricks
Radiant arch	Pyroblock module, fiber blanket, fiber board, insulating castables, insulating firebrick
Convection section	Superwool block, fiber module block, fiber blanket blocks, insulating firebrick
Tubeseals	Fiberwool
Corbels	Ceramic fiber block module, insulating castables, insulating firebricks
Stack	Ceramic fiber block module, dense castables, insulating castables

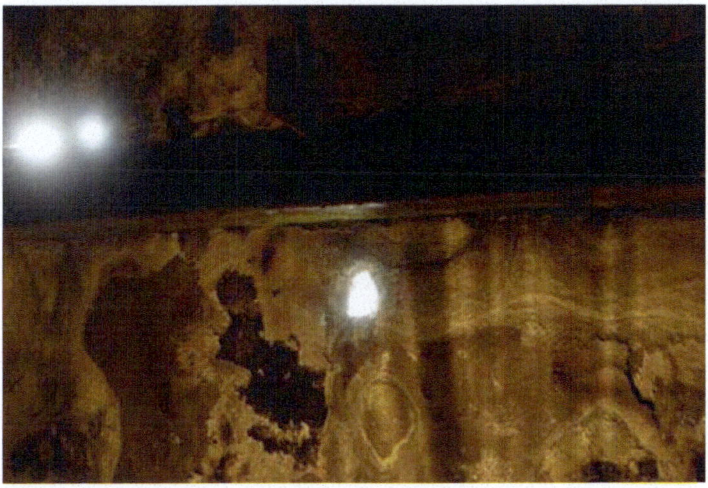

Fig. 9.6 Hole formation in steel shell, because of acid attack

the brick or castable, being the junction of dissimilar metals, are found to suffer more corrosion. It has been found after dismantling of the existing lining that where such condensation occurs, the shell gets a hole (Fig. 9.6) and the fibers become hard like a rock losing its insulation property. The dense sticky material formed on the casing was analyzed and was found to contain 42% H_2SO_4 and 12% Fe_2O_3 [2].

If the backup insulation is used behind the castable lining, the insulation layer must be sealed to prevent the moisture of the castable to migrate from the castable to the insulation material.

The temperature of the outside casing of the radiant and convection sections and the hot ductwork should not exceed to 82 °C with an ambient temperature of 27 °C and zero wind velocity.

Anchoring norms – For radiant and convection section roof, the anchor spacing shall be 1.5 times the lining thickness up to a maximum of 300 mm. For the walls and the breeching, the anchor spacing should be two times the lining thickness with maximum up to 300 mm. For lining thickness of more than 75 mm, the anchor rod diameter should be minimum 6 mm.

9.3 Fluid Catalytic Cracking Unit

This is a type of secondary unit operation primarily used to produce additional gasoline in the refining process. Catalytic cracking is the most important and most widely used process in a refinery to convert heavy oils or feedstock (with high molecular weight) into products with lighter molecular weight, like gasoline, LPG, etc. It is the most critical and highest revenue producer unit. Many processes in the refinery are dependent on the availability and output of the FCCU. It is the largest single unit in the refinery and requires more than 1000 tons of refractory for a new unit. Catalytic cracking is a thermal conversion process to speed up the chemical reaction kinetics in the presence of a catalyst. Figure 9.7 shows the schematic diagram of the FCCU.

9.3.1 Process Description

- In this process, the hot oil feed is contacted with the catalyst in either the feed riser line or the reactor.
- As the cracking reaction progresses, the catalyst is progressively deactivated by the deposition of coke on the surface of the catalyst.
- The catalyst and hydrocarbon vapors are separated mechanically, and the oil remaining in the catalyst is removed by steam stripping before the catalyst enters the regenerator.
- The oil vapors are taken to a fractionation tower for separation into streams having the desired boiling ranges.
- The spent catalyst flows into the regenerator and is reactivated by burning off the coke deposits with air. Regenerator temperatures are carefully controlled to prevent catalyst deactivation by overheating and to provide the desired amount of carbon burn-off.
- This is done by controlling the airflow to get a desired CO_2/CO ratio in the exit flue gases or the desired temperature in the regenerator.
- The flue gas and catalyst are separated by cyclone separators.
- The catalyst, as it leaves the regenerator to remove adsorbed oxygen before the catalyst is contacted with the oil feed, is steam stripped in some units.

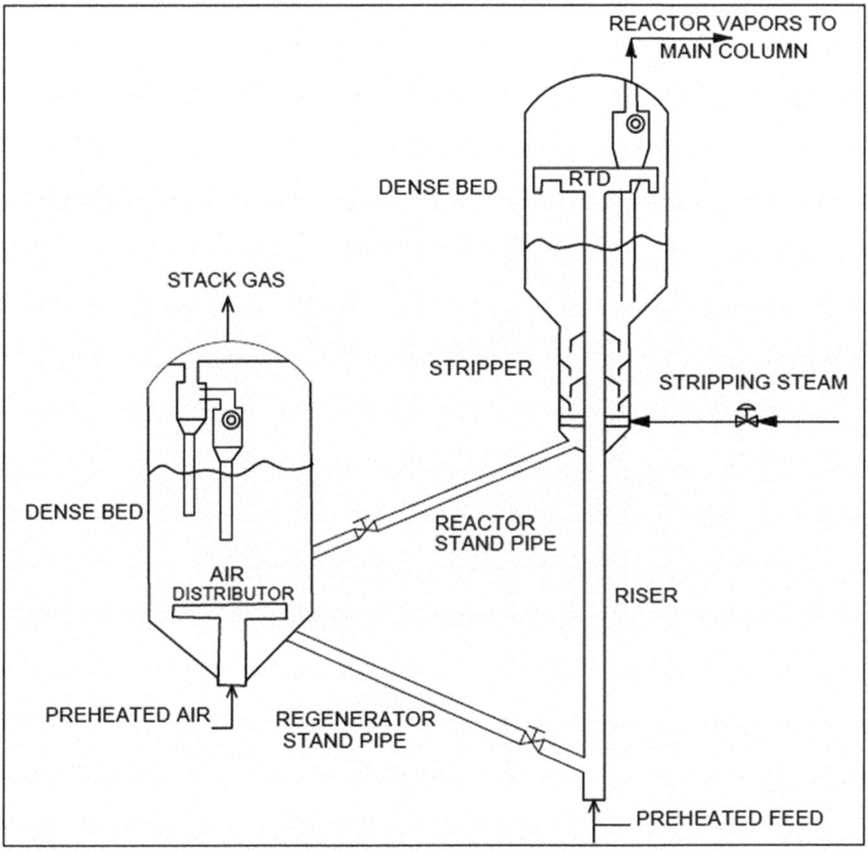

Fig. 9.7 FCCU unit

The detail of operational parameters of a FCCU reactor with 2500 MT of feed/day is shown in Table. 9.2.

Main Elements The main parts of the FCCU where refractories are used to protect those from different kinds of abuses like high temperature, abrasion, and chemical corrosion shown in Fig. 9.8 [5] are as follows:

- Riser pipe
- Reactor/regenerator walls
- Transfer lines that connect the vessels
- Cyclones
- Flue gas lines/CO boiler

The service conditions exist in the different parts mentioned above, and the solutions that can be provided to protect the parts from abusive condition are discussed next.

Table 9.2 Operational parameters of a typical FCCU unit of 2500 MT/day capacity

Regenerator	
Temperature (°C)	650–682
Dry air rate (ton/day)	1360
Flue gas analysis (vol %)	
CO_2	9.0
CO	12.5
O_2	1.5
N_2	77
Reactor	
Temperature (°C)	498
Fresh feed flow (MT/day)	2001
Slurry + CRO recycle (MT/day)	335
Total feed (MT/day)	2336
Combined feed temperature (°C)	311
Catalyst/oil ratio	6.4
Catalyst circulation rate (MT/min)	10

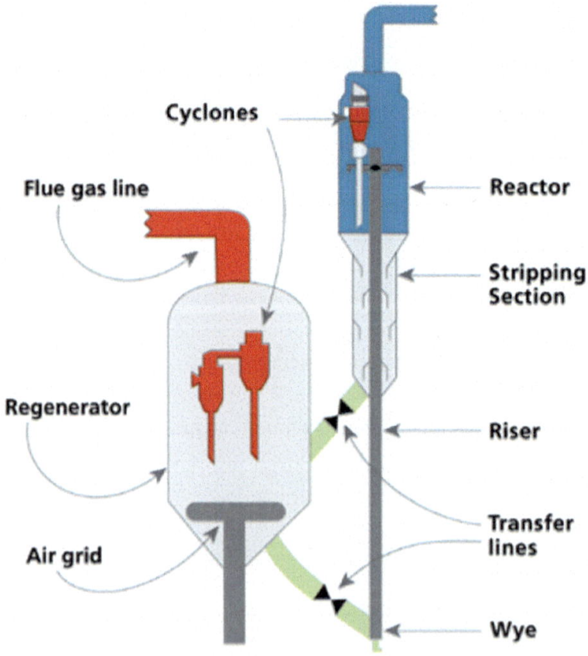

Fig. 9.8 Schematic diagram of FCCU showing the parts required for refractory lining

9.3.2 Refractories for Riser Pipe

Hot gases laden with catalyst particles move in the riser pipe at a very high velocity of 15–23 meter/sec, upwards to the reactor. The temperature inside the riser pipe remains in the range of 600–800 °C [6]. All the reactions in which the heavier hydrocarbons are converted into lighter molecule of higher economic value occur in that short amount of time. It poses a very high abrasion on the refractory lining on the wall of the riser pipe. Lining of riser pipe wall is usually done by dense abrasion-resistant castable of about 150 mm thickness installed by gunning.

The castable should have a maximum abrasion loss of 10 cc as per the ASTM C-704 test method. The abrasion is not the same everywhere; in Wye section corners and the injector nozzle, the abrasion is still higher, and castable with abrasion loss below 6 cc is preferred there for lining. Besides the abrasion resistance, thermal conductivity and PLC of the refractory at about 850 °C are very important, and it should be as low as possible and should not be lower than 0.3%, because the shrinkage in refractory at higher temperature leads to development of cracks in the lining.

Besides this, the environment inside the riser pipe is reducing, and under such condition, the carbon gets deposited inside the pores causing microcracks in the castable. As a consequence, the castable behaves as a castable carbon composite material with different properties than the original material. The use of the castable for long duration develops crack and the castable spall. Therefore, it is important to know the thermo-mechanical property of the castable before its use and also under the altered condition.

The determination of Young's modulus of the castable at various temperatures gives a good idea about the mechanical behavior of the castable (Fig. 9.9). It has been found that there is a reduction of Young's modulus when castable is heat treated at elevated temperature. When castable is heated and cooled in a cycle, it is found that the modulus of elasticity reduces with the number of cycles. This is explained by the polymorphic transformation of quartz particle and microstructural damage with repeated heating and cooling due to the thermal stresses generated because of the anisotropy of different phases.

Expansion Joint There are many pipelines in the whole FCCU with different diameters. When hot gases travel through those pipelines, the shell gets heated and tries to expand; thus, it is very important to design the piping to accommodate the expansion and related different kinds of movements like axial, lateral, angular, etc. which otherwise can cause a disastrous failure. The overhead line exits the top of the regenerator, and it enters the third stage of the separator, the power recovery turbine. This is typically the largest and the heaviest joints used in the refinery (Fig. 9.10).

To take care of the expansions in the pipeline in service, the expansion joints are provided which are subjected to high temperature (about 730 °C), high abrasion by the recirculating catalyst particles, and large movements of different types (axial, lateral, angular, etc.). The flexibility of the piping and resultant load on equipment

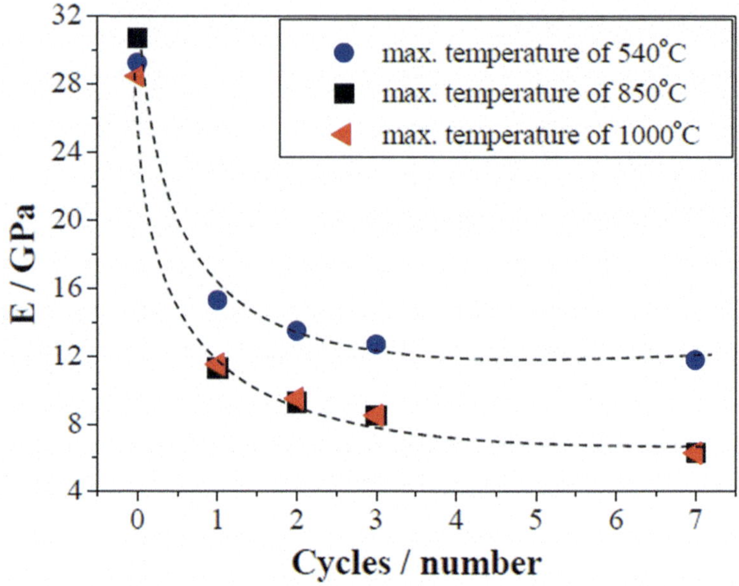

Fig. 9.9 Change in the E value of the castable at different temperatures

Fig. 9.10 Expansion joint in installation

are affected by the presence of the refractory lining inside the pipe. The flexural stiffness of the refractory-lined pipe is significantly higher than that of unlined pipe. The main concern in the design of the refractory-lined pipe is that the end reaction is significantly higher than the unlined pipe [7].

Expansion joints for FCC can be divided into three categories, namely, cold wall, hot wall, and unlined.

Cold Wall This kind of expansion joints are designed with a refractory lining to ensure the outside shell temperature does not exceed the limiting temperature (normally 340 °C). The shell may be constructed with low alloy or carbon steel. Lining

a

b

Fig. 9.11 (**a**) Stainless steel anchors prior to vibration casting of the refractory in the cold wall expansion joint, (**b**) expansion joint after casting and drying of the castable

is made with self-flow-type high abrasion-resistant castable. Stainless steel anchors are used for casting. Lining thickness ranges from 100 to 200 mm (Fig. 9.11). The lining is dried properly to remove the total water of casting before it is made ready for use. The refractory lining is placed inside the expansion joint, and behind the dense refractory lining, a layer of insulation is used to protect the shell from attaining higher temperature. It is a very critical application and requires a lot of experience to design the material. It has been found that refractory using very selective raw materials only performs well in this application.

Hot Wall In the hot wall design, the refractory is used to protect the steel shell from high abrasion of the recirculating catalyst particles. It is not used as a thermal barrier, and because of that, the shell outside may attain a high temperature and is constructed of chrome moly alloy or stainless steel. For placement of refractory, normally hexmesh is welded on the shell, and phosphate-bonded high alumina plastic material is hand packed over the hexmesh to provide a thin but abrasion-resistant lining (Fig. 9.12).

9.3.3 Refractories for Reactor and Regenerator wall

The temperature inside the reactor and regenerator remains at about 500–650 °C and 600–800 °C, respectively. The thermography of regenerator is shown in Fig. 9.13 [6]. There is mild abrasion of the gas on the wall and the environment inside is reducing in nature. Castable refractory having a density of 1.2–1.4 gm/cc and having a compressive strength of minimum 8.0 N/mm^2 at 800 °C should be used for lining the wall. Lightweight castable is recommended because of its low thermal conductivity which can retain the heat inside the reactor or the regenerator.

a b

Fig. 9.12 (a) Hexmesh preparation before installation of castable and (b) after installation in hot wall expansion joint

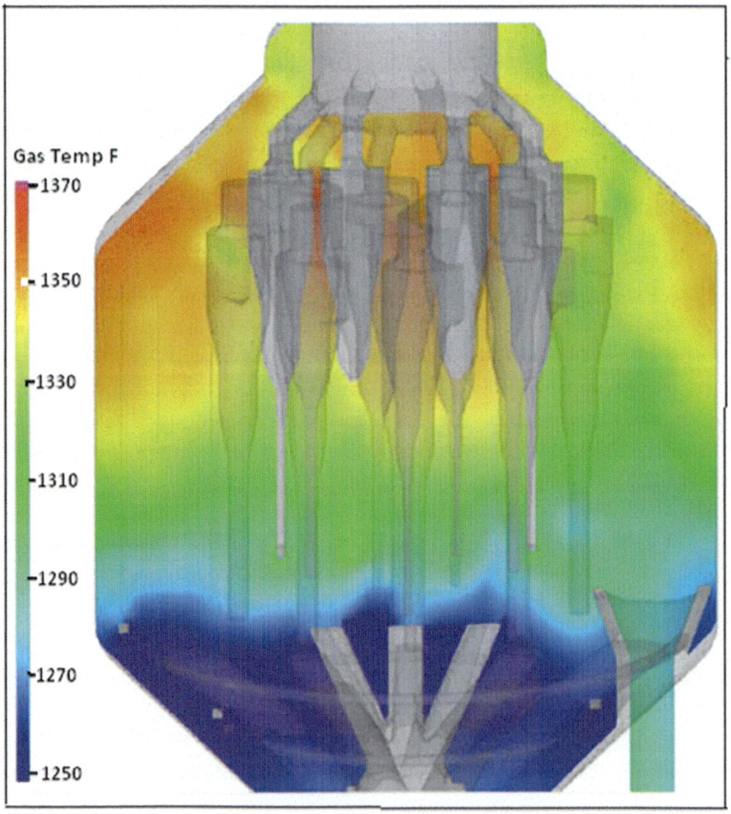

Fig. 9.13 Thermography shows the temperature profile inside the regenerator

The catalyst carried by the gas stream falls out the end of the cyclones through tubes called dip legs into a collector at the bottom of the vessels, where more abrasion takes place than that of the wall. Therefore, in the bottom region, a castable with higher density in the range of 1.7–2.0 gm/cc and abrasion loss of 5 cc as per ASTM C-704 test is used for better performance. The details of the properties of the castables are shown in Table 9.2.

9.3.4 *Refractories for Transfer Lines*

They connect all the vessels together. These lines contain and control the flow of catalyst. The refractory used in the transfer line is the same that is used at the bottom of the regenerator, i.e., a castable with a density of 1.7–2.0 gm/cc. The flow of the catalyst in the transfer line is controlled by operation of slide gates. The refractory used in this area is plastic refractory same as that used in cyclone wall by ramming over Hexmesh.

9.3.5 *Refractories for Cyclones*

The riser feeds the gas stream carrying catalysts in two sets of cyclones in the reactor to separate the catalyst from the hydrocarbon; similarly cyclones in the regenerator separate the catalyst from the flue gas stream and send it back to the riser pipe. The separation process of the catalyst, from the gas, is done by injecting the high-velocity catalyst-laden gas into the cyclone. The velocity of the gas at the entry of cyclone is about 20 meters/sec [8, 9]. Due to the centrifugal force, the higher density solid catalyst particles lose their kinetic energy by colliding with the cyclone wall and fall down. An FCC unit on average circulates about 50 tons of catalyst per minute between the reactor and the regenerator and run continuously for 4–5 years, which means a movement of 100 million ton of catalyst. Therefore, the cyclone wall experiences a constant impingement of the high-velocity catalyst particles and needs to be very highly abrasion resistant. Phosphate-bonded high alumina plastic material is the most common refractory used in the cyclone wall. The thickness of the plastic material maintained is about 25–30 mm and is rammed over hexmesh (Fig. 9.14a). After installation of plastic, it is heat treated at 350 °C before starting the FCCU operation, because the steam at higher temperature may damage the plastic unless it is heat treated prior. The plastic used has a density of 2.8–2.9 gm/cc and the abrasion loss by ASTM C-304 test is 2–3 cc.

Barracuda's erosion model can be used to quantify the impingement of the particles on the refractory lining and has proposed an equation to quantify the erosion as given below [6]:

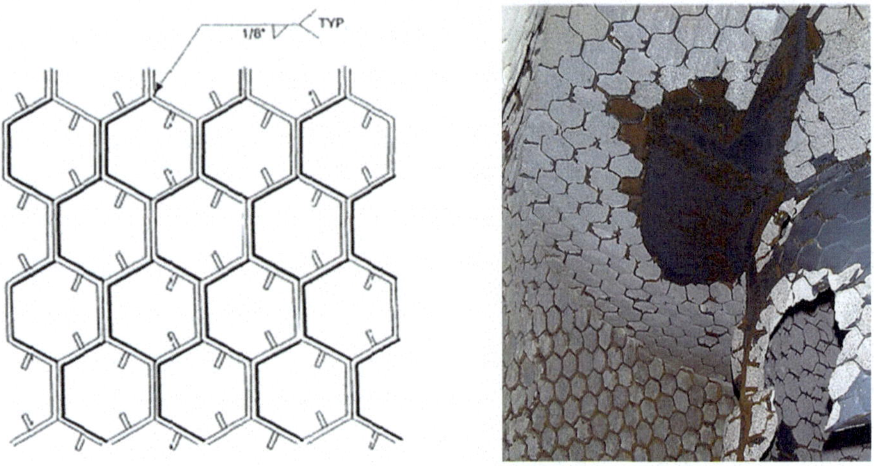

a b

Fig. 9.14 (**a**)The hexmesh anchor to fix plastics in the cyclone. (**b**) Severe erosion in the cyclone, FCCU

$$\text{Erosion index} = C_{\alpha}.m^{1.5}.v^{3.5} \tag{9.1}$$

where C_{α} = a coefficient as a function of angle α, m = particle mass, and v = velocity of the particle.

The regions with erosion exceeding a tolerable limit can be plotted and compared for various designs of the equipment.

Phosphate-bonded products are used here because it has excellent strength at medium range of temperature that exists in the cyclone. Here the insulating effect of the refractory is irrelevant, because the cyclones are situated inside either of the reactor or the regenerator and the temperature exists both outside and inside of the cyclone wall [10]. In fact, phosphate-bonded plastic material can be used to get a good performance in many other applications in FCCU (Fig. 9.15). The phosphate-bonded materials are hand packed on hexmesh and then rammed to finish. The effective thickness can be 25–30 mm. Properties of a typical phosphate-bonded plastic suitable for the above application are given in Table 9.2.

In the reactor, the coke as a reaction product gets deposited and accumulated underneath the hexmesh in the crevices of refractories pushing the refractory apart and causing damage in the lining (Fig. 9.14b) [8, 11]. This gap is formed due to the difference of thermal expansion of stainless steel and the refractory. The thermal expansion of stainless steel is about two times that of the refractory, and that difference in thermal expansion creates the crevices which are big enough for the catalyst particles to go inside. During reaction, the coke is deposited into the catalyst particles, and the reducing condition is created locally. This reducing condition damages the protective chromium layer on the steel and initiates the carburization and sulfurization of the steel which is most severe at the bottom and ultimately leads to

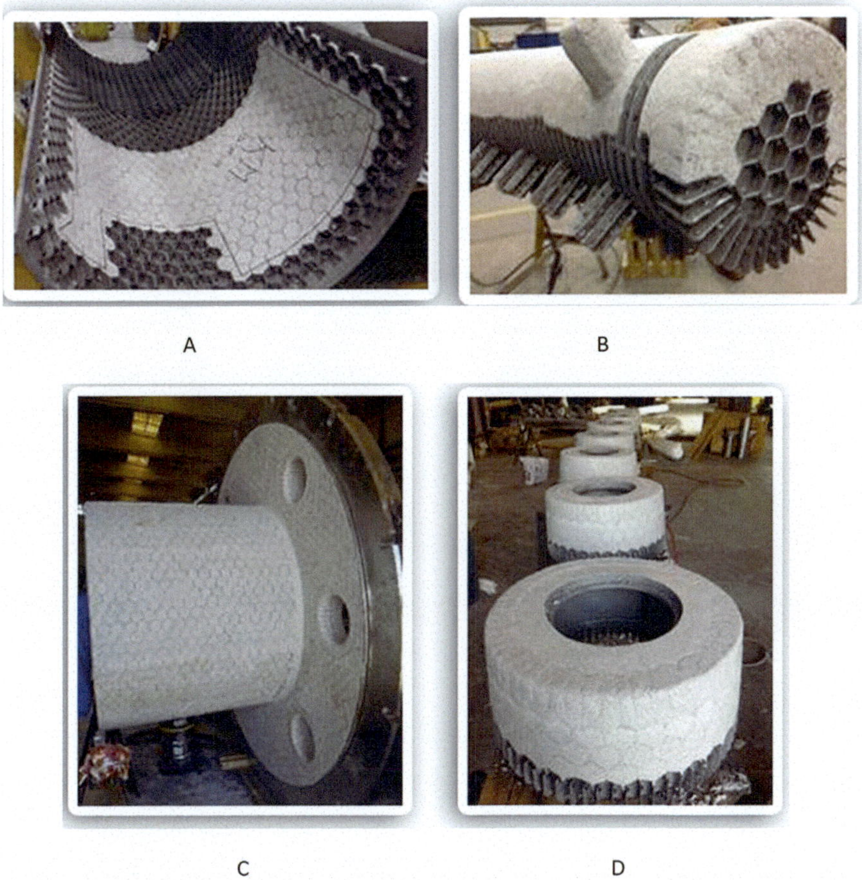

Fig. 9.15 Lining of phosphate-bonded plastics on (**a**) regenerator cyclone, (**b**) air distributor, (**c**) slide valve, (**d**) lampshades. (Courtesy Thermbond Refractory Products, USA)

the failure of anchors and subsequently the refractory lining itself [12, 13]. Hexmesh anchors in the reactor should be fully welded along the seam, and all the cracks should be repaired during turnarounds.

Two different types of cyclones are, primary and secondary (Fig. 9.16), in the FCCU. There is a fundamental difference in the erosion patterns of the lining in primary and secondary cyclones. Highly loaded primary cyclones normally experience much less erosion compared to lightly loaded secondary cyclones. The key difference in the erosion pattern of these two types of cyclone lies in the solid flow pattern in the cyclones (Fig. 9.17) [9].

There are more numbers of turns of the vortex in the secondary cyclone, and the spinning continues to the top portion of the dipleg, below the cone. The concentrated solid stream rotates at high velocity in the lower cone and causes significant erosion there and in the dipleg [9].

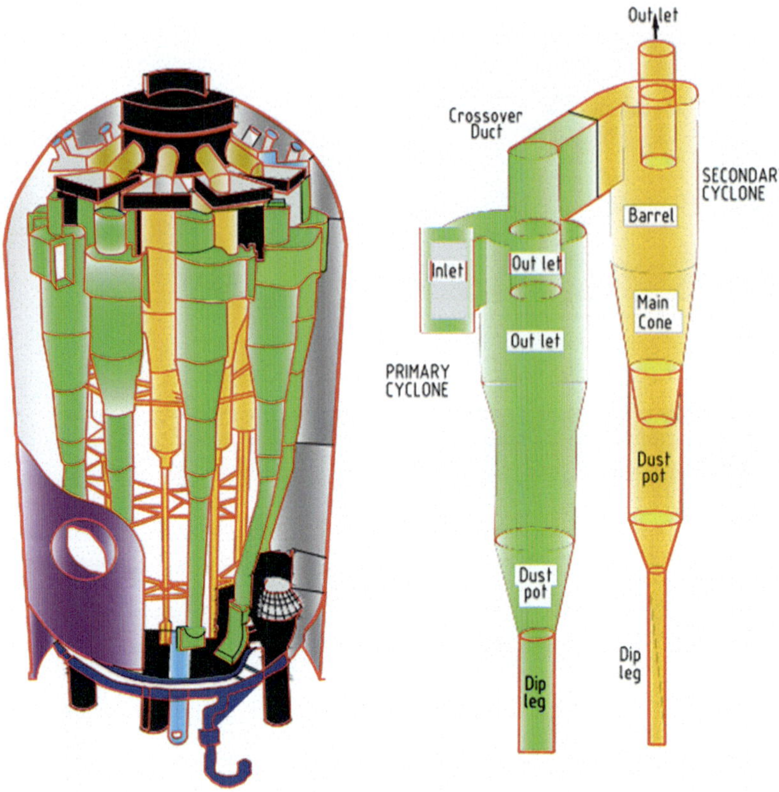

Fig. 9.16 A typical two-stage cyclone, inside regenerator, showing primary (green) and secondary (yellow) cyclones

Normally, both conventional and phosphate-bonded plastics are used, and the three types of unshaped refractories used in FCCU are shown in Table 9.3. Most of these requirements are specified by the oil companies and licensors worldwide [14]. The castable refractories can be installed by gunning also, with little modification in their rheological behavior. It has been found that the extent of abrasion also depends upon the geometry of the cyclone, and it has got a direct bearing on the L/D ratio of the cyclone. It has been found that the measured erosion rate at an L/D of 5.1 was about 70% of the erosion rate of a cyclone with L/D of 3.1 [9].

9.3.6 Refractories for Flue Gas Line

The temperature that exists in the flue line is 500–580 °C. Only mild erosion takes place in this line carrying flue gas from the regenerator. Some thermal shock may take place depending upon the specific operating practices. Normally, a single layer of castable of medium density is used.

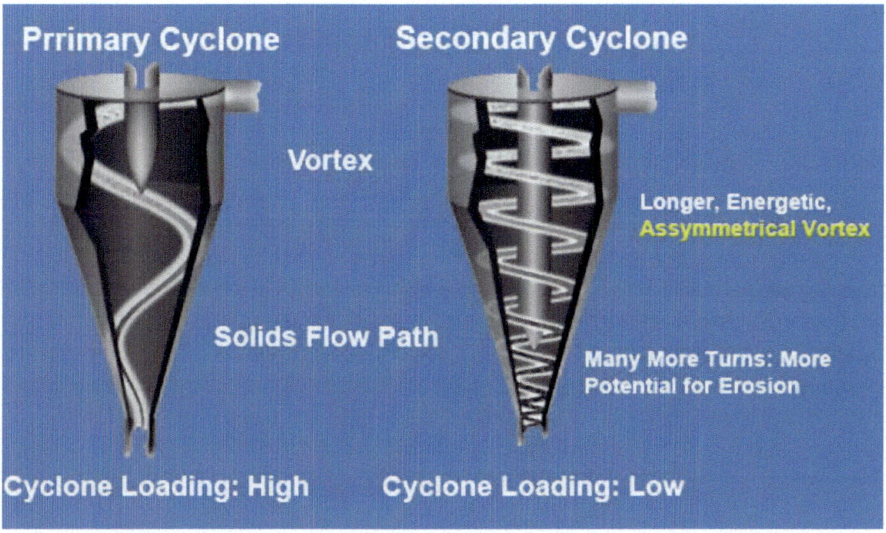

Fig. 9.17 Difference in flow pattern in primary and secondary cyclones

Table 9.3 Refractories used in FCCU

	Products		
Properties	Wall of reactor and regenerator, transfer line	Flue gas line, reactor/ regenerator bottom	Cyclone wall
Type of product	Conventional castable	Conventional castable	Phosphate-bonded castable
Service temp. °C	1350–1400	1300–1350	1250
A12O3%	45–50	50–55	78–80
SiO_2%	–	–	8–9
Fe2O3%	0.5–0.8	0.8–1.0	0.8–1.0
CaO%	10–12	6–7	P_2O_5-4.0%
B.D. (gm/cc)	1.6–1.9	2.1–2.3	2.80–2.90
CCS (MPa)			
/110 °C	50–60	60–100	60
/1000 °C	45–50	50–70	–
CMOR (MPa)	0.4 MPa	90–130	17
PLC/1000C (%)	0.2	0.2	0.1–0.2
Th. cond (W/mK)			
400C	0.8–1.0	1.3–1.4	2.1
800C	1.0–1.05	1.5–1.6	2.2
Abrasion resistance (CC) (ASTM C704)	10–12	10–11	<3.0
Application method	Casting, gunning	Casting	Ramming

9.4 Sulfur Recovery Unit

9.4.1 Introduction

Sulfur recovery units are used to remove sulfur compounds (mainly H_2S) from gases produced by sweetening of refinery gases or sour field production gases. SRU converts the H_2S gas to elemental sulfur and condenses the sulfur in liquid form. The feed gas most commonly comes from amine regenerators and sour water strippers located in refinery units and gas processing plants. The chemical reaction that occurs in the SRU converter is very complicated in nature and forms sulfur in various species. In Claus reactor, 50–70% of the sulfur compounds can be converted to elemental sulfur. Further conversion of the sulfur-containing gases must be carried out in the presence of catalyst in catalytic reactors in multiple stages. Up to 93% of sulfur is removed in the two-stage convertor and up to 96% in the three-stage convertor.

9.4.2 Process

Claus sulfur recovery unit is composed of two different sections (Fig. 9.18):

1. Thermal section in which the H_2S is converted to elemental sulfur at high temperature (>980 °C) without the presence of catalyst
2. Catalytic section in which sulfur is formed at much lower temperature (200–350 °C)

Thermal Section H_2S bearing acid gas is fed to the burner of the reactor, and enough air is injected to burn all the impurities like NH_3 and hydrocarbons to N_2 and CO_2 and to oxidize 1/3 of the H_2S to SO_2 as per the reaction shown:

$$H_2S + 3/2O_2 = SO_2 + H_2O \qquad (9.2)$$

The reaction is highly exothermic.

In the reaction furnace, the unburned H_2S (remaining 1/3) in the acid gas reacts with SO_2 already formed to yield elemental sulfur and water as per reaction 9.2:

$$2H_2S + SO_2 = 3/2S_2 + 2H_2O \qquad (9.3)$$

This reaction is endothermic and is controlled by chemical equilibrium.

The flame temperature of combustion of H_2S when the acid gas contains about 80–90% H_2S is over 1000 °C, and presence of hydrocarbons and other impurities can shoot up the temperature up to 1400 °C [15].

Reaction Furnace Reaction furnace is the heart of the SRU. The furnace is a horizontal refractory-lined vessel equipped with a choke ring and/or checker wall for

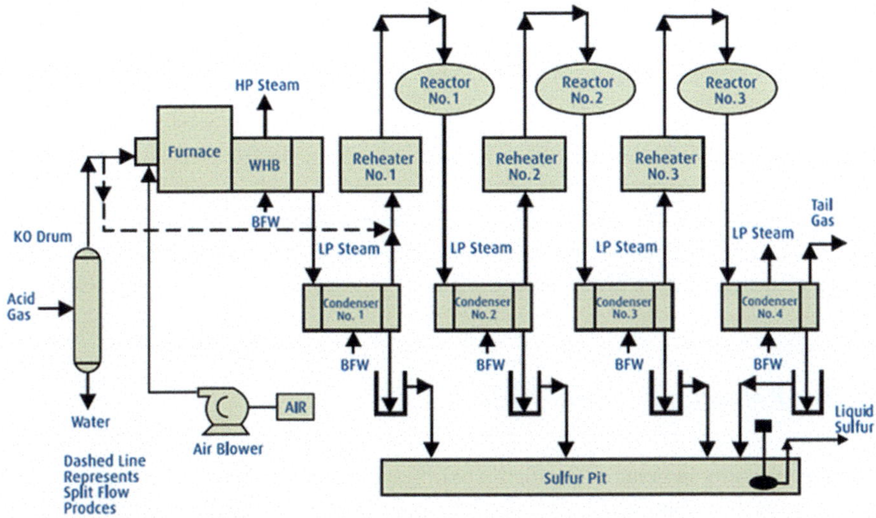

Fig. 9.18 Lay out of Claus process of sulfur recovery unit. (Courtesy Linde)

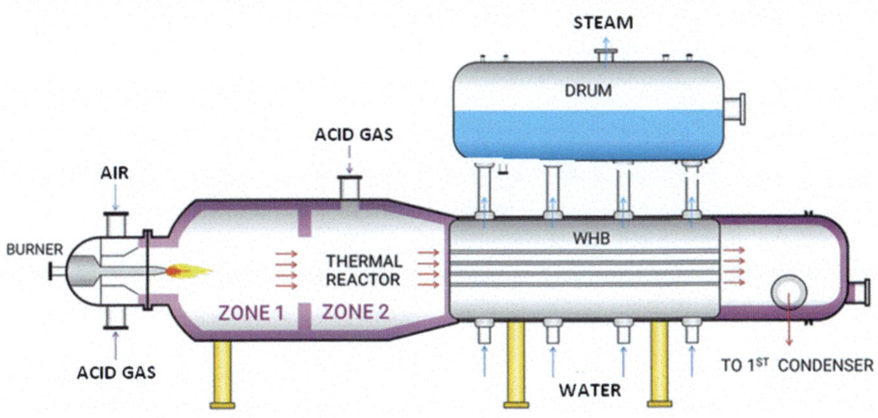

Fig. 9.19 Schematic figure of construction of refractory lining in SRU reactor

improving the mixing of the reactant gases in the furnace (Fig. 9.19). This also acts as a reflector to radiate heat to the front of the furnace and to shield the waste heat boiler tubes from the direct heat of the furnace. WHB is used to cool the hot gas leaving the thermal reactor. A typical feed composition is given in Table 9.4 [15].

More damage occurs in the SRU refractory lining during start-up and shutdown time. During start-up of SRU thermal reactor, the natural gas, in the main burner, is burned with stoichiometric air quantity to heat it up. The flame temperature of the natural gas and stoichiometric air is about 1925 °C and needs to be controlled carefully. In some cases, additional oxygen is also used to increase the capacity by

Table 9.4 A typical feed composition

Components	H_2S	H_2O	NH_3	Hydrocarbons	H_2
Percentage	97.6	2.2	0.0	0.1	0.00067

reducing nitrogen. Therefore, the refractories to be used for lining need to have, besides high-temperature corrosion to the sulfur oxide gases, high thermo-mechanical stability at high temperature. It has been found that the high-temperature resistance to deformation is one of the very important criteria to decide the performance of the refractory in the SRU.

9.4.3 Refractory Lining Pattern

Refractory lining is generally constructed in two different layers. The backup layer is insulating in nature of 75 mm thickness, and the front lining is made of dense alumina bricks of 90–95% Al_2O_3 purity and 230–350 mm thickness depending upon the furnace diameter. Fe_2O_3 must be low because it will be used in the acidic atmosphere. Refractory lining thickness is designed to keep the shell temperature between 180 and 325°*C* and to keep the sulfur in vapor state and not to allow the condensation of sulfur or sulfur oxides to avoid severe corrosion of the shell by the acidic gases. The suitable refractories for the lining of the SRU reaction furnace are given in Table 9.5.

The refractories to be used in the working lining of the SRU must be resistant to chemical attack of sulfur oxide gases at high temperature, should have good thermal shock resistance, and must have very good resistance to deformation under load at high temperature [16]. Besides the chemistry of the refractory material used, the microstructure of the refractory is also very important to decide its load bearing capacity, and the quantity of glassy phase and its viscosity at high temperature, its pore volume and the pore size distribution are all very important parameters.

The test methods ASTM C832, that determines the creep under load and ASTM C-16-03, the standard load testing of shaped refractories at high temperature, are often prescribed by the licensors to determine the suitability of the refractories. The standard test method is to determine the deformation in the direction of load when 0.172 MPa load is applied on a rectangular sample at a certain temperature for 90 min. But some licensors have modified the standard methods to provide more soaking time to ensure minimal deformation of the refractory at high temperature of 1650 °C. The load bearing capacity of the refractory, suitable for the lining of SRU as given by some licensors, is given in Table 9.6 [16].

The brick shape used to line the Claus reactor is in the side arch shape as shown in Fig. 9.20, and the refractory-lined reactor is shown in Fig. 9.21.

The refractory shapes are laid dry without any mortar, and the bricks remain tight in their position because of the key effect and the thermal expansion of the bricks during operation. But during long and continuous use of the refractory in the reactor, due to thermal shrinkage, vibration, etc., the bricks tend to come down from the

Table 9.5 Properties of the refractories for the SRU

	Product		
Properties	Alumina brick	Low cement castable	Conventional HA castable
Al2O3	90–95	40–50	95
Fe2O3	0.1–0.2	0.5–1.0	0.1
Na2O	0.3	–	0.1
CaO	–	1.5–3.5	SiO2–0.3%
BD	2.95–3.0	2.2–2.35	2.7
Porosity	15–16	–	–
CCS/110 °C	100–120	50–80	60
/1000 °C	–	100–110	70
RUL	>1700	–	–
Th. expnsn (%)/1400 °C	1.07%	0.6% at 1000C	0.75/1000C
Th. conduct (W/mK)			
/200 °C	2.0	1–1.5	2.05
/1000 °C	2.2	1.1–1.6	2.1
PLC/1000C (%)	–	−0.25 max	0.1
Area of application	Working lining of reaction zone and burner area	Backup lining of reaction zone	Lining of the ports

Table 9.6 Modification of ASTM C-16-03 standard method to ensure minimal deformation in service

Licensor	Soak time at 1650 °C	Maximum deformation allowed
1	90 min	+/− 0.5
2	25 h	−0.5
3	50 h	−0.5
4	100 h	−0.4

ring slowly and sometime fall down which requires repairing of the existing lining or new lining (Fig. 9.22).

To prevent the slippage of the brick from the lining, the design of the bricks is modified with the tongue and groove to impart more stability in the lining as shown in Fig. 9.23.

9.4.4 Checker Wall

Inside the Claus reactor vessel, there is a refractory checker wall (Figs. 9.19 and 9.24) at some point (or points) between the burner and the waste heat boiler to serve a number of functions such as (a) enhancing mixing of gases, (b) increasing residence time of the gases and (c) protecting the tube sheet from the radiant heat.

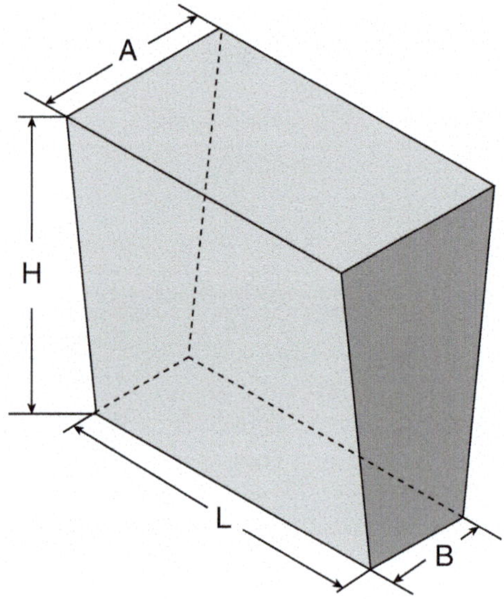

Fig. 9.20 Side arch brick (A,H,L,B defines the brick)

Fig. 9.21 Newly refractory-lined Claus reactor

Different designs of refractory wall were tried over since the beginning of the use of the Claus Reactor in the industrial scale. A lot of research has been done to arrive at the optimum design to control the three Ts, i.e., temperature, time, and turbulence, which are the factors to control the kinetics of reactions 9.2 and 9.3.

The initial design was made with alumina brick lining with high strength and volume stability at high temperature. Cylindrical tubes are also tried to construct the wall, but both designs are found not stable over a long duration. In another design,

Fig. 9.22 Slippage of the bricks from the ring of the lining of the Claus reactor during its use

Fig. 9.23 The tongue and groove arrangement of the brick lining

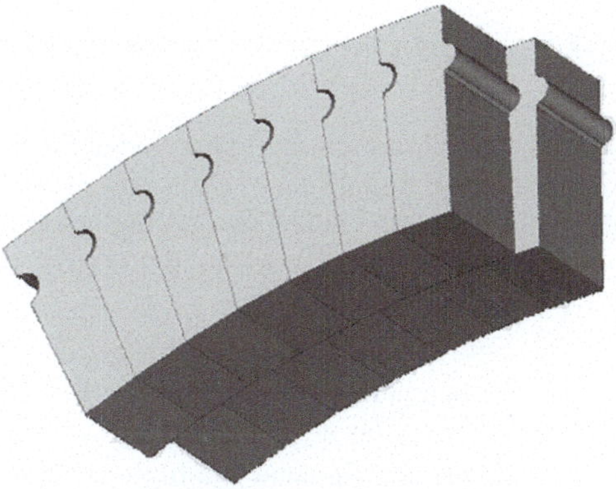

Fig. 9.24 Refractory checker wall

the ceramic ferrules are implanted in the wall made of castable. The main causes of failure are thermal cycling during heat up and shutdown of the unit [17]. Moreover, the carryover of liquid in the combustion chamber loosens the brick lining system, and running the unit above the design capacity causes vibration to damage the lining (Fig. 9.25).

Most modern design using the computational fluid dynamics (CFD) modeling has made the holes not straight but a hood on the holes at different angles to divert the gases in different direction to ensure better efficiency of the reactor to optimize the three Ts and at the same time ensure least pressure drop as shown in Fig. 9.26 [18].

Orifice throat and checker blocks can be made of high alumina precast and pre-fired (PCPF) quality of the properties shown in Table 9.7. The blocks are made by casting of low cement castable based on tabular alumina.

9.4.5 *Refractory or Ceramic Ferrule*

This is a small but an important element in the SRU system and used to protect the tube sheet and boiler tubes of the waste heat boiler (WHB) where the exit gas from the SRU enters into WHB for cooling down and to convert the heat in the exit gas to generate steam as shown in Fig. 9.27 [19]. The tube sheet and the tubes of the WHB are exposed to high-temperature gases at a temperature of about 1100 °C containing H_2S and sulfur oxides and need protection from sulfidation at high temperature.

The refractory ceramic ferrules are used for different designs (Fig. 9.28), according to the particular design of the equipment [20], but irrespective of the design, it must be able to withstand the high temperature and the temperature fluctuations during start-up and shutdown because it has been found that 60% of failure of ferrule occurs during start-up and shutdown of the SRU. The ferrules must also be able to accommodate the thermo-mechanical stresses generated due to mismatch of properties of tube sheets and the ferrules. The arrangement of fixing of the ferrule on the tube sheet is shown in Fig. 9.28.

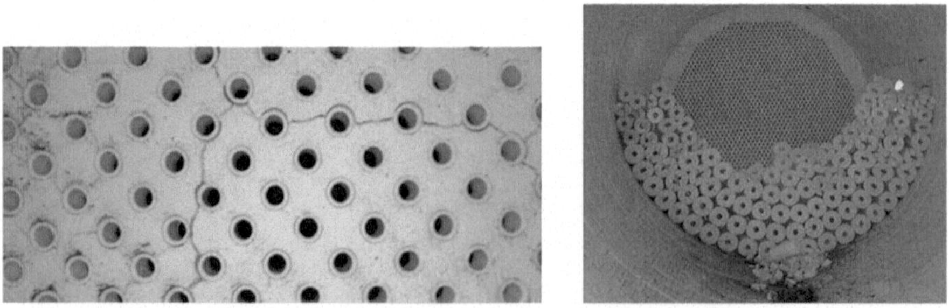

Fig. 9.25 Damage of the castable wall due to thermal cycling and the falling of the checker wall constructed with cylindrical tubular shape

Fig. 9.26 The design of the vector wall. (Courtesy Blasch Ceramics)

Table 9.7 Properties of PCPF checker blocks

Properties	Values
Al2O3%	94–95
Fe2O3%	0.3–0.4
CaO%	1.5
BD (gm/cc)	3.1
CCS (MPa)/ 110 °C	90–100
/1100 °C	100–110
/1550 °C	120–130
PLC (%)/1100 °C	+/– 0.3
/1550 °C	+/– 0.3
PCE	38 Orton cone

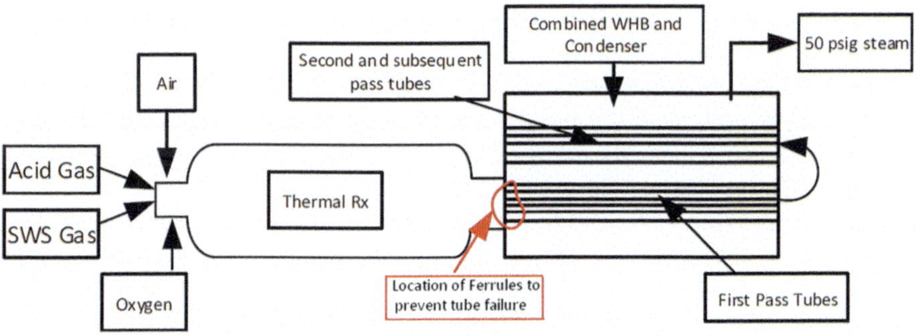

Fig. 9.27 Location of use of ceramic ferrules

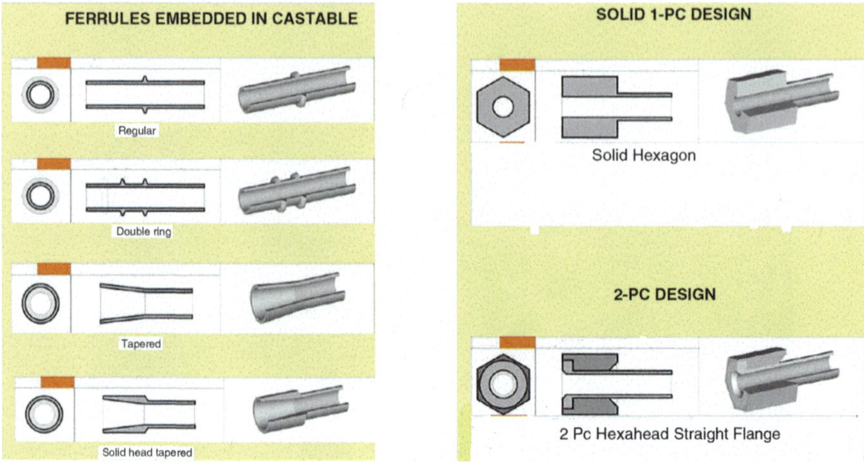

Fig. 9.28 Different designs of the ferrules

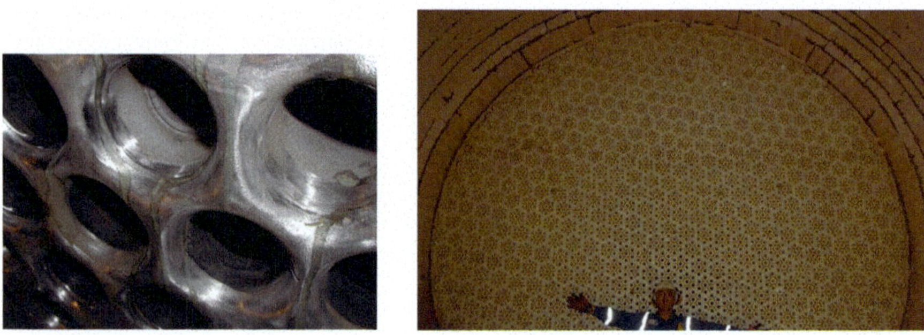

a) Tubesheet

b) 4200 number Ferrules are placed over the

Fig. 9.29 (**a**) Tube sheet, (**b**) 4200 number ferrules are placed over the tube sheet

The actual tube sheet and the ferrules after installation over the tube sheet in a SRU are shown in Fig. 9.29 [21].

The temperature distribution across the ferrule during service is shown in Fig. 9.30 [18]. Normally, the ferrules are made of 96–99% alumina by injection molding or casting methods followed by high-temperature firing to get sintered, volume-stable product. Alumina has all the desired properties like high refractoriness, high volume stability, and high thermal shock resistance and corrosion resistance to most chemicals. Ferrule design, quality, and installation quality are the major factor to decide the performance of the ferrule. The design of the ferrule must take care of the stress generated due to the constrained expansion at high temperature. Figure 9.31 shows the actual ferrule design and how the stress, generated due to thermal expansion, is mitigated [22].

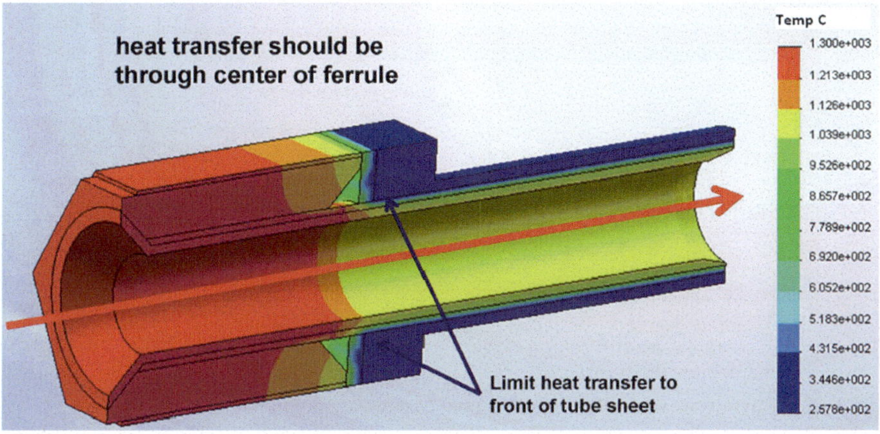

Fig. 9.30 Temperature distribution across a ferrule in service

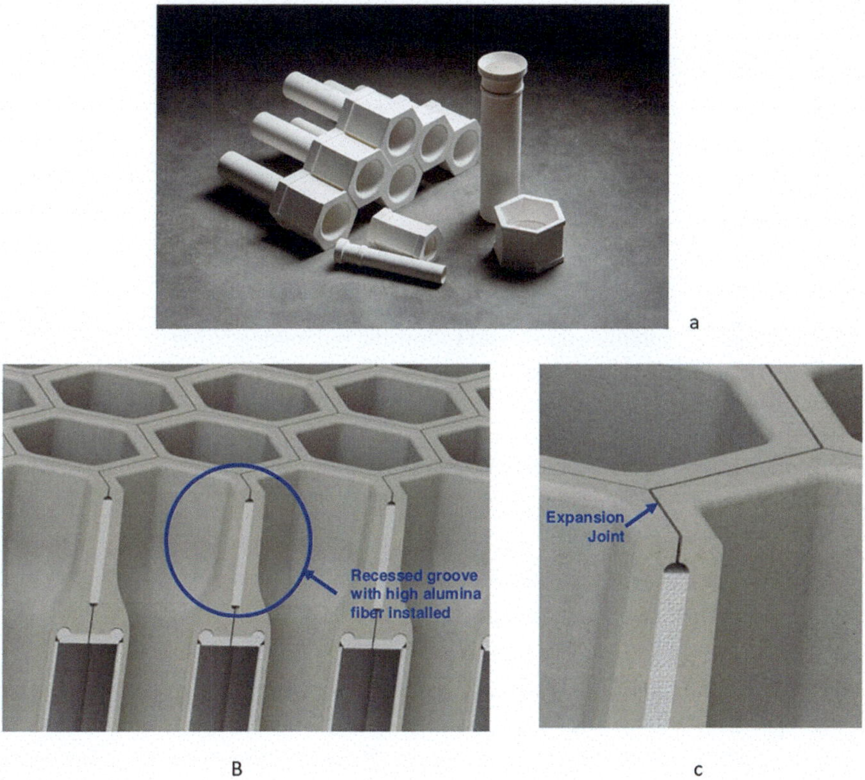

Fig. 9.31 Actual ferrules and its design to counter the stress due to thermal expansion. (Courtesy Blasch Ceramics)

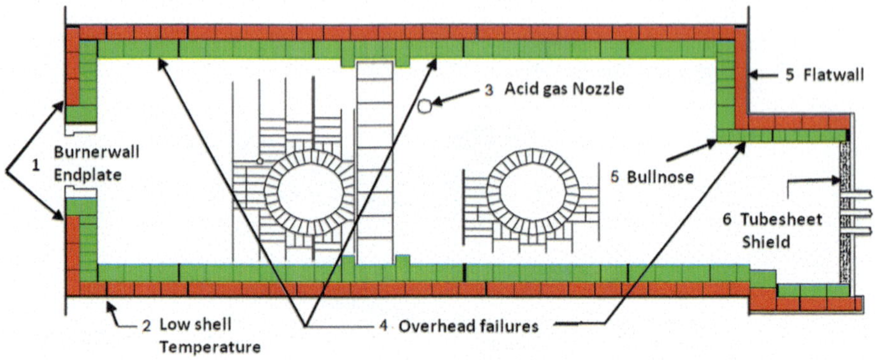

Fig. 9.32 Different areas of SRU furnace prone to refractory failure

Table 9.8 Recommendation of refractories for the SRU reactor

Plant type	Gas plant	Refinery	Refinery
SRU process	Air only	Air only	O_2 enriched
SRU feed	Amine acid gas	Amine acid gas	Amine acid gas
Operating temperature	1100 °C	1200–1430	1540–1650
Backup lining Quality Thickness(mm)	Fireclay brick 115	Fireclay brick 115	Fireclay brick 115
Hot face lining Quality Thickness (mm)	70% alumina 115	90% alumina 115	90% alumina 115

The performance of the refractory lining in SRU reactor depends upon the design of the reactor also, and the user should follow the recommendation of the refractory supplier to use the refractory quality, suitable for different locations inside the Ortloff reactor (Fig. 9.32). Table 9.8 shows the recommendation of the refractories in the Ortloff furnace and gives a quantitative idea about the problems and the locations of occurrence in the furnace, related to refractory, that causes the outage of the Ortloff reactor, off the service [23] (Table 9.9).

9.4.6 *Tail Gas Incinerator and Waste Heat Recovery Boiler*

Tail gas incinerator and waste heat boil are the other important accessory equipment in SRU process. Tail gas treating unit converts the small amount of unconverted sulfur (<5%) from the Claus reactor exit gas to H_2S and recycles it to the Claus reactor for further processing [24]. The thermal oxidizers typically work in the temperature range of 650–900 °C, and the large quantum of heat energy produced in the thermal oxidizer can be utilized in waste heat boiler to generate steam for further

Table 9.9 Problem areas in SRU furnace of the Ortloff design

Area reference number shown in Fig. 9.32	Area of refractory failure	The proportion of total refractory damage incidents
1	Reaction furnace endplate (burner end)	5%
2	Shell area adjacent to burner endplate (low shell temperature)	10%
3	Acid gas bypass nozzle – refractory, ferrule, and shell	30%
4	Reaction furnace overhead	15%
5	Transition piece bullnose and flat wall	25%
6	Tube sheet shield	15%

Table 9.10 Properties of refractories used in incinerator and waste heat boiler

	Product				
Properties	High alumina brick	High alumina brick	High alumina plastics	Bubble alumina castable	Insulating castable
Al_2O_3%	90	70	80	94	50
Fe_2O_3%	< 0.2	<0.5	1.0	0.3	<2
BD (gm/cc)	3.3	2.7	2.7	1.65	1.5
CCS/110 C(MPa)	120	100	–	11	15
Th. conductivity (W/mK)/800C	–	–	–	0.85	0.5
Max. service temperature °C	1750	1650	1600	1400	1100
	Incinerator floor and lower wall	Upper side wall	Roof	Backup lining	Waste heat boiler

use. High-grade alumina refractories are used to line the incinerator, and the properties of the refractories are given in Table 9.10. The iron oxide content in the refractory must be low because of the acidic gas environment.

The waste heat boiler is used to cool down the gas that comes out from the SRU reactor and to utilize the heat energy produced to form steam, and in a similar way, the thermal energy produced in the incinerator of tail gas is also used in waste heat boiler. The refractories used in waste heat boiler are shown in Table 9.7.

References

1. Hassan Al Haj Ibrahim. Fired Process Heaters, Matlab, www.intechopen.com
2. Saha P et al, Study of Refractory lining performance of heaters fired with high Sulfur fuel oil. In: UNITECR Proceedings

3. Zink J, Combustion Process
4. Fired heaters for general Refinery service API standard 560, fourth edition August, 2007
5. Karns S (2019) Delicate balance in a tough environment: refractory selection for Fluid catalytic cracking units. Am Ceram Soc Bull 98(2)
6. Cabrelon MD (2013) Thermo-mechanical study of a refractory castable for Petrochemical Riser. UNITECR
7. Truong KT (1998) Improve FCCU refractory lined piping design. In: Hydrocarbon processing, July 1998
8. Crowley MS, Schacht CA (1997) A comparison of the thermo-mechanical properties of 'As installed' Vs 'Coke filled' Refractories for fluid catalytic cracking unit. In: UNITECR Proceedings
9. Ye Mon Chen (2011) Evolution of FCC- past, present and future,- and the challenges of operating a high temperature CFB system. In: 10th International conference on circulating fluidized bed and fluidization technology, CFB-10, May 1–5, USA
10. API Technical report 979, 1st edition Oct 2018
11. Kuo R, Galveston R (2016) Troubleshooting catalyst losses in the FCC Unit
12. Dries H, Mc Auley R (2001) FCC Cyclones – a vital element in profitability, Shell Subbed, 5/7/2001
13. Bujaski M et al (2005) New multifunctional Refractory concretes for FCCU in Refineries. In: Proceedings UNITECR
14. Goar BG, Fenderson S (1998) Fundamental of Sulfur recovery by Claus Process. In: Laurence Reid Gas Conditioning Conference, Engg and Geosciences College of Continuing Education, University of Oklahoma March, 1–4, 1998
15. Chidambaram S (2018) Thermal damage of Sulfur processed chamber under Claus operating reaction condition- a case study- case studies in construction materials. Science 8:517–529
16. Brueck J, Stimpfl C (2015) Alumina Mullite Bricks for Sulfur recovery application, RHI Bulleting, Feb, 2015
17. Bolebruch J et al (2014) Field experience with Claus Furnace checker wall, www.digitalrefining.com/article/1000892, PTQ, Q1 2014 1
18. Reliability in recovery- Matthew Allen and Jeffrey Bolebruch
19. Clinton J. Schulz (2017) Improved ferrule and tube design to prevent the film boiling in SRU waste heat boiler. In: Vail Brimstone Conference, Sept. 2017
20. Lyttle DM, Barrow D (2015) Improving SRU reaction furnace reliability: boiler tube protection, Middle east Sulfur plants operation network forum –October 20, 2015
21. Muhammaed Bashir Khan Sulfur recovery unit reaction furnaces waste heat boiler Ferrule failure- root cause analysis. Al Hosn gas, Abu Dhabi Gas Development Company Ltd
22. Gicewitz L (2015) Engineered Ferrules for high temperature waste heat boiler, Blasch Precision Ceramics
23. Grigson S et al (2010) A case history of Refractory lining improvements for O_2 enriched furnace service, Sulfur unit.com conference, April 14–16, 2010, League city, Texas
24. Pickard P, McQuigg K (2013) Understanding thermal oxidizer, PTQ, Q2, 2013,1

Chapter 10
Refractories for Glass Manufacturing

10.1 Introduction

One of the most important materials, intimately associated with the human civilization, is glass. It is used in every walk of life as a versatile material for building construction, for solar panels, and for communication as glass fiber, as optical material, etc. Present world production of glass is close to 120 million tons per year.

Various types of glasses are produced based on the different use, but almost 90% of total glass production is comprised of container glass, flat glass, and fiberglass. The distribution of the production of different kinds of glass is given in Fig. 10.1 [1]. The production line for all the bulk glass product is continuous. The specialty glass products are made normally in batch process. The heart of the glass plant is the furnace where the glass batch is melted and then processed further to give its desired shape.

The chemistry of glass can vary widely, and depending upon the chemistry, the physical properties also vary widely. The main ingredients of glass are soda (Na_2O), lime (CaO), and silica (SiO_2), along with other minor additions made for refining, decolorization and to incorporate special properties. For commercial production, a glass has certain prerequisites, e.g.:

(a) It can be melted and refined below 1550 °C, and the viscosity at the melting temperature should be in the range of 50–100 poise.
(b) Rise in viscosity during cooling should be such so that glass can be worked by drawing or pressing or rolling, etc.
(c) It should not recrystallize during cooling.

The flow sheet of glass manufacturing process is shown in Fig. 10.2 [2].

Chemical compositions of different types of glasses are shown in Table 10.1.

P. Sengupta, *Refractories for the Chemical Industries*,
https://doi.org/10.1007/978-3-030-61240-5_10

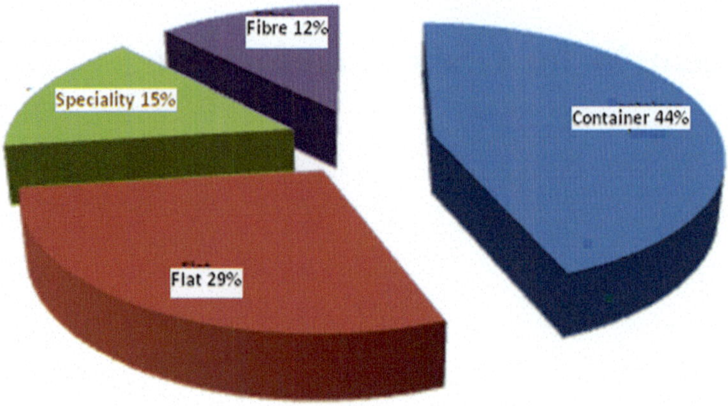

Fig. 10.1 Distribution of production of different kinds of glass products

10.2 Container Glass Manufacturing Process

The basic glass batch consists of silica sand (SiO_2), lime ($CaCO_3$), and soda ash(Na_2CO_3), cullet (broken glasses, which melt first and enhance the glass batch reaction), and other minor ingredients like BaO, $NaNO_3$, MnO_2, Al_2O_3, litharge (PbO), spodumene (Li_2O), arsenious oxide, etc. to impart special properties or for refining and decoloration of glass melt. All the ingredients are fed to glass melting furnace.

The furnace can be of batch type or continuous type. Batch-type furnaces are used for melting special glasses in small crucibles. The continuous-type furnaces are used for manufacturing of bulk glass products, like container glass, float glass, or fiberglass. We shall discuss here the refractory requirements for continuous glass production furnace which is also called glass tank furnace. In these types of furnaces, glass batch is fed continuously at a certain rate at one end of the furnace, and the molten glass is drawn continuously at a certain rate from the other end, depending on the design of the furnace. Glass melting furnaces and regenerator consume most of the refractories required for the glass plant and small quantities of special feeder expendable refractories like orifice ring, spouts, tubes, plungers, etc. are used in the forming end of the glass furnace. Here we shall focus our attention on the refractories for the glass tank furnace and regenerator.

10.2.1 Glass Melting Furnace

The furnace can have various designs depending upon the capacity, burner positions, etc. Two most common designs for furnaces are cross port firing and end port firing system (Fig. 10.3) [3]. The fuel is either natural gas or furnace oil, and for combustion preheated air or oxygen-enriched air is used. Normally 4–6 burners are

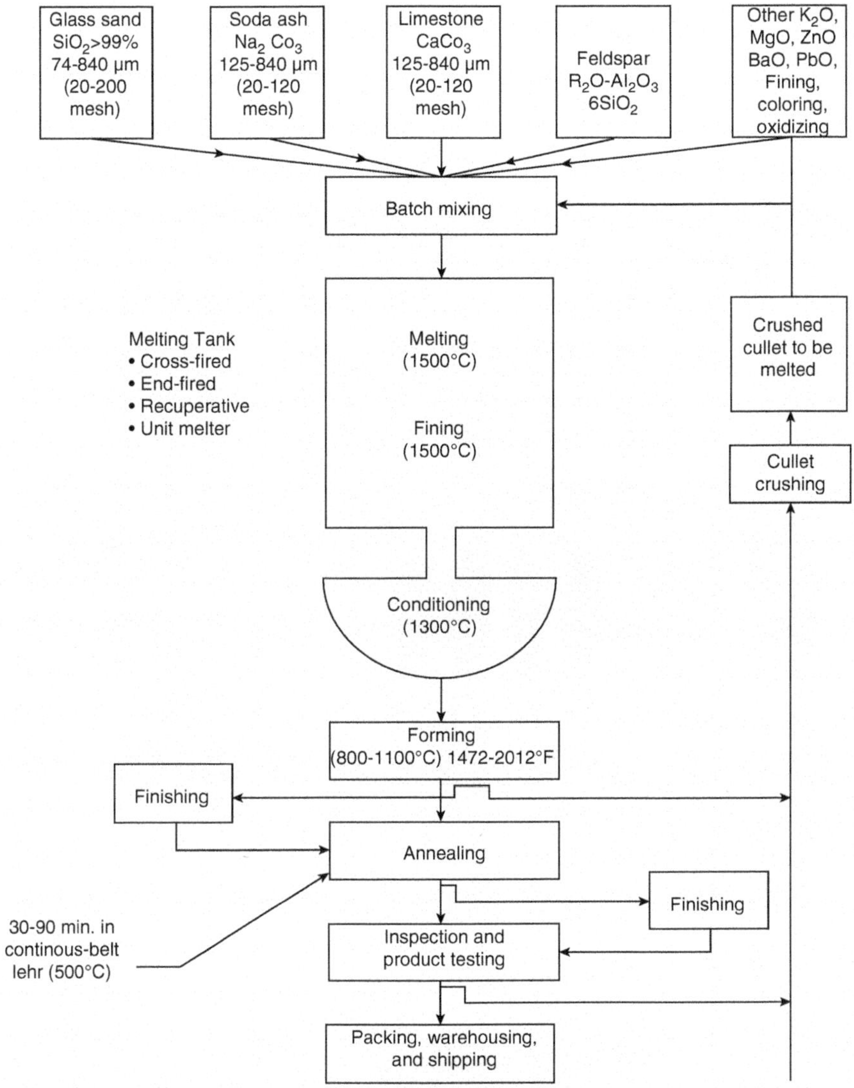

Fig. 10.2 Flow sheet of manufacturing of continuous glass melting plant

used on each sidewall. Burners in opposite walls are not placed in one line and are normally staggered. The furnace melting capacity is expressed as MT of glass melted per 24 h, and depending upon the furnace type and product type, it can vary from 20 to 700 tons/day.

The main elements of a glass tank furnace are as follows, shown in Fig. 10.4:

(i) Melting Tank – Bottom.
(ii) Super structures – Combustion chamber.
(iii) Bottom sidewall.

Table 10.1 Compositions of different glasses

Component wt %	Container glass	Flat glass	Fiberglass
SiO_2	73.0	72.7	55.2
Na_2O	13.6	1.0	0.3
CaO	10.6	8.3	18.7
MgO	0.3	4.0	3.3
Al_2O_3	1.3	1.0	14.8
B_2O_3	–	–	7.3

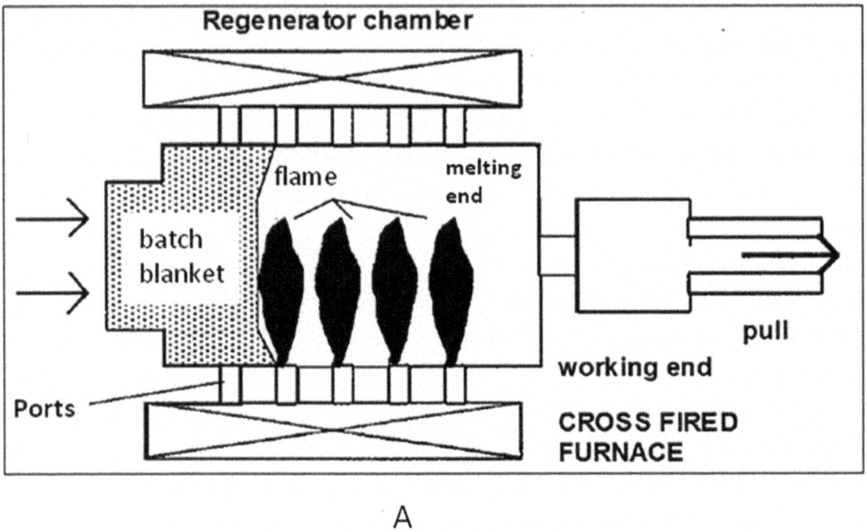

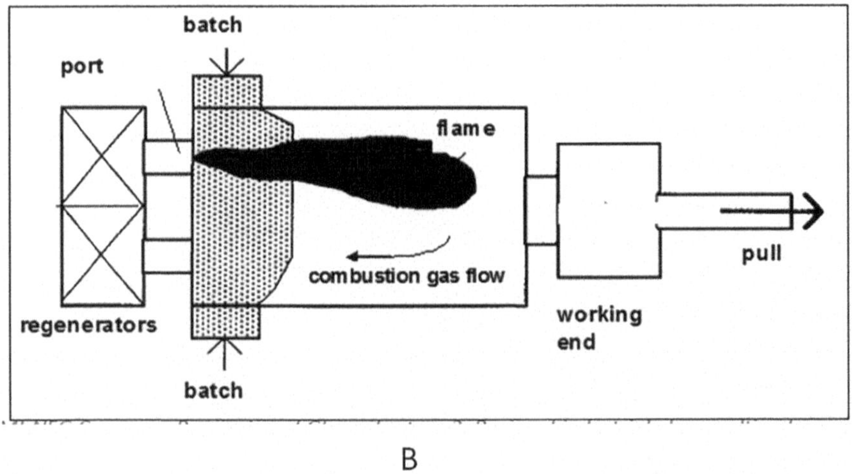

Fig. 10.3 (**a**) Cross-fired furnace, (**b**) end-fired furnace

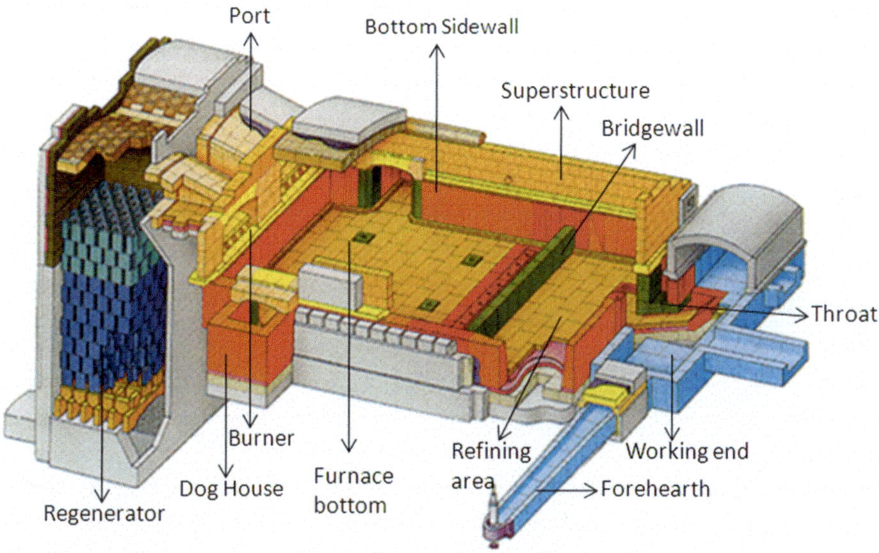

Fig. 10.4 Major elements in a glass tank furnace with end firing system

(iv) Throat – Connection between the melt end and the riser that brings the glass to the working end.
(v) Working end.
(vi) Regenerator.
(vii) Crown – Not shown in this figure, but it is the arch roof constructed over the superstructure to cover the total melting and reefing area of the furnace.

10.3 Refractories

Good quality Refractories are absolutely necessary for the smooth operation of all the elements of a glass tank furnace, which is the heart of a glass plant. The glass furnaces are connected to regenerators, which utilize the waste heat of the furnace to preheat the combustion air to provide the fuel economy to the process. The performance of refractory in the regenerator also is very important for smooth operation of the glass tank furnace. The world average specific consumption of refractory, per ton of glass produced, is about 4 kg. Performance of refractory in a glass tank furnace is very important to decide the production cost and the quality of glass manufactured. Different areas in the glass tank furnace have different operating conditions prevailed there, and, therefore, different types of refractories are used in different zones, in a glass tank furnace, which will be discussed in more detail.

Table 10.2 Operating parameters of a glass tank furnace

Parameters	Value
Type of glass	Container
Pull of the glass	100–300 tons/day
Melting rate	2–3 tons/m2/day
Energy requirement	4000–7000 KJ/kg
Temperature in melting chamber	1600 °C
Temperature in working chamber	1350 °C
Glass depth in melting chamber	1.1 meter
Glass depth in working chamber	0.8 meter
Throat – height/width	1:2
Cross section	0.1–0.3 m^2
Flow rate of glass	5–20 m/hr
Combustion chamber heat input	60–210 KW/M^3 of space per second
Velocity of air at port mouth	8–12 m/sec
Velocity of flue gas	10–14 m/sec

The typical design and operational parameters of a standard glass tank furnace are given in Table 10.2, to understand the operating parameters of the furnace.

10.3.1 Bottom

Bottom refractories in the furnace always remain in contact with molten glass, but the temperature at the bottom remains lesser, and the convection current of molten glass is slower, compared to the glass at the surface. This results in slower corrosion of the bottom refractories. The insulation of the hearth increases the temperature and the current of the glass stream at the bottom and aggravates the corrosion. Characteristically the undersurfaces of the paving bricks suffer more upward corrosion than the top surfaces by molten glass penetrating through joints. If the cullet used in the glass batch is not properly sorted and contains metal pieces (e.g. bottle lids, metal containers, etc.), the metal will get deposited on bottom, and downward corrosion is observed. This phenomenon is called metal penetration or downward drilling and is very prominent along the joints and can increase the corrosion of bottom refractory by tenfold [4]. The thermal shock is not experienced by the bottom refractories. Therefore the refractories selected for the bottom must have the following characteristics:

1. Good corrosion resistance to molten glass
2. Low porosity
3. Sufficient resistance to metal penetration
4. Low contaminability to molten glass

Table 10.3 Properties of fused cast and zircon refractory

Properties	Quality				
	Standard AZS	High zirconia AZS	β- Al_2O_3	Al_2O_3-Cr_2O_3 fused cast	Zircon brick
Chemistry (%)					
Al_2O_3	48–53	44–48	92–94	56–60	0.5–2.5
ZrO_2	32–36	40–44	–	Cr_2O_3 - 26–28	64–68
SiO_2	11–17	10–13	0–0.5	1–3	30–35
Fe_2O_3	0–0.3	0–0.2	0–0.1	5–7	0.1–0.3
Na_2O	0–0.3	0–0.2	5–7	–	0.5–2.0
Mineral phases (%)					
Corundum	45–55	42–48	–	Al_2O_3-Cr_2O_3 solid solution 60–70	–
Baddeleyite	20–30	27–33	β-Al_2O_3	Compound spinel – 30–40	–
Glassy phase	20–30	20–25	0–1.0	0–3	–
Physical property					
True sp. gravity	3.8–4.0	4.05–4.25	3.15–3.35	3.95–4.10	BD- 3.4–3.8 gm/cc
Apparent porosity %	0–2	0–2.0	0.5–5	3–6	15–25
CCS(MPa)	250–500	250–500	150–250	–	70–120
RUL °C	1700–1750	+1730	1750	–	1620–1700
Coefficient of th. Expansion (%)					
At 1000 °C	0.6–0.85	0.6–0.9	0.75–0.85	0.7–0.9	0.35–0.4
At 1500 °C	0.7–0.9	0.7–0.9	1.1–1.3	1.2–1.4	–
Th. Conductivity (W/mK)					
600 °C	3.0–3.5	3.0–3.5	2.8–3.4	3.0–3.4	1.6–1.8
1000 °C	3.2–3.7	3.2–3.7	3.4–3.8	2.6–3.0	1.8–2.2

Fused cast alumina-zirconia-silica (AZS) refractories have been found best as the paving block for the bottom of glass tank furnace, and the properties of the blocks are shown in Table 10.3. In the lining with multilayer insulation backup, the temperature of the molten glass increases, and viscosity gets reduced making the glass more fluid, and chance of penetration through brick joints increases. Therefore the refractories selected for the layer below the paving bricks needs to be corrosion resistant to molten glass. Zircon bricks are found as a good candidate for this purpose. It has also been found that a layer of ramming mass below the dense brick layer provides a joint less monolithic lining, which helps in arresting the flow of the glass, if it gets penetrated. A typical lining pattern of the bottom is shown in Fig. 10.5.

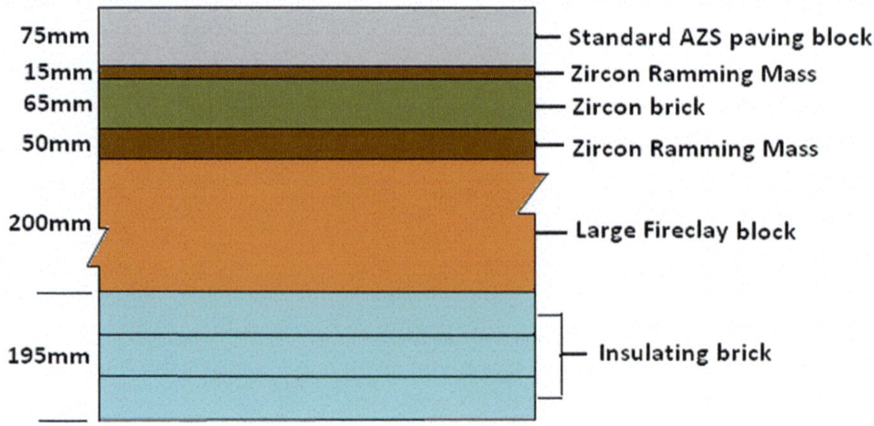

Fig. 10.5 Multilayer insulation lining of the glass tank bottom

10.3.2 Side Wall Lining

In the melter, the molten glass reaches a temperature of 1550–1600 °C, and convectional current in the glass bath remains strong. Therefore both chemical corrosion in contact with molten glass and physical erosion by the moving glass stream are the most prominent damaging factors to the refractory lining. Thermal shock is not an important parameter here because once the glass tank furnace started, it runs continuously for years together.

Looking into the operating conditions of the refractory for sidewall, the best selected refractory is fused cast AZS with 33–35% ZrO_2. AZS fusion cast refractories have three major phases alumina, zirconia, and glass and are differentiated by their ZrO_2 content. The different qualities are 32–34%, 36% and 38–41% ZrO_2. The higher the ZrO_2 content, the more resistant it is towards glass corrosion. Common thickness of the blocks used are about 250 mm and outside is insulated by fireclay brick. The depth of the glass bath in the tank normally remains about 1000–1500 mm. If the total bath height is covered by blocks with horizontal joints, then the joints are found to corrode very severely by upward corrosion (Fig. 10.6). Therefore single blocks are used to cover the total depth of the glass bath. Figure 10.7 shows the lining configuration of bottom and sidewall of the furnace.

The AZS blocks can have chemical segregation in it depending upon the manufacturing process. This results due to the difference in densities of different phases present and their different solidification temperatures. For example ZrO_2 has the highest density, and it solidifies first; therefore it tends to settle at the bottom. In addition to chemical segregation, there can be large variation in grain sizes and porosity, and the blocks should be tested before use [5, 6].

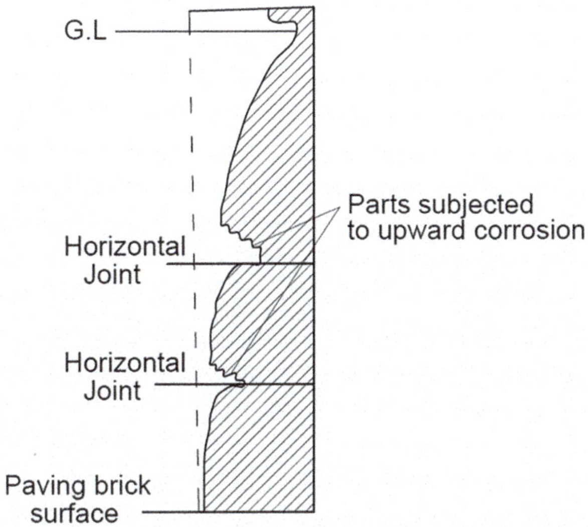

Fig. 10.6 Upward corrosion of the horizontal joints of the glass tank refractory lining

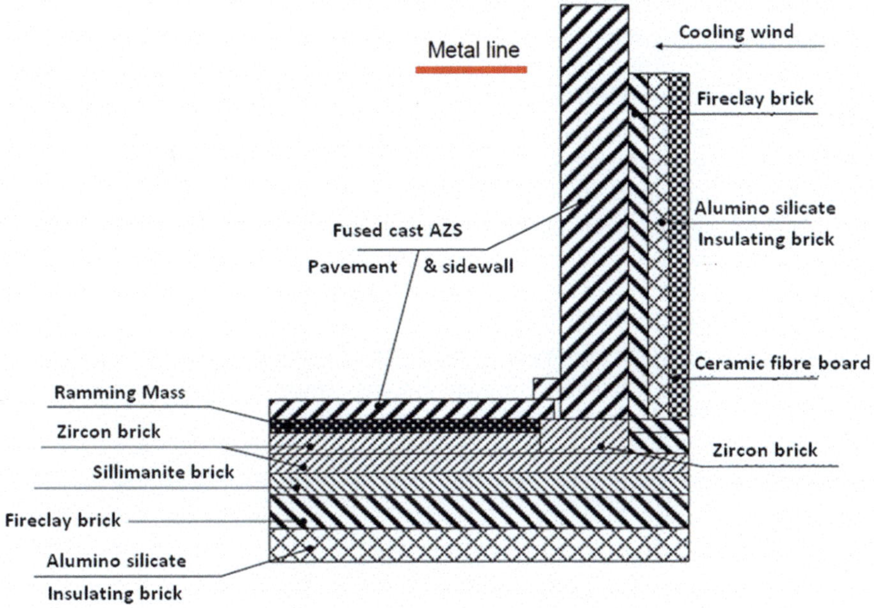

Fig. 10.7 Refractory lining configuration of bottom and side wall of the furnace

10.3.3 *Throat*

In this area of the furnace, the path of the glass stream gets narrowed, and therefore the velocity of the stream increases, and physical erosion becomes the major cause of damage to the refractory lining. Besides this, the corrosion is enhanced by the higher velocity of the glass stream that removes the product of corrosion from the surface of the refractory, favoring the forward reaction.

AZS fused cast refractories are most suitable refractory to line here. In some cases high zirconia electrocast blocks and alumina-chrome fused cast refractory also are used. The properties of some refractories used, are given in Table 10.3.

For extending the service life of the refractories in throat area, the design of the refractory shapes and its mechanism of cooling are also very important over and above the refractory quality.

10.3.4 *Superstructure*

This consists of wall above the glass bath, port and bridge wall, and doghouse sidewall. The factors responsible here for damage of the refractories are almost the same, with little variation from upstream to downstream. The refractories here are exposed to an atmosphere of combustion gases and alkali vapors at high temperature 1550–1600 °C. The combustion gases, laden with dust particles, move at high velocity inside the furnace. The dusts are formed of size 0.03–0.05 micron by evaporation process from the glass bath (Fig. 10.8) [7, 8].

In the side port or cross-fired furnace, the severity of condition for refractories differs greatly from upper stream to downstream. The doghouse suffers a good deal of batch carryover but temperature is low and here the temperature fluctuation is more due to the contact with air from outside. AZS fused cast blocks are best for this area. In the doghouse where the temperature fluctuation is too much, fused silica or fireclay refractories are preferred.

In the common regenerator-type furnace, the combustion gases and combustion air are sucked in the regenerator through port alternatively and temperature varies from 1000 to 1500 °C. Intensity of these factors depends on the furnace size, the furnace type, and working conditions. In the end port furnace, the whole melter is subjected to severe condition because the furnace is rather small and the flame traverses in U-shape after hitting the bridge wall. Therefore the severity of condition to which the refractories are exposed to is almost the same over the whole melting area.

The end wall and breast wall, near the doghouse, are greatly affected by batch carryover, but the temperature there is comparatively low. At the breast wall and port near the center, the temperature is high, and the most corrosive environment is created by the batch carryover (Fig. 10.8) [7]. The different parts of the furnace, attacked by the batch carryover, are shown in Fig. 10.9 [8]. In all these places, standard AZS refractories perform best.

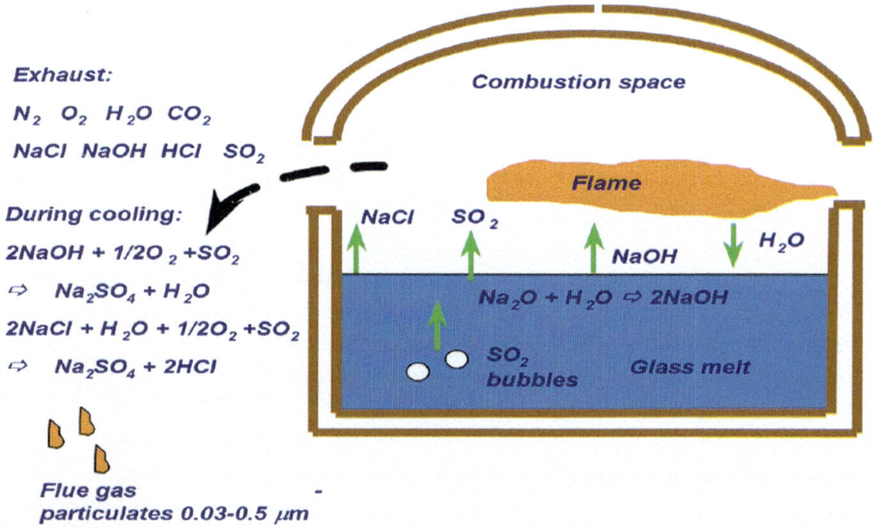

Fig. 10.8 Mechanism of fine dust emission from the glass bath

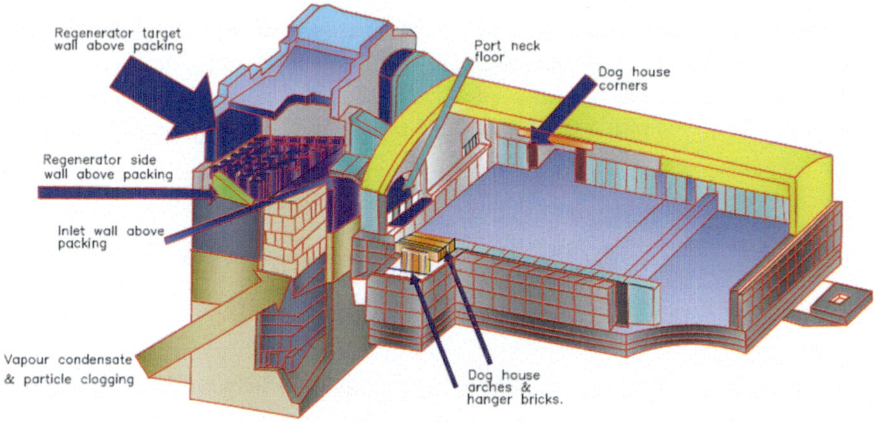

Fig. 10.9 Different parts of the furnace attacked by batch carryover (the width of the arrow indicates the intensity of the attack)

The suitable refractory lining for these areas of application needs to have the following characteristics for its best performance:

1. Excellent corrosion resistance to alkali vapors
2. Spalling resistance at higher temperature range
3. Low contaminability to glass batch

β-Alumina fused cast refractories can be used in the port closed to downstream, breast wall, and the bridge wall. In the area closer to the downstream, the temperature

Table 10.4 Comparison of chemical analysis of β-alumina before and after use

β-Alumina fused cast block		
Chemical analysis (%)	New	Corroded
Al_2O_3	89.8	72.7
SiO_2	0.46	4.02
MgO	5.31	12.1
CaO	0.09	9.19
Na_2O	3.73	0.2
K_2O	< 0.02	<0.02

is lower and is decreasingly affected by batch carryover. β-Alumina fused cast refractory is essentially a single-phase refractory, saturated with alkali, and has good corrosion resistance to alkali vapor and imparts least contamination to molten glass. Service experience has shown that a 1/9.2 ratio of Na_2O/Al_2O_3 in fused cast β-alumina is most effective to ensure high thermal shock resistance and stability towards alkali and alkaline earth environment [9]. It is not stable in the presence of SiO_2 at higher temperature because of the release of soda in the bonding phase when it gets converted to α-alumina and the resulting shrinkage causes crack in the refractory body [10]. The attack of the carryover on the β-alumina is shown by the comparative chemical analysis of unused and used refractory corroded by high silica carryover (Table 10.4) [11].

Corrosion resistance to batch carryover is more important for high temperature area near doghouse, but low contaminability is more important for the area closer to refiner. Therefore in refining area, feeder and forehearth area, it is the best refractory to line with [12]. The property of β-alumina is shown in Table 10.3.

10.3.5 Crown

Refractory lining in crown is subjected to be in operation at high temperature for long term and chemical corrosion by alkali vapor. The refractories here are attacked by alkali vapor and must be resistant to that. Refractories here must have:

1. High PCE
2. High resistance to alkali vapor at high temperature
3. Resistance to thermal shock
4. High volume stability at high temperature and
5. High creep resistance

Silica and mullite bricks are mostly used in the crown. Alumina fused cast refractory with 40% α-alumina and 60% β-alumina is also found to work very good in the crown and is a stable refractory for the crown [12].

Silica refractory is used most commonly in the crown of regenerative and recuperative glass furnaces. Presence of impurities and fluxing oxides like Al_2O_3, TiO_2, and alkalis has very prominent effect on the properties of silica refractories. The

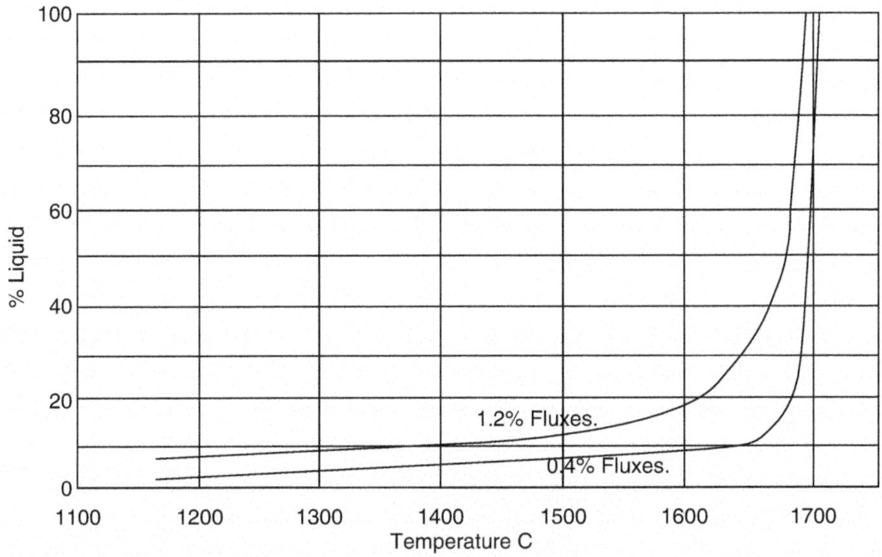

Fig. 10.10 The liquid formation in the silica refractory as a function of flux content

calculation of percent liquid content in the silica refractories is shown in Fig. 10.10. It shows the liquid content in two bricks containing 0.4% and 1.2% of fluxing oxides. It clearly shows the presence of more liquid in the silica refractories at high temperature, and the difference in the content of liquid starts increasing from 1600 °C, and the liquid content increases abruptly after 1600 °C in case of brick with 1.2% flux and 1650 °C in case of brick containing 0.4% fluxing oxides [13]. This also indicates that the service temperature of the silica refractories should not exceed 1600 °C.

Another interesting finding has been reported [13] that there is a migration of the fluxing oxides towards the cold face increasing the silica percent more in the hot face. The used silica bricks from the crown were chemically analyzed from hot face to cold face, drawing samples from different distances from cold face Silica and the combined flux oxides were analyzed and the result is shown in Fig. 10.11.

The corrosion starts at the bonding phase of traditional silica bricks with wollastonite as the bonding phase, which is attacked by NaOH followed by formation of liquid phase that dissolves the silica grain and destroys the refractory. The corrosion process is enhanced by high temperature, high alkali vapor pressure, and CaO in the brick. Looking into the explained corrosion mechanism, lime-free silica refractory has been developed and found to have superior behavior in crown. The used refractory bricks were taken from the crown, and the samples were drawn, from hot face towards cold face, at certain distances and analyzed. Lime-free silica bricks have shown much less Na_2O infiltration (Fig. 10.12) [14].

The properties of silica and mullite refractories suitable for crown are given in Table 10.5. Dense silica brick can be made of porosity 13% [15] with high abrasion resistance (Table 10.5).

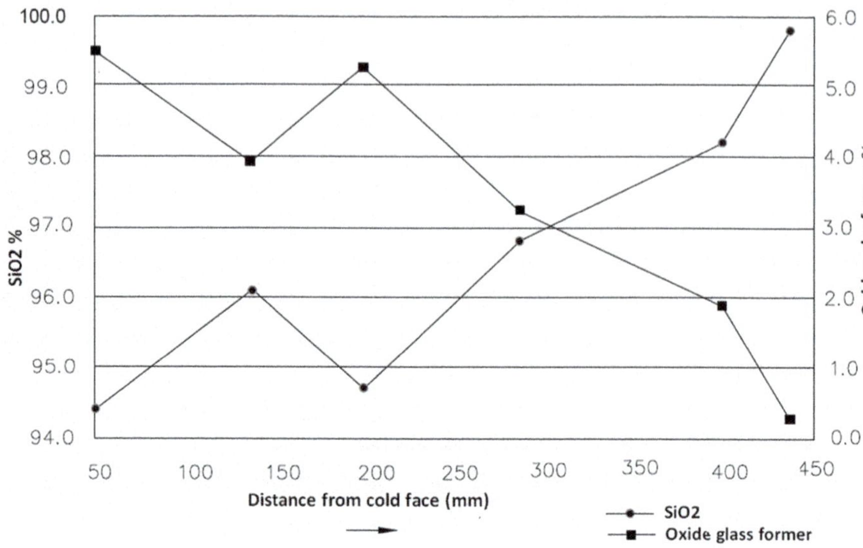

Fig. 10.11 Migration of oxides in the silica refractory during its use

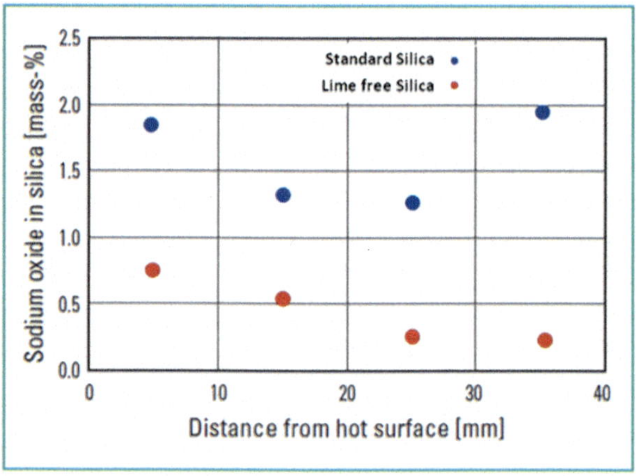

Fig. 10.12 The Na_2O penetration behavior of different types of silica refractories

10.4 Float Glass Process

This is used for manufacturing of flat glass plates of different thickness. In this process the glass batch is melted in a furnace similar to that used for container glass, and the molten glass from the furnace is poured on a bath of molten tin at about 1050 °C (Fig. 10.13a). The bath temperature of molten tin varies from 1050 to

Table 10.5 Properties of refractories for the crown

Properties	Quality			
	Standard high-duty silica refractory	Lime-free silica refractory	Dense silica	Mullite refractory
SiO_2 %	96–97	99	97.50	70
Al2O3 %	0.1–0.3	0.1	0.3	28
CaO %	1.8–2.5	0.1	1.5	–
Fe2O3 %	0.4–1.0	0.1	0.45	0.7
Na2O + K2O %	0–0.06	–	0.14	–
Physical property				
Sp. gravity %	2.25–2.35	–	2.30	–
BD (gm/cc) %	1.8–1.95	1.90	1.99	2.66
AP %	17–23	17	12.7	16
CCS (MPa) %	35–50	40	113	
RUL (C) %	1640–1660	1690	1677	1700
Th. expansion %				
1000 C %	1.2–1.4	1.60		0.66
Creep %	–	–	0.055/1500C For 5–25 hr	<0.15/1600 C For 5–25 hr.

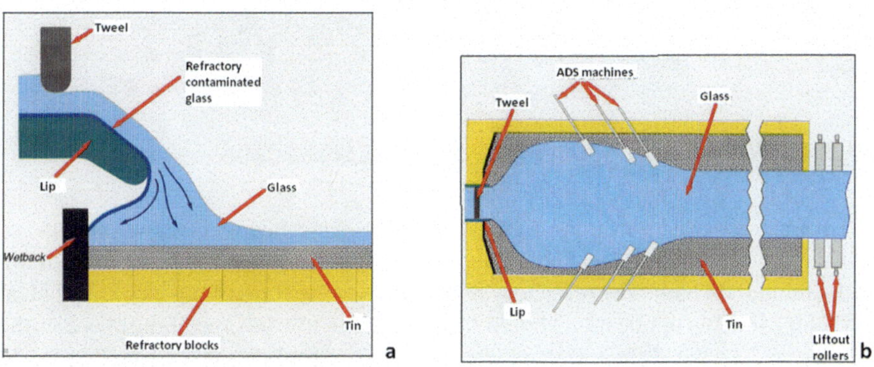

Fig. 10.13 (**a**) Side view of glass delivery system, (**b**) plan view of glass ribbon on molten tin

600 °C over its length across the furnace, and the molten glass spreads itself over the molten tin bath and travels slowly to the colder end (Fig. 10.13b). Finally the solidified glass ribbon is continuously drawn over a series of rollers and goes inside annealing lehr for stress removal (Fig. 10.14).

Atmosphere of hydrogen and nitrogen is maintained in the molten tin bath to prevent the oxidation of molten tin.

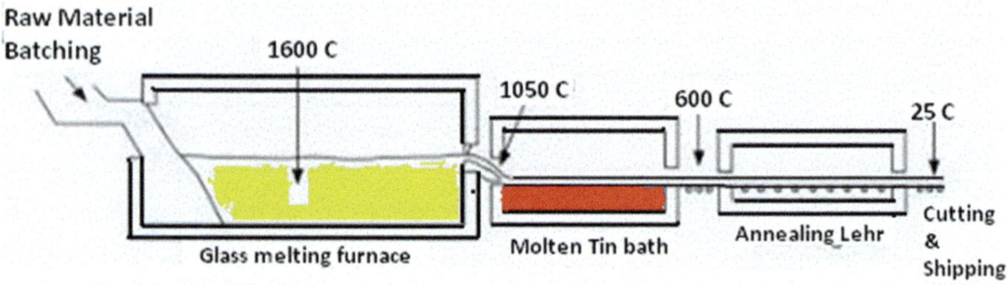

Fig. 10.14 The sectional views of the float glass process

Table 10.6 Properties of tin bath blocks

Physical properties	Fireclay refractory block	Calcium aluminate block
Bulk density(gm/cc)	2.10	2.25
Apparent porosity (volume %)	19	23
CCS (N/mm^2)	60	55
RUL (°C)	1400	1350
Young's modulus (N/mm^2)	13,500	12,000
H_2 – Diffusivity (mm/WG)	10	10
Gas permeability (nPm)	3	7
Thermal expansion/1000 °C (%)	0.6	0.65
Thermal conductivity at 1000°C (W/mK)	1.3	1.0
Al_2O_3%	38–40	67–68
SiO_2%	55–56	5
CaO + MgO %	–	25

10.4.1 Refractory Selection

The refractory selection in the furnace is in the same line as is shown in case of furnace for container glass. The different and major requirement here is the lining of the molten tin bath. The refractory used here is in the form of big block. The ideal refractory for this application should have inertness towards the atmosphere of the furnace and to the diffusion of elements from the molten glass to molten tin to refractory. The thermal conductivity should be low around 2.0 W/mK. Thermal expansion should be around 0.55% at 1000 °C and the pore size distribution should be optimum to maintain low hydrogen permeability [16].

Aluminosilicate blocks with Al_2O_3 40–45% are mostly used in this area (Table 10.6). The bulk density of the block is 2.15–2.2 gm/cc, and the density of the molten tin is 6.50–7.20 gm/cc. Therefore the blocks will tend to float. Therefore it needs the fixing of the blocks on the steel shell by studs through predesigned holes in the block (Fig. 10.15).

The Na_2O from the glass gets dissolved in molten tin and gets diffused into the bottom block and reacts and forms nepheline, when about 8% Na_2O is absorbed,

Fig. 10.15 Bottom block for lining of tin bath, with predesigned hole

Fig. 10.16 Flake containing nepheline peels off from the bottom block of tin bath

which is associated with a volume increase of 20% as per the reaction shown below [12]:

$$Al_2O_3 + 2SiO_2 + Na_2O \rightarrow Na_2O.Al_2O_3.2SiO_2\,(\text{Nepheline}) + 20\%\,\text{Volume change} \quad (10.1)$$

This nepheline peels off as flakes (Fig. 10.16) and causes problem in operation of the float glass process.

It has been found that calcium aluminate is a very good material for the tin bath application. It is inert to alkali and meets all other conditions that an ideal refractory demands there. The properties of the calcium aluminate brick are shown in Table 10.6. The comparison of alkali penetration of calcium aluminate brick compared to the traditional aluminosilicate bricks is shown in Fig. 10.17.

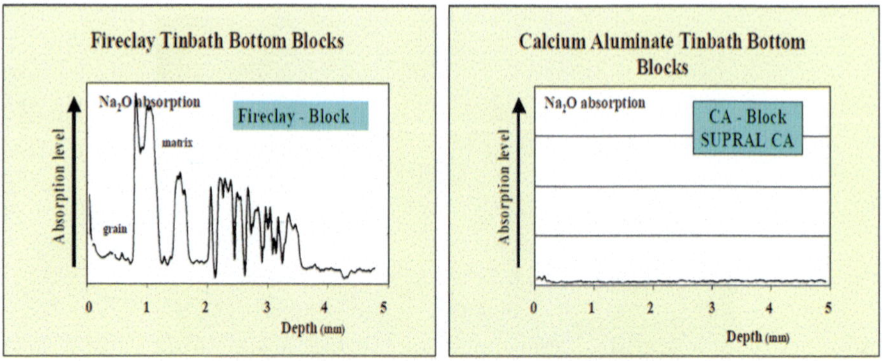

Fig. 10.17 Comparison of alkali penetration (fired at 1000 °C for 100 hr in reducing condition)

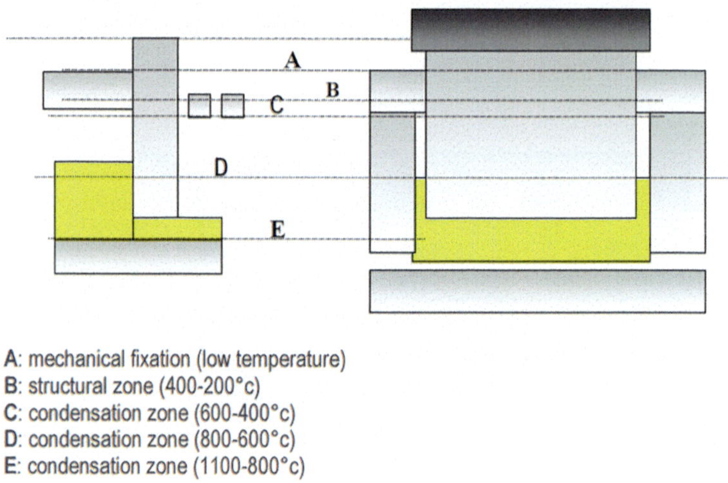

Fig. 10.18 Temperature service condition of Tweel

10.4.2 *Tweels*

Tweels (Fig. 10.13) are used to regulate the flow of molten glass to the tin bath. Its approximate size is 1100 × 800 × 110 mm. A good-quality Tweel comes in contact with molten glass without preheating and needs very high thermal shock resistance. Figure 10.18 shows the temperature condition in the various sections of a Tweel [17].

It is very important component in the float glass process because any impurity in it may impart bubble to the molten glass, and for automobile quality glass, the specification is very stringent to accommodate any bubble.

Fused silica meets all the requirements and is an ideal candidate for making this product.

The stress-relieving furnace, operating at around 600 °C, uses series of rollers over which the glass plate moves from entrance to exit end. The rollers have diameters of 150–300 mm. Fused silica rollers are found to work very well because of its low thermal expansion and its special character that molten glass does not stick on it.

10.5 Refractory Corrosion by Molten Glass

The knowledge of mechanism of corrosion is very important to control the glass defects. The wear of refractories in glass melting furnace is a complex phenomenon, and it is a combination of both chemical and physical processes, and the simultaneous involvement of both aggravates the ultimate wear. For example, the penetration of the glass inside the pores of the refractory is a physical process, but it aggravates the chemical corrosion because more surface area is available for chemical reaction. Similarly when the chemical corrosion loosened the grain to matrix bond in a refractory, the turbulence or convectional flow of molten glass dislodged the particles of refractory and aggravates the wear.

The penetration of molten glass is a physical process and is governed by Poiseulli's law:

$$l^2 = \left[r.\sigma.\text{Cos}\theta / 2\eta\right]t \tag{10.2}$$

where l = depth of penetration in time t, r = capillary or pore radius, σ = surface tension of molten glass, η = viscosity of molten glass, and θ = wetting or contact angle of glass on refractory

Equation 10.2 indicates that the molten glass penetration inside the refractory can be suppressed by increasing viscosity and decreasing surface tension of the molten glass. As the molten glass penetrates inside the refractory and travels from hot end to cool end, the glass viscosity increases, which hinders the further penetration. The steeper the temperature gradient across the refractory wall, the lesser will be the penetration of the molten glass.

Viscosity is also an important measure of the rate of diffusive mass transport within the molten glass after penetration into the refractory. According to the Stokes-Einstein equation:

$$\text{D} = \text{kT} / 6\Pi\eta r \tag{10.3}$$

where r = radius of the diffusing species and D = ionic diffusivity.

Equation 10.3 indicates that if the viscosity of the glass changes due to dissolution of the solid, the ionic diffusivity also changes. The refractory corrosion will be enhanced if the dissolution of the solid in molten glass decreases the viscosity of the glass.

Corrosion is the chemical attack of the molten glass on refractory. The reaction products may directly dissolve in the glass which is called direct dissolution, or the solid reaction product can form a layer on the refractory-glass interface (boundary layer) and then dissolve in the molten glass, called indirect dissolution [18–22]. When the diffusivity of the reaction products is faster than the rate of chemical reaction at the interface, it is termed as reaction controlled or the interface controlled, because the slower of the two processes will ultimately determine the rate of the process. The dissolution rate then can be expressed mathematically as

$$J = K\left(A_c / A_o\right)C_m \tag{10.4}$$

where J = dissolution rate (gmcm^{-1} sec^{-1}), K = reaction rate constant, A_c = actual area of the refractory (cm^2), A_o = apparent area of the refractory (cm^2), and C_m = concentration of the reactant in the glass melt (g/cm^{-1}).

In the case of indirect dissolution, the rate of removal of the reaction products by diffusion is smaller than the rate of the chemical reaction; a boundary layer, rich in solute, is built up whose interface with the refractory is saturated with reaction products. The diffusion of the reactant products, to the boundary layer and away from the boundary layer, govern the dissolution process. Here the corrosion rate can be expressed by Nernst equation as

$$J = D\left(C_s - C_a\right)/\delta \tag{10.5}$$

where D = diffusion coefficient (cm^{2}sec^{-1}), C_s = the concentration of the component at which it is saturated in the glass melt (gm/cm^3), C_a = the concentration of that component in bulk glass (gm/cm^3), and δ = the thickness of the boundary layer (cm).

The corrosion rate is a function of many variables which include temperature, composition of refractory, composition of the glass and their interface layer, glass density, viscosity, diffusivity and the degree of turbulence. Let us discuss the effect of different variables on the refractory corrosion.

Temperature The dissolution of the refractory in the molten glass is diffusion controlled, and the diffusion coefficient D is related to temperature by the Arrhenius equation

$$D = A\exp\left(\Delta E / RT\right) \tag{10.6}$$

where A, ΔE, R, and T are frequency factors for diffusion, activation energy, universal gas constant, and temperature.

The thermal radiation at the boundary of refractory in contact with molten glass surface plays an important role in deciding the refractory corrosion. Figure 10.19 shows the effect of temperature on the rate of corrosion [21]. The temperature difference between the surface and bottom of the glass in the tank or temperature difference of glass in different areas of glass in the furnace sets up the convective current which enhances corrosion rate.

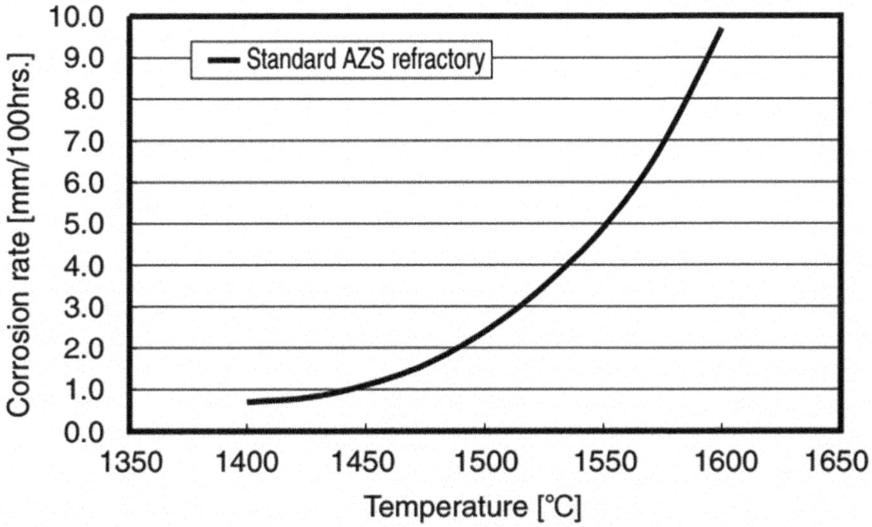

Fig. 10.19 Effect of temperature on the corrosion rate of AZS refractory

Convection Flows Convection plays a major role in the corrosion process, and the quantitative investigation of corrosion needs a sound knowledge of the influence of various convection processes active in the glass melting furnace. There are three types of convection in the glass melting tanks, namely, forced convection, free laminar convection due to density variation, and interfacial convection driven by surface tension differences of glass melts.

Forced convection in glass tank is caused by the bubblers, mechanical stirrers, and melting rate. The melting rate caused high velocity of flow in the throat area.

Free laminar density convection is caused by the temperature gradient in the tank and by concentration gradient of the dissolved components from the refractory wall of the tank.

The free interfacial convection is a very important factor for corrosion of refractories in glass tank. The surface tension of the molten glass changes from that of the bulk glass melt when the refractory components get dissolved in it at the refractory-glass interface. The tendency of heat and mass to travel to areas of higher surface tension in the liquid (Marangoni effect) sets up a convection current at the glass refractory interface.

The influences of different types of convection, on the rate of corrosion, are tabulated [19] in Table 10.7.

In case of AZS lining in the glass tank melting soda-lime glass, the study of the corrosion shows that the velocity of the chemical solution of refractory is much faster than the convective-diffusive transportation of the dissolved ions. That is why a boundary layer is formed on the refractory-glass interface saturated with the dissolved ions and the composition of the boundary layer is shown in Table 10.8 [17] .

Table 10.7 Influence of different factors on the corrosion rates for different types of convection

Convection flow (general)	R = Function of (driving force, geometry, kinetics, solution)
Forced convection	$R_{fc} = f(\omega^{1/2} - D^{2/3}, \nu^{-1/6} \Delta C)$
Density convection	$R_{dc} = f(g.\Delta\rho/\rho_a)^{1/4} y^{-1/4} D^{3/4}, \nu^{-1/4} \Delta C)$
Interfacial convection	$R_{ic} = f(g.\Delta\sigma/\sigma_a)^{1/3} B D^{2/3}, \nu^{-1/3} \Delta C)$

ω = angular velocity of stirrer, D = diffusion coefficient, ν = kinematic viscosity, $\Delta C = C_s - C_a$, C_s = saturation concentration of the dissolved component in the interphase layer, C_a = concentration of the component in the bulk molten glass, g = acceleration due to gravity, $\Delta\rho = (\rho_s - \rho_a)/\rho_a$, ρ_s = density of the saturated glass melt at the interface layer, ρ_a = density of the host glass melt, y = current coordinate along the wall, $\Delta\sigma = (\sigma_s - \sigma_a)/\sigma_a$, σ_s = surface tension of the saturated glass melt in the interface layer, σ_a = surface tension of the host glass melt, B = factor dependent on the geometrical feature of the formed fluxline.

Table 10.8 This shows the compositions of boundary layer along with that of the container quality soda-lime glass and the AZS 41 refractory in contact with glass

Component wt %	AZS 41 refractory	Container glass	Boundary layer
SiO_2	12	74	53
Al_2O_3	46	1.4	27
ZrO_2	41	–	<2
$Na_2O + K_2O$	<1	12	14
CaO	–	11	4
Others	0.2	2	–

The large difference in the composition of pure container glass and the glass of saturated concentration in the boundary layer leads to different physical properties which are the driving force behind the refractory corrosion.

Mathematical model was developed to predict the refractory corrosion of the AZS blocks in glass tank furnace, and the calculated value tallied fairly accurately to the actual values determined experimentally [23]. The effect of temperature on corrosion rate of 250-mm-thick AZS refractory block was calculated and is shown in Fig. 10.20. It shows that the corrosion rate is very high initially and falls with time, and the effect is more prominent at higher temperatures, above 1500 °C. It also indicates that refractory corrosion rate is getting lesser as its thickness is gradually reducing and after the thickness reaches 50 mm, the rate comes down drastically, which is more prominent when the temperature is high.

10.6 Defects in Glass due to Refractories

Refractories are corroded by interaction with molten glass, and the reaction products, ultimately, form different types of defects in glass. The defects in glass, originating from refractory at the different parts of the furnace, are shown in Table 10.9 [24]. The different types of defects found in the glasses [25] are given below:

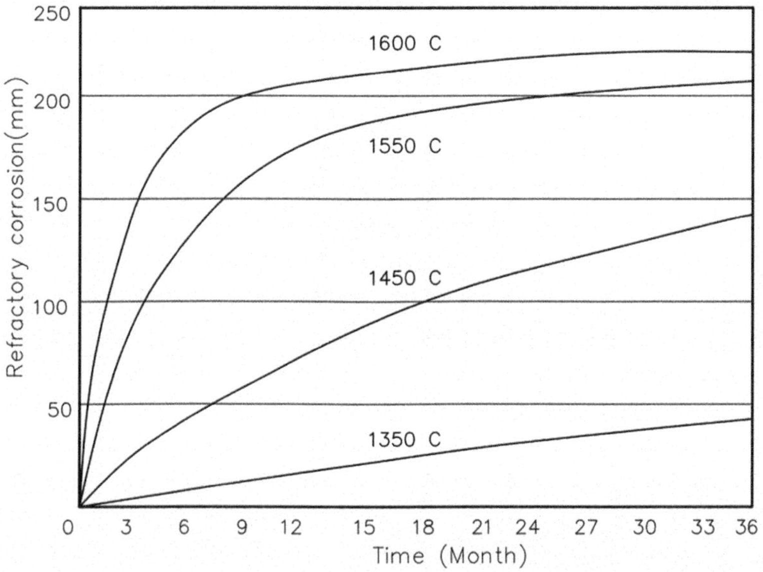

Fig. 10.20 Variation of refractory corrosion rate with time

Table 10.9 Different types of defects in glass from originating from refractory

Refractory type	Location in furnace	Defect potential
Silica	Melter and working end crown. Melter refining and working end superstructure	Silica refractory stone, Silica rundown/ drip, frost
Alumina	Melter superstructures (float), Wall (float), waist crown (float), forehearth canal/superstructure (containers, tablewares)	Aluminous stone, cord
Mullite	Regenerator walls, waist crown	Aluminous stone
Sillimanite	F/H cover block	Aluminous stone/ cord
Zircon	Isolation between alumina and silica mortar, Monolithics	Zircon stone
Fused cast AZS	Melter sideblocks, melter tuckstones, melter pavings. Melter superstructure	AZS stone, knots, cords
Magnesia	Regenerator walls, crowns, checkers	Spinel

Bubbles or blisters, which are gaseous inclusion in glass
Stones, which are solid unmolten or recrystallized material inclusion
Knots are vitreous particle of deviating composition of bulk glass. The different refractive index of it produces a lens effect

Fused cast AZS product with 33% ZrO_2 is most commonly used in the superstructure and side wall of the glass tank furnace. This product contains a significant

Fig. 10.21 Exudation of glass from the AZS sample

Primary AZS stone | Secondary zirconia stone | AZS knot

Fig. 10.22 Defects in the glass from ZrO_2 crystal

quantity of glassy phase in it. This glassy phase tends to be expelled (exudation) from the brick at high temperature (Fig. 10.21). From this exuded glassy phase, ZrO_2 crystal can enter the glass melt and forms defects, e.g., stone, cord, and knot (Fig. 10.22) [24, 25]. The exudation of glassy phase from fused cast AZS depends upon the temperature, the oxidation state, and the amount of impurities. The relation of volume of exuded glass on temperature is shown in Fig. 10.23.

It has been found that if the composition of the glass is adjusted to augment the viscosity of the glassy phase in the fused cast refractory, the exudation can be minimized to a great extent. Development of such fused cast refractory, with different chemistry and almost same physical properties as normal AZS, but with lesser exudation behavior (Table 10.10), was reported by RHI [26].

The electrochemical reaction takes place at the AZS refractory-glass interface. The diffusion of cations from the glass melt to refractory and subsequent formation of O_2 (as per the reaction 10.2 and 10.3) is responsible for bubbles and blisters [25, 27]. It has also been proved by the fact that application of an external voltage (−ve pole at the outer wall) reduces the formation of bubbles in AZS/glass interface:

$$4Fe^{3+} + 4e \leftrightarrow 4Fe^{2+} \tag{10.2}$$

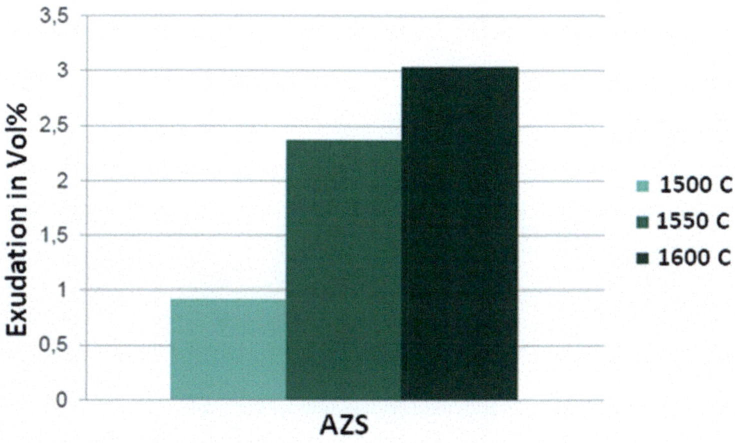

Fig. 10.23 Exudation behavior of AZS with temperature

Table 10.10 Properties of special fused cast in comparison to normal fused cast block

	Special fused cast block	Standard fused cast block
Al_2O_3%	64.2	48–53
ZrO_2%	16.9	32–36
SiO_2%	16.5	11–17
Na_2O %	2.4	0–0.3
BD (gm/cc)	3.55	3.8–4.0
CCS (N/mm^2)	195	250–500
Th. expansion %	1.25 at 1645 °C	1.0 at 1600 °C
RUL (°C)	>1700	>1700
Glass phase (%)	25	20–25

$$2O^{-2} \leftrightarrow O_2 + 4e \tag{10.3}$$

The more is the amount of polyvalent impurities like Fe^{2+}, Fe^{3+} and Ti^{2+}, Ti^{3+}, the more is the redox reaction and gas formation. Total Fe_2O_3 + TiO_2 should be below 0.2%. Presence of B_2O_3 and high content of carbon from the electrode eases the dissolution of AZS. The presence of nitride is also unwanted.

Silica drip is very common in float glass process to impart defect in tridymite form. The crown or the superstructure silica refractories are attacked by alkali vapor and bring the melting point down, and the melt runs down in the form of finger (Fig. 10.24). The melt drips into the glass bath and forms stone as defect.

It is clear from the above discussions that refractory corrosion generates different kinds of defect in the glass, during its production. The improvement in the properties like thermal shock resistance and corrosion resistance in refractories can reduce the defects in glass. Table 10.11 has compiled all these properties of different refractory qualities at a glance.

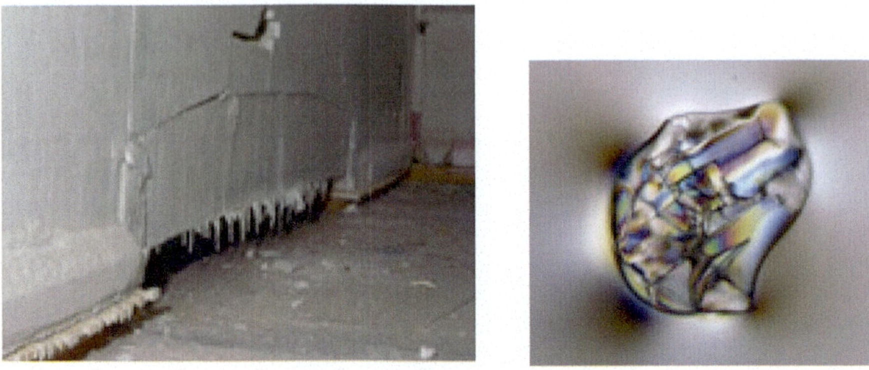

Fig. 10.24 The silica melting and tridymite crystal as defect in glass

Table 10.11 Behavior and recommendation of refractories for lining of different areas of glass melting furnace

Properties		Refractory type					
				Fused Al_2O_3			
		Fused mullite	Fused AZS	α	αβ	β	Dense silica
Corrosion resistance		P	G	M	M	M	P
Spalling resistance		M	G	M	P	E	G
Colorant contamination		E	E	E	E	E	E
Stoning potential		G	G	E	E	E	M
Melter	Upper structure	Yes	Yes			Yes	
	Glass contact		Yes				
	Crown		Yes			Yes	Yes
	Pavement		Yes	Yes			
	Shadow wall		Yes			Yes	
Bridge wall cover							
Throat assembly			Yes				
Refiner	Crown	Yes	Yes			Yes	Yes
	Glass contact		Yes		Yes		
Feeder			Yes		Yes		

P poor, *M* medium, *G* good, *E* excellent

10.7 Regenerator

The regenerator plays a very important role in the thermal efficiency of the furnace. The combustion gases discharged from the melter pass through the port into the regenerator chamber, stacked with checker bricks. It transfers its sensible heat to the checker refractories and then discharged to the flue. During the next changeover after 15–20 min, the air, required for combustion, is preheated by passing through these checker bricks before it is discharged in the melter.

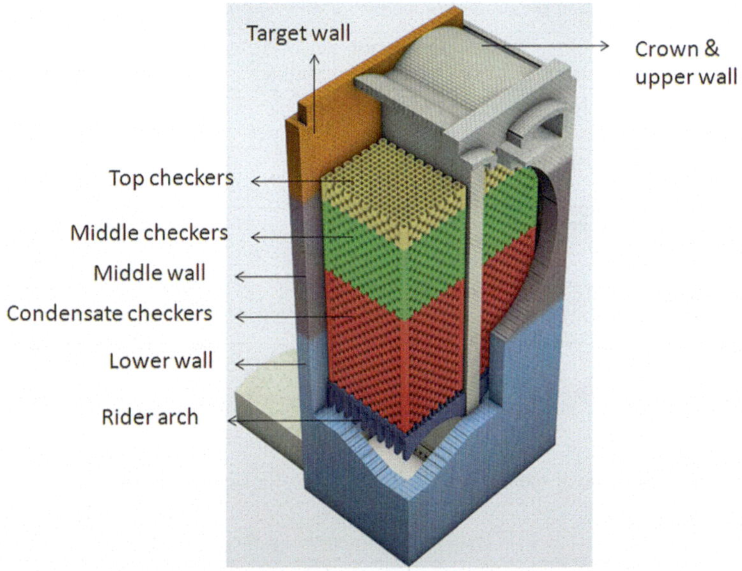

Fig. 10.25 Glass furnace regenerator (Courtesy of Harbison-Walker)

The regenerator is roughly divided into the housing which is the outer shell and a checker blocks stacked inside as shown in Fig. 10.25 [28].

Regenerator housing consists of side walls, bottom, and ceiling. The superstructure (crown and the upper wall) and casing refractories of the regenerator has two functions considering the waste heat recovery from flue gas:

1. Heat transfer from the flue gas to the combustion air stream
2. To thermally insulate the whole regenerator to disallow heat flux to go out to surrounding environment and to prevent ingress of cold air from surrounding to inside

Therefore the selected refractories should have lower thermal conductivity and good thermo-mechanical and thermo-chemical stability. Table 10.12 shows the characteristics of different refractories and their suitability for construction of superstructure [29].

10.7.1 Ceiling

In this area, temperature is lower than melter and goes up to 1300–1400$^{\circ}C$. Because all the combustion gases passes through regenerator, it is exposed to both high temperature alkali vapor and batch carryover. The alternate cycle of combustion gases and combustion air put the ceiling under thermal stress also. The regenerator ceiling is built in arch form, but the span of the arch is much smaller. The refractory to be

Table 10.12 Comparison of the desired properties of different refractories for superstructure

Desired properties	Basic	Fused cast AZS	Mullite	Silica	Fireclay
Low thermal Conductivity	No	No	Yes	Yes	Yes
Resistance to batch and alkali	Yes	Yes	Reacts to form stable layer	No	No
Low thermal expansion	No	Yes	Yes	Yes	Yes
Creep resistance	Depends on grade	Yes	Yes	Yes	No
Resistance to redox change	Variable	Yes	Yes	Yes	Depends on Fe_2O_3%
Cost/unit volume	High	Very high	Moderate	Low	Very low

Table 10.13 Comparison of mullite and magnesite bricks for superstructure

Brick quality Properties	90% Mullite	97% Mullite	Direct-bonded magnesite
Bulk density(gm/cc)	2.55	2.52	2.98–3.02
AP (%)	14.1	18.5	15–16
CCS(MPa)	75	90	50–60
TSR(950 °C/ water)	30	16	2
Creep% @ 1550C, (25-50 Hr)	0.3	0.07	0–0.85
Th. expansion %, 1500 °C	0.91	0.92	1.5–2.0
Th. conductivity/1000 °C	1.35	1.10	4.6

used here must have very high alkali resistance and excellent creep resistance at high temperature. The best refractories to be used here are of fused cast AZS, fused cast mullite, silica, and magnesite-chrome quality. But when the insulation is used over the ceiling, the temperature of the roof goes high, and the effect of alkali corrosion gets intensified. Under such condition silica refractory is not a good choice, and AZS, fused mullite, or magnesite-chrome are better choices [7]. But magnesite-chrome refractory has got a severe drawback that chromite, used in magnesite-chrome brick, reacts with alkali at high temperature and forms alkali chromates containing hexavalent Cr+6 which are soluble in water and are carcinogenic in nature. This bricks after use, if is disposed in open area, the alkali chromate salts may get washed by rainwater and contaminates the ground water posing serious threat to health hazards [10]. Therefore, as an alternative, magnesia-zircon refractory has been developed with very good performance [30]. The selection actually depends upon many other factors like operation conditions, degree of insulation, running period, etc.

Properties of fused mullite based bonded refractories containing 96–97% mullite compared to magnesite refractory is given in Table 10.13. It shows clearly the advantage of fused mullite over magnesite in respect to weight, thermal shock resistance, thermal expansion, and creep. In magnesite refractory, the creep depends upon the CaO/SiO_2 ratio and crystal size of periclase in base raw material. Refractory with fused magnesia having large crystals and fired at 1700 °C can maintain the creep at 1550 °C. Mullite refractory is not so sensitive to the presence of its minor constituents. Reaction with alkali forms a glassy impervious layer on the mullite refractory and inhibits further reactions [29].

10.7.2 Side Wall

The side wall can be divided into three parts depending on the temperature. The upper part has temperature of 1300 °C, and the lower part has temperature of 500–700 °C, and in the middle part, the temperature is in between.

10.7.3 Upper side wall

Upper side wall is exposed to the same condition as the ceiling of the regenerator, and the refractories here are mostly attacked by the alkali vapor and the batch carryover material. The use of insulation bricks at backup aggravates the condition. The most used refractories here are magnesite chrome. In the middle and lower wall because of much lower temperature, low porosity fireclay bricks can be used.

10.7.4 Target Wall

It is opposite to the ports and exposed to the impingement of the exhaust gas-borne vapor and particulates. Depending on the distance between target wall and the port, the wear of the target wall can vary significantly. Fused cast AZS and fused cast mullite are the best refractories to construct target wall. Bonded AZS refractory is not suitable as it reacts with alkali, forming β-alumina and nepheline ($NaAlSiO_4$) as the major phase. Formation of these compounds results in about 17% expansion and causes the spalling of the refractory [7].

Mullite under attack of alkali gets converted to corundum with needle-shaped crystals surrounded by glass composition. This glass layer provides a sealed layer on the surface and prevents further attack of alkali [8].

10.7.5 Checker Refractory

Glass melting is an energy-intensive process, and its economy largely depends upon efficient energy management. Checkerwork inside the regenerator casing serves the purpose of heat exchanger. It stores heat when flue gas is passed over it and dispenses the heat to the combustion air in the next cycle. A good checkerwork must fulfill the following requirements:

1. High resistance to thermal shock
2. High chemical resistance towards alkali salts
3. High mechanical stability
4. High efficiency to recover the heat from flue gas
5. Low tendency to clogging

Different types of design are available for the checker bricks as shown in Fig. 10.26. Besides the refractory quality, the shapes of the checker blocks are very important to decide the efficiency of the regenerator. Among all these designs, the chimney blocks are found to be most efficient for heat transfer. Chimney blocks also have different designs to create turbulence in the gas stream for better heat transfer.

The efficiency of the regenerator depends on two dimensionless factors [31], namely, reduced length (Λ) and reduced period (Π) defined as

$$\Lambda = hS_p / mC_{pg} \tag{10.4}$$

$$\text{and}\,\Pi = hS_p\tau / MC_{ps} \tag{10.5}$$

where h = heat transfer coefficient in W/m²/k, S_p = specific surface area in m²/m³, m = mass flow rate in Kg/sec, C_{pg} = heat capacity of gas (J/Kg/K), C_{ps} = heat capacity of the solid (J/Kg/K), M = the mass of the checkerwork in kg, and τ = time between the flow inversion.

Both reduced length and reduced period are different for flow of air and flue gas. If it is considered that for an ideal regenerator $\Lambda_{air} = \Lambda_{flue}$ and $\Pi_{air} = \Pi_{flue}$, then the efficiency can be found out from a graph given in Fig. 10.27 in which the efficiency is plotted against different values of reduced length for different reduced period. For a typical regenerator, reduced period (Π) is always less than unity because of short duration of air or flue gas flow and large thermal mass of the checker refractories.

It is clear from Eqs. 10.3 and 10.4 that the efficiency of regenerator depends on:

1. The surface area available, for heat transfer of the checker installation. It should be largest, for a particular tonnage of refractory.
2. The thermal conductivity, sp. heat, and density of the refractory, which are important design considerations.

Different attempts have been made to increase the surface area by designing different shapes of checker bricks, and it has been found that 13% increase in efficiency of heat recovery could be achieved by changing design of checker bricks [31, 32].

The specific checker volume should be between 60 and 70 m³ /Nm³ air, and large heating surface with high heat transfer coefficient is a standard requirement [33]. Typical parameters of different factors and the corresponding efficiency of a regenerator system are shown in Table 10.14.

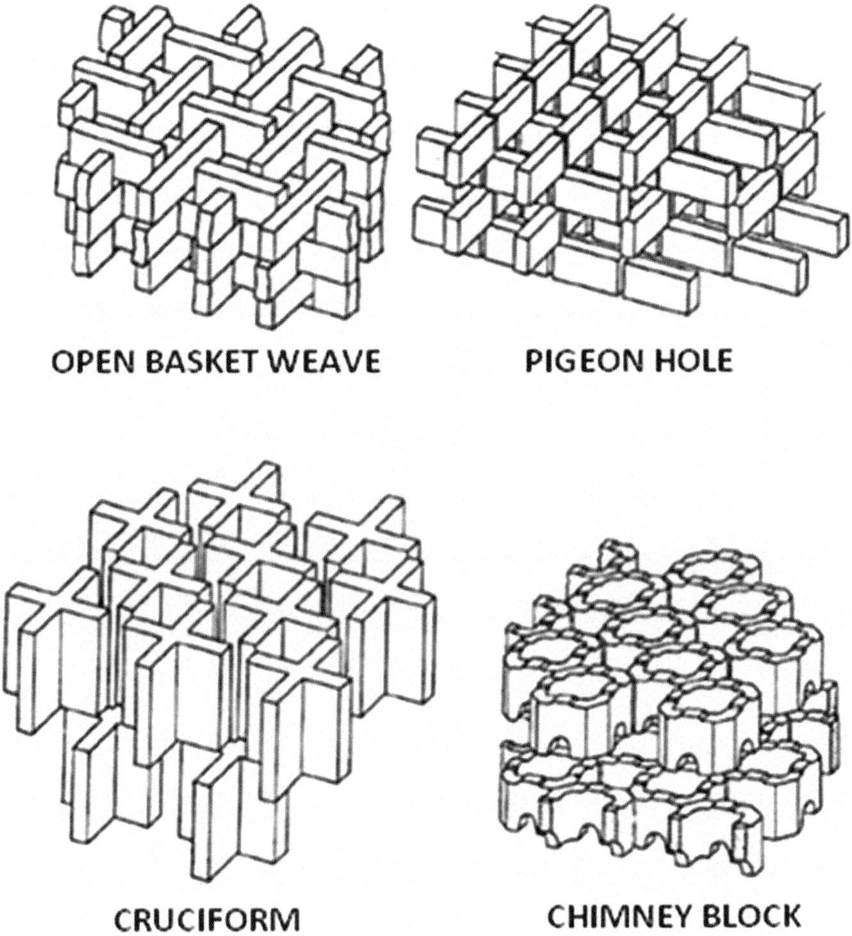

Fig. 10.26 Different designs of checker bricks and blocks

For designing the checker bricks, the conditions prevailing in different levels of checker refractory from top to bottom must be taken into consideration. The different factors responsible, to decide the refractory quality, are temperature, load, alkali attack, and sodium sulfate attack. The checker conditions at different levels during operation are shown in Table 10.15.

It shows the temperature is highest on the top course and lowest at the bottom. The static load of refractories is least on top and maximum on the bottom course. Alkali attack is severe on top and gradually becomes less towards bottom. Sodium sulfate attack is mild at top, severe in the middle course, and mild again at lower course. Its effect is most at the temperature range 800–900 °C, where the reversible solid to liquid transformation of sodium sulfate takes place.

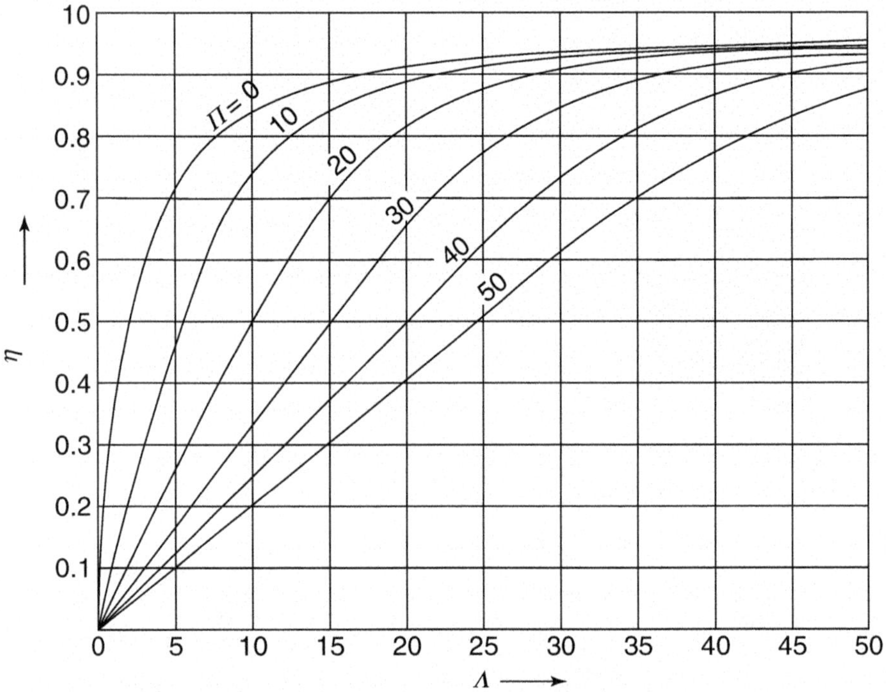

Fig. 10.27 Curves showing reduced height Vs regenerator efficiency

Table 10.14 Different parameters and efficiency of a typical regenerator system

Weight of checkerwork	5.17 tons
Surface area of the checkerwork	17.2 meter^2
Air flow	866 Nm^3/hr
Flue gas flow	1205 Nm^3/hr
Air inlet temperature	153 °C
Air outlet temperature	934 °C
Flue gas inlet temperature	1179 °C
Flue gas outlet temperature	658 °C
Energy efficiency	44%

Table 10.15 Checker conditions at different levels during operation

Location	Temperature	Load	Alkali attack	Na_2SO_4 attack
Top course	High	Mild	Severe	Mild
Middle course	↓	↓	↓	↓
Bottom course	Low	Severe	Mild	Severe ↓ Mild

Top Course Checkers High-purity magnesite bricks are found to work satisfactorily in this area. The chemistry of the magnesite brick is very important. The total CaO/SiO_2 ratio determines at what temperature the first liquid phase will appear, and the $CaO + SiO_2$ combined quantity decides the content of liquid phase in the $MgO\text{-}CaO\text{-}SiO_2$ phase system (Fig. 10.28). Table 10.16 shows the formation of mineral phases and the liquid phases formed at different temperatures in magnesite refractory [34].

A good product demands MgO > 96% and $CaO/SiO_2 > 2.0$. Besides this Fe_2O_3 in the bricks should be less because the phase 2CaO. Fe_2O_3 forms liquid at 1449 °C. Fe oxides also take part in redox reaction and form magnesiowustite (MgO.FeO) in reducing condition and magnesioferrite ($MgO.Fe_2O_3$) in oxidizing condition and disturb the refractory stability. Therefore, lower Fe_2O_3content (below 0.7%) is preferred. The property of a magnesite refractory for checker upper course is given in Table 10.17.

Besides the chemistry of the refractory, the manufacturing process and the development of microstructure are also very important. It has been found that firing at

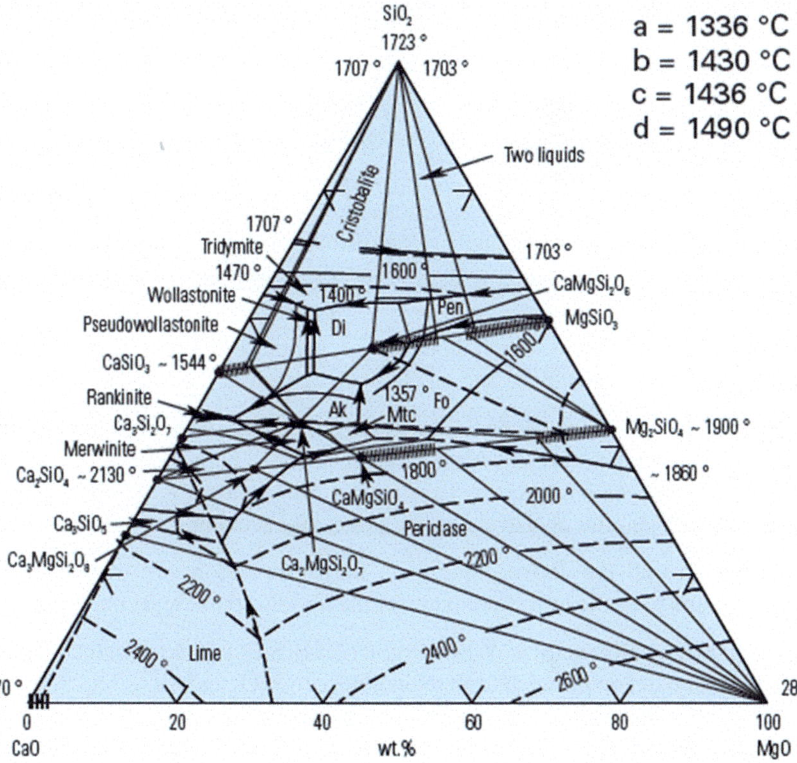

Fig. 10.28 MgO-CaO-SiO2 phase system

Table 10.16 Relation between CaO/SiO_2 ratio and formation of mineral phases in magnesite refractories

C/S ratio	Mineral formation in paragenesis with MgO (M) in the C-M-S system	Formation of first melts (Invariant points) °C
< 0.93	M, M_2S	1860
	M,M_2S,CMS	1502
0.93–1.4	M,CMS,C_3MS_2	1490
1.4–1.87	M, C_3MS_2, C_2S	1575
1.87–2.8	M, C_2S	1800
	M, C_2S, C_3S	1790

Table 10.17 Properties of basic refractories for regenerator checker

Quality Properties	Magnesite	Magnesia-zircon	Magnesia-spinel
Chemical analysis %			
MgO	97	75	73-75
CaO	1.5-1.7	1.2	0.9-1.0
SiO_2	0.5-0.7	8-10	1.2-1.5
Fe_2O_3	0.5	0.5	0.5
Al_2O_3	–	–	23-24
ZrO_2	–	12-13	–
Physical properties			
Bulk density (gm/cc)	2.96	3.10	2.95
App. porosity (%)	15-16	14-15	15-16
CCS (MPa)	70-80	80-90	55-65
RUL (°C)	1700	1570	1590
Th. expansion (%)			1.4/1400 °C

high temperature develops direct bonding of periclase crystals which can bear higher load at high temperature and has lower creep value.

Middle Course The refractories for the middle zone must have high thermal shock resistance, high temperature strength, creep resistance, high thermal conductivity, high specific heat and resistance to alkali, sulfur oxides, and redox reaction. Environmental friendliness also should be taken into account.

Looking into all the above conditions, magnesia-zircon and magnesia-spinel bricks are found to be the best candidates for the middle course and condensate zone.

Magnesia-Zircon Refractory When zircon (ZrO_2. SiO_2) is added in the magnesite brick, the zircon decomposed at high temperature to ZrO_2 and SiO_2. The SiO_2 reacts with MgO to form forsterite ($2MgO.SiO_2$) in the matrix which acts as bond between the periclase grains. It has been found that a rim of forsterite with small ZrO_2 crystals forms, which protects the MgO against corrosive acidic sulfates. The property of magnesia-zircon refractory is shown in Table 10.11 [35].

The other chrome-free refractory solution for the checker is magnesia-spinel. The field trial of magnesia-spinel refractory in the condensation zone under oxidizing condition shows very good result, and the bricks are found quite resistant to Na_2SO_4 and SO_3 [23]. The properties of the spinel-bonded magnesia refractory is shown in Table 10.13.

The following reactions occur in the top- and middle-level checkers [12]:

Reactions with SiO_2, $2MgO + SiO_2 = 2\ MgO.\ SiO_2$ (10.6)

$2CaO.\ SiO_2 + SiO_2 + MgO = CaO.MgO.SiO_2(CMS) + 3CaO.MgO.2\ SiO_2(C_3MS_2) + 2MgO.SiO_2(M_2S)$ (10.7)

Reaction with CaO, $2MgO.SiO_2(M2S) + CaO = CaO.MgO.SiO_2(CMS) + 3CaO.MgO.2SiO_2(C_3MS_2)$ (10.8)

Reactions with V_2O_5 from oil, C_2S, C_3MS_2, $M_2S + V_2O_5$ = Ca-Vandate and Ca-Vandite (10.9)

Reactions occur in the condensation zone under *oxidizing condition* are as follows:

$$2SO_2 + O_2 = 2SO_3 \tag{10.10}$$

$$2NaOH + SO_3 = Na_2SO_4 + H_2O \tag{10.11}$$

$$MgO + SO_3 = MgSO_4 \tag{10.12}$$

$$CaO + SO_3 = CaSO_4 \tag{10.13}$$

$$MgO + Na_2SO_4 = \text{Na-Mg-Sulfates} \tag{10.14}$$

Reactions occur in the condensation zone under *reducing condition* are as follows:

Alkali attack on forsterite, $M_2S + NaOH$ = Na-Mg- Silicate + H_2O (10.15)

Alkali attack on aluminosilicate refractories, AxSx + NaOH = Nepheline + Al_2O_3 (Volume Expansion 6-36%) (10.16)

Alkali attack on alumina, Al_2O_3 (α – Alumina) + NaOH = β-Alumina (Vol. Expansion 28%) (10.17)

Magnesia-zircon refractory is corroded in reducing condition or by furnace carryover with high CaO. The comparison of chemistry of the used and unused block is shown in Table 10.18.

The reducing condition in regenerator reduces the performance of the refractory lining. It has been found that CO content up to 1000 ppm in the exhaust gas does not harm the magnesia-zircon refractory, but higher CO content attacks the forsterite bond and deteriorates the refractory performance [36]. Similarly the use of fused cast AZS and the β-alumina refractory in the regenerator with reducing condition is not recommended [35, 36].

Table 10.18 Composition of unused and used magnesia-zircon bricks from regenerator

Chemical analysis (%)	Unused magnesia-zircon	Used magnesia-zircon
MgO	77	78.8
ZrO_2	11.7	7.7
CaO	1.04	2.78
SiO_2	8.84	8.64
Na_2O	0.24	0.29
Major phases	Periclase, ZrO_2	Periclase, ZrO_2
Minor phases	Forsterite	Forsterite (M_2S) Merwinite (C_3MS_2) Monticellite(CMS)

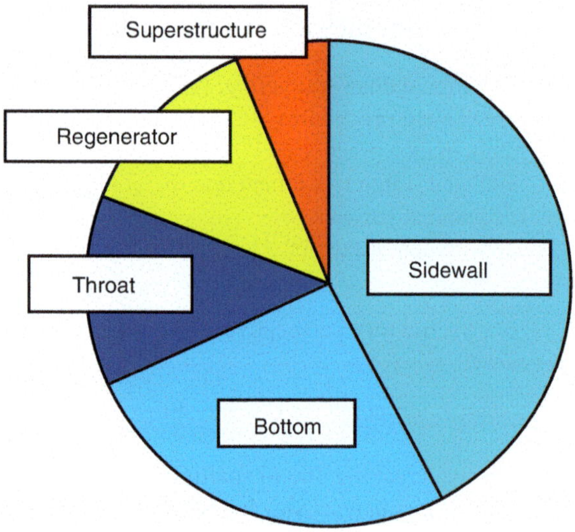

Fig. 10.29 Weak spots in the furnaces and failure probability is proportional to the area

10.8 Monitoring and Maintenance

Premature failure or unscheduled maintenance of the refractory lining of the glass melting furnace is very expensive, and therefore ability to monitor the furnace refractory lining can contribute to commercial success of a glass making unit. The modern container glass furnace once starting operation is expected to run continuously for 11–12 years, minimum. In case of float glass furnace, this figure will be 20% higher, in spite of operation at higher temperature, because in float glass furnace, the critical components like throat, etc. are not there and the chance of metal contamination from cullet is not there, which ensures a good bottom life. Although adequate site holes are provided in the furnace to monitor the lining condition inside, some areas cannot be visible due to the furnace geometry, for which the water-cooled periscope can be provided. Moreover the areas of the furnace, in contact with glass, are more prone to erosion and failure (Fig. 10.29), which are not visible during the operation of the furnace.

Therefore remote sensing techniques are adopted to monitor the furnace lining condition especially in critical areas. The following are the main techniques used:

(a) Thermography – It is the infrared photography of the furnace wall. When the thermography of a wall is done from outside, it gives a colored pattern of the wall, and from the color pattern, one can know which area of the furnace is overheated at outside. The overheating means the refractory thickness gets reduced in that area (Fig. 10.30). But this is a comparative test, and it can only tell where the lining thickness is more and where is less, if the same quality of refractory (same thermal conductivity) and same lining configuration is used. It cannot tell about the lining thickness at any point.

(b) Ultrasonic method – In this method a supersonic sound wave is allowed to pass through the refractory; the attenuation of signal will be there, depending on the softening of the brick at higher temperature.

(c) Resistance monitoring – The electrical resistance of AZS refractory depends upon its temperature. For AZS refractory an increase of temperature of 100 $^{\circ}C$ decreases the electrical resistivity by 80%. In actual process a low voltage is applied to different refractory points during the furnace operation and the resistances are measured. By analyzing of these results and using computer algorithm, the thickness can be estimated. Actual measurement by this method is found to be +/– 10% accurate [36].

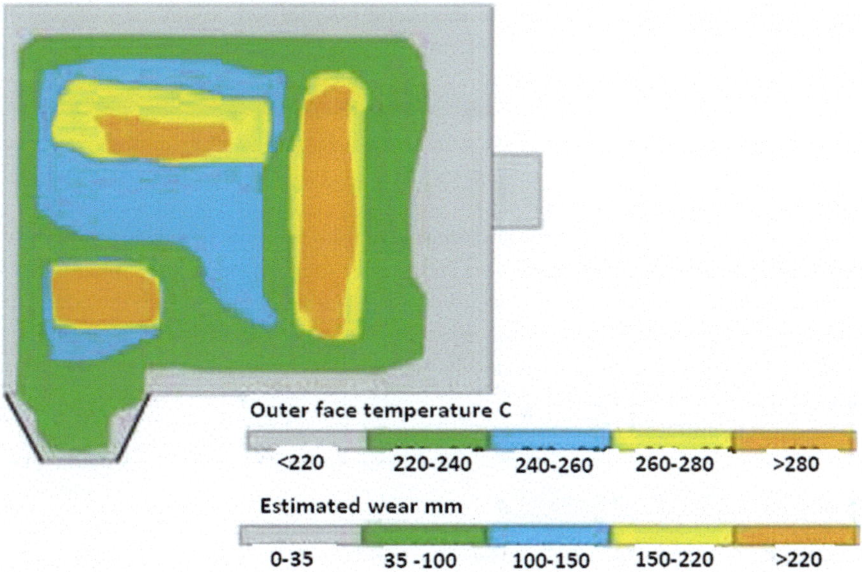

Fig. 10.30 Thermometry of the bottom of glass furnace showing temperature Vs estimated wear

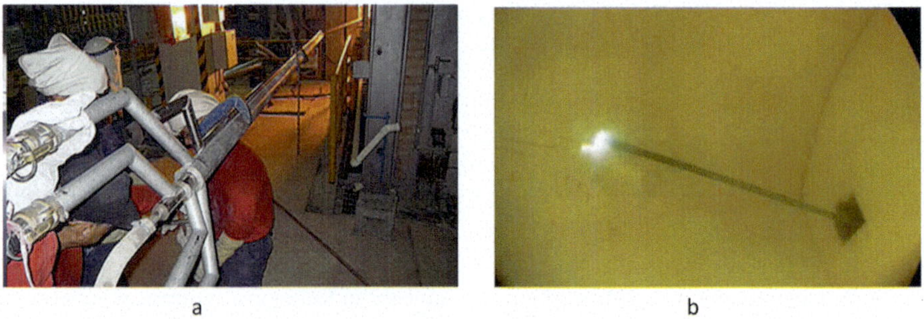

Fig. 10.31 (a) The lance used for welding. (b) Welding of the crown of regenerator

10.8.1 Ceramic Welding

Once a glass melting furnace starts operation, it should run continuously for long time without any major shutdown. The ceramic welding process has been developed and standardized for the repairing and maintenance of the furnace refractory lining in any part without taking any shutdown.

In this process the furnace lining is inspected first. For inspection of the remote corners of the furnace, the endoscopic method or periscopic methods are utilized. The powdered mixture of ceramic material and metal powders (corresponding to the oxides present in the refractory) of definite proportion are conveyed through stainless steel pipe by means of oxygen-enriched air to the spot to be welded. The pipes may be water cooled or air cooled to protect it from high furnace temperature (Fig. 10.31). At furnace temperature the metal powder gets oxidized and evolves further heat which melts the total mass and sprayed on the damaged wall to fill up the cavity or to restore the contour of the furnace.

Ceramic welding method deposits a mass having very similar physical and chemical properties to the substrate refractory material so that bonding becomes strong. Figure 10.32 shows the refractory lining before and after repairing to understand the efficacy of the welding repair. In actual process, the stainless steel pipe is inserted inside the furnace through some available opening to reach the point of welding. If the place of damage is inaccessible, then an opening is made in the opposite wall or in a convenient place, with water-cooled chainsaw, to introduce the pipe to access that place and welding repairing is done. After the welding is over, the newly opened hole is either permanently closed by refractory or temporarily closed to serve as a peephole.

The composition of the ceramic weld material can have SiO_2 95–99% which is suitable to repair the crown having silica brick lining. For repairing of the AZS refractory lining, the composition is available which can deposit a mass with composition of SiO_2-38%, Al_2O_3-37%, and ZrO_2-20%. Similarly for welding of fused mullite corundum bricks, the material is available with 69% Al_2O_3 and 27% SiO_2.

The welding deposit can be made about 100 kg per hour, and the oxygen consumption for the welding is about 50m^3/hour.

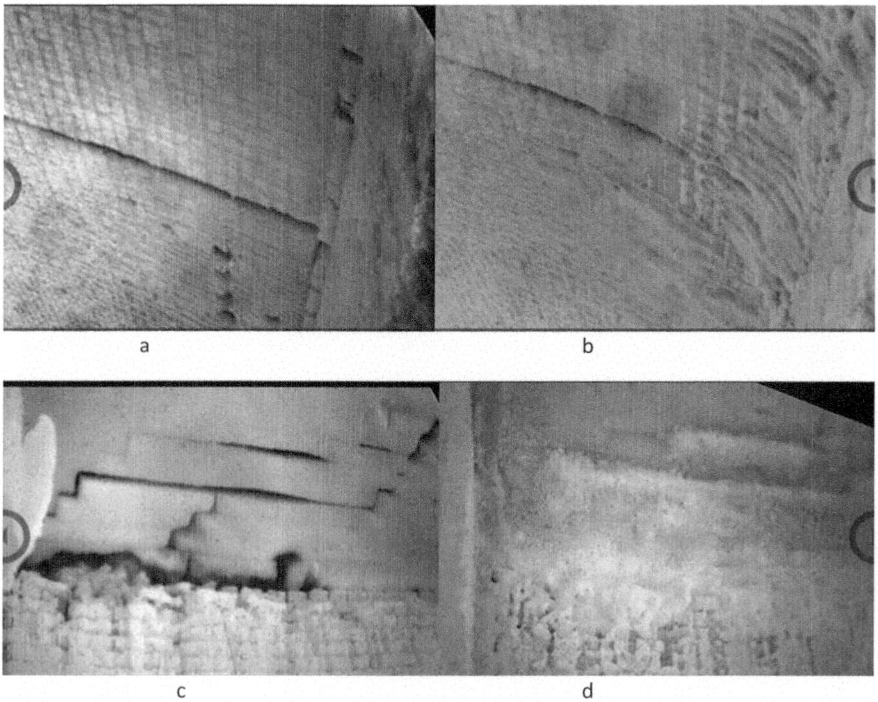

Fig. 10.32 (**a** & **b**) Crown expansion joint, above 3–port before and after repairing, (**c** & **d**) regenerator division wall, before and after repairing (Courtesy of Nosco Asia)

References

1. Jarvis A Glass Industry Refractories and their increasing role in energy conversion
2. Glass the industry of future- Energy and Environment profile of US Glass industry, 2002, DOE/GO-102002-1590, pp 26
3. Beerkens RGC, Van der Schaff J (2002) Advanced heating techniques for glass melting, Technische Universiteit, Eindhoven, March 5, 2002
4. Glass furnaces- Monitoring refractory wear- Geof Evans, The Refractories Engineer, Nov, 2006 issue
5. Gupta A, Silkregg K (2009) Testing and performance of the fuse cast Refractories. In: 70th conference on glass problems, Ohio State University, Oct-11–12, 2009
6. Fleischmann B (2010) Fused cast AZS Refractory materials-Structure and structural changes during application, Refractories world forum, 2, 2
7. IMI-NFG course on processing in Glass, Spring 2015, Lecture-3 : Basics of Industrial Glass melting furnaces- Mathieu Hubert, Celsian Glass and Solar, Endhoven, The Netherlands
8. Wilson T (2020) A Pernicious problem-Batch carryover attack of regenerator Refractories, The Refractories Engineer, March 2020
9. Postrach S et al (2009) State of the art Refractories for the float glass furnace, Refractories Worldforum, 1, [2]
10. Ohnaka H (1981) Present status of Refractories for the Glass industry. Takabutsu Overseas 1(1)

11. Meynkeins JP, Cherdon B Environmental impact for the regenerator materials selection in the soda lime flat glasses furnace. Adv Mater Res 39–40:619–624, Transtech Publications, Switzerland
12. Schmitt GJ (2007) Development of Refractories to extend Glass melting furnace life time. In: 31st Asean Glass Conference, Chiang Mai, Thailand, Oct 4th, 2007
13. Curtis W (2009) Postmortem of a conventional Silica brick from the soda-lime float glass furnace crown. In: 70th Conference on Glass problem, Ohio State University, Oct 11–12, 2009, pp 161–172
14. Heidrich R, Postrach S (2013) Refractory solutions for the new challenges in glass furnace construction. In: Conference paper, 2013, Research gate Publication No 272085677
15. Kotoucek M et al (2013) Dense Silica- Properties, production and perspectives. World Refractory forum 5 [2]
16. Strohm H et al (2007) Refractory materials for the production of flat glass by Float process, Interceram, Refactories manual
17. Rancoule G, Tweels in Sodium Silicate Glass chemistry, Vesuvius, http://www.vesuvius.de/fileadmin/user_upload/Tweels_in_sodium_silicate_glass_chemistry.pdf
18. Lee WE, Zhang S (2004) Direct and indirect corrosion of oxide and oxide-C refractories. In: VII International conference on Molten slag, Fluxes and Salts, The south African Institute of Mining and Metallurgy
19. Glass-melt Refractory interaction- Michael Dunki, UNITECR Proceedings, 1991
20. Craig AG (2016) The study of molten liquid-Refractory interaction – It is all about phases. In: 10th International conference on Advances in Molten slag, Fluxes and Salts, Molten 16, 2016
21. Vadasz P et al, Corrosion profile on boundary solid phase- Slag melt, Interceram, 06-07/16, pp 232–236
22. Lee WE (2004) Theory, experiment and practice of slag attack on Refractories, Tehran International conference on Refractories, 4–6 May, 2004
23. Nishikawa T et al (2005) Development of numerical prediction of Refractory corrosion for glass melting tank, Reports Research Lab Asahi Glass Company, (55)
24. Izmirlioglu B, Ilmaz S (2015) Glass melting furnace Refractories and Refractory related defects. J Chem Technol Metall 50:4
25. Van Dijk FAG (Franciscus Arnoldus Gerardus) (1994) Glass defects originating from glass melt/fused cat refractory interaction, Thesis, Eindhoven University of Technology
26. Heinrich R et al (2014) REFEL 1616ULX- An ultra low exudation Fused cast product for the superstructure of glass melting furnaces. RHI Bull 2:16–17
27. Canaguier J (2016) How to reduce Glass defects generated by Fused cast AZS. In: 16th Conference on Electrical Glass Melting, Sept 14, 2016, Prague, Czech Republic
28. Harbison Walker International, Refractory solution for regenerators
29. Lechevalier D et al, New Refractory cruciform for improved energy efficiency of regenerative glass furnace. https://pdfs.semanticscholar.org/04f9/4af06aeda42830e6f248a02824792d7c42ca.pdf
30. Lechevalier D, Citti O, Bourdonnais S, New refractory cruciform for improved energy efficiency of regenerator glass furnace
31. Bei R et al (2014) Postmortem analysis of Magnesia and Magnesia checkers after one campaign in Soda- Lime glass melting furnace. RHI Bull 2
32. Dzyuzer VY (2013) Boundary condition for calculating the regenerator and checker brickwork in a glass making furnace. Glas Ceram 70(7–8)
33. Bei R et al (2012) Industrial experience with Spinel bonded Magnesia checker brick in the condensation zone. World Refractories forum, 4 [2]
34. Schmalenbach B et al (2003) New solutions for the regenerator of the Glass melting furnace. RHI Bull 1
35. Schmalenbach B, Reiepl K, Impact of the reducing atmosphere on the corrosion of Refractories in the regenerator of the glass melting tanks
36. Evans G (2006) Glass furnace- monitoring of refractory wear, The Refracories Engineer, November, 2006

Chapter 11
Refractories for Portland Cement Manufacturing

11.1 Introduction

Portland cement is one of the most important products which has contributed to the journey of human civilization from cave to skyscraper. Road, bridge, tunnel, skyscraper, ports, and airports – nothing could be constructed with an affordable cost without concrete. Today, cement and concrete is very intimately associated in all walks of life after almost 200 years since it was first invented and patented by Joseph Aspdin in Leeds, UK. The present global production of Portland cement is nearly about 2.8 billion tons.

Portland cement manufacturing is considered as a heavy chemical industry, and new plants are being built with higher capacities with progresses of time to get better energy and cost efficiency.

11.2 Process of Manufacturing

The raw materials required for manufacturing Portland cement are naturally occurring limestone or chalk, as the source of CaO, and clay or shale as the source of silica (SiO_2) and other small ingredients to adjust the chemistry of the final product. Limestone and clay contain other compounds, like Al_2O_3, MgO, Fe_2O_3, Na_2O, and K_2O which have a profound influence on the performance of refractories.

The flow sheet of the manufacturing process is shown in Fig. 11.1 [1].

In brief, the process of manufacturing is to grind limestone and clay together in a definite proportion along with other ingredients and is heated to about 1400–1500 °C in a huge horizontal cylindrical steel, tubular vessel, called rotary kiln. Kilns are about 3–7 meters in diameter and about 60–75 meters in length, and it rotates at a very slow RPM (2–4 rotations per minute). The kilns are lined inside with refractory

P. Sengupta, *Refractories for the Chemical Industries*,
https://doi.org/10.1007/978-3-030-61240-5_11

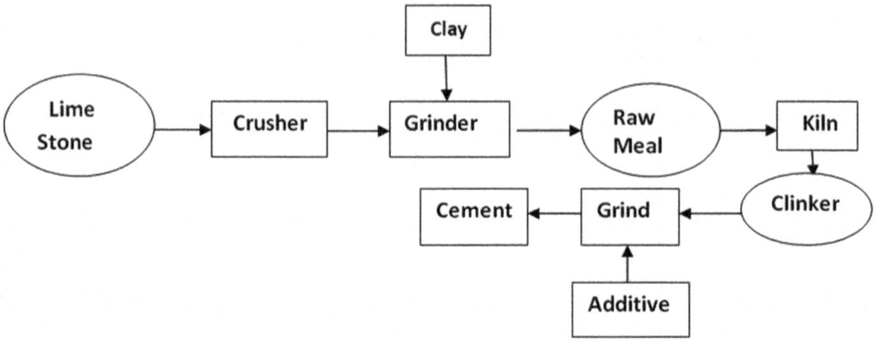

Fig. 11.1 Flow sheet of cement manufacturing process

Table 11.1 Chemical reactions in the cement manufacturing process

Temperature (°C)	Reactions that occur
100	Evaporation of physically bonded water
>500	Dehydroxylation of clay minerals and evolution of chemically bonded water
>900	Decomposition of limestone as per the reaction, $CaCO_3 = CaO + CO_2$
>900	Reaction started between CaO and SiO_2, Al_2O_3, and Fe_2O_3
>1200	Liquid formation started
>1280	$3CaO.SiO_2$ (C_3S) is formed and total lime is reacted

bricks to protect the steel shell from high temperature. The kilns are mounted on two or three supports with the axis, inclined slightly from the horizontal. The finely ground raw material is fed into the higher end. The lower end is fitted with a concentric burner producing a strong flame, by controlled burning of powdered coal, oil, alternative fuels, or gas under forced draft. Due to slow rotation and gravitational pull, the material inside flows slowly from the higher end to the lower end. As the material moves through the kiln, the materials react among themselves at high temperature. Low melting liquids are formed in the material to the extent of 20–28% which helps in forming nodules of average size of 10–100 mm which is called clinker. Clinker comes out of the kiln as dark gray balls which are cooled fast to arrest some desirable phases and then is ground fine along with some additives to control its setting and flow behavior and bagged as commercial Portland cement.

The chemical reactions taking place at different temperatures, throughout the firing process, are shown in Table 11.1

11.3 Fuel

Different fuels like coal, oil, petcoke, gas, etc. are used as the sources of heat energy required to carry out the reactions inside the cement rotary kiln. The fuel cost is about 50% of the total energy cost and 10% of the total cost for cement production. Therefore, a continuous effort is going around the globe in search of cheaper source of fuels. As a result, a lot of other fuels are being found and used more and more to replace the conventional fuels, and these are called alternate fuels. The cement kilns are being used in developed countries as an incinerator, and all kinds of industrial and domestic wastes are used as sources of energy. The common alternate fuels used are petroleum coke, shredded tires, plastics, waste oil and solvents, animal meat, bones and fats, municipal wastes, agricultural wastes, etc. Some of these materials contain high percentage of sulfur and chlorine, for which some special cares are to be taken in the manufacturing processes to maintain smooth operation, and these impurities again affect the refractory performances to a great extent. The acidic gases formed, because of burning of alternate fuels, permeate through the refractory and corrode the kiln shell and anchors which hold the monolithic lining.

11.4 Process Design

Energy is the major cost in clinker manufacturing process. The energy means the combination of thermal energy from the fuel and the electrical energy required to rotate the kiln and to run the cooler and the number of ID fans.

In the cement plant, the preheater system functions as the heat exchanger. The hot gas generated in the kiln to burn the clinker leaves the kiln and moves up the preheater to heat up the raw meal falling vertically through the preheater cyclone system (Fig. 11.2). The thermal energy cost of the cement manufacturing depends upon the heat exchange efficiency of the system. The electrical energy is required mainly to rotate the kiln and its requirement depends on the kiln diameter and length. Good amount of electrical power also is required to run the ID fans and the cooler.

The operating temperatures of different equipment like preheater cyclones, kiln, tertiary air duct, and cooler are shown in Fig. 11.2. Among all these equipment, the rotary kiln is the heart of the cement plant, and most of the major shutdowns in the cement plants are because of the outage of the refractory lining of the rotary kiln.

The rotary kiln can be divided into several zones depending upon the operational conditions inside, and these zones are (Fig. 11.3):

1. Inlet cone – 1000 °C
2. Calcining zone – 1000–1200 °C
3. Upper transition zone –1200–1350 °C
4. Burning zone – 1350–1450 °C
5. Lower transition zone – 1250–1350 °C
6. Outlet area (nose ring, retainer ring) – 1000–1250 °C

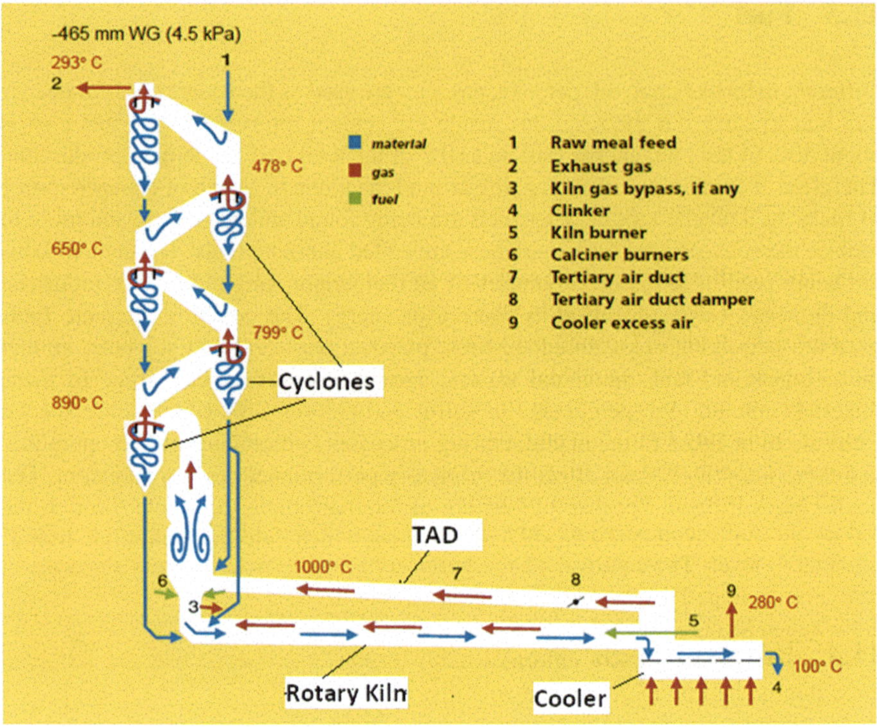

Fig. 11.2 Flow of solids and gases in an ILC kiln. (Courtesy FLS, preheater calciner system)

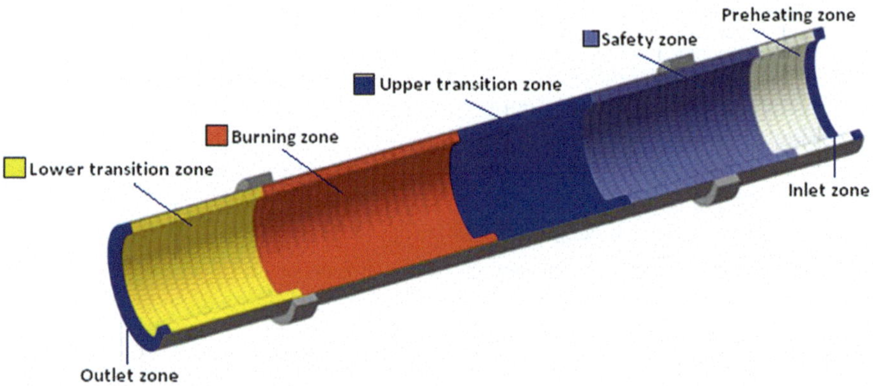

Fig. 11.3 Different zones inside the cement rotary kiln

11.5 Selection of Refractories

Refractories should always be designed and installed to provide, wherever possible, a balanced and predictable economic life. There are many reasons for the shutdown of the kiln, but the reparation, if due to refractory problem, needs the total cooldown of the kiln which is very problematic and expensive. It is therefore important to get a refractory life which coincides with the scheduled maintenance of the plant, and normally, it is 12 months or 24 months, rather than 14 months or 26 months [2].

Because of the wide variation in the size, capacity, raw materials, fuels used, and operational practices, it is not possible to provide a standard recommendation for refractory use, which any plant can adhere to. But it is desirable to follow a general guideline which will improve the refractory performance in each zone of the kiln and in all other ancillary equipment with minimum cost and minimum failure of refractory lining.

Refractories are mostly tailor-made products. A refractory, performing well in any particular equipment, in a particular plant, does not always guarantee the similar performance in the same equipment in another plant. Therefore, it demands a careful and thorough study of the conditions prevailed in individual plant before selecting the suitable quality of refractory for the lining.

A hypothetical relationship, although qualitative rather than quantitative, can be proposed (Table 11.2), indicating which are the physical and chemical properties that exert an influence on the rate of wear by common mechanical, thermal, and chemical factors [3].

Table 11.2 Relation between the refractory characteristics and wear factors

	Wear factors						
	Thermo-mechanical		Thermal		Thermo-chemical		
Refractory characteristics	Stress	Abrasion	Overload	Shock	Infiltration	Alkali attack	SO_2 and CO_2 attack
Dimensional accuracy	VI						
Density		VI					
Porosity				I	VI	VI	VI
CCS		I					
MOR		I					
Chemistry						I	I
Refractoriness			VI				
HMOR	I						
Th. expansion	VI			VI			
PLC	VI						
Th. conductivity	VI			VI	I		
MOE	VI			VI			

VI very important, *I* important

This table can be utilized to compare the suitability of a refractory to others, for a certain application, where the factors responsible for the damage of the refractories are identified to a certain extent.

The overall use and consumption pattern of shaped, unshaped, and insulation refractory in different areas of application in a cement plant is shown in Fig. 11.4

11.6 Refractories for Different Equipment

Let us now discuss the operating conditions prevailing in different equipment and the refractory quality to suit those conditions.

11.6.1 Cyclones and Ducts

The cyclone is used to separate the raw meal fed from the top to separate from the flue gas moving in opposite direction. The cyclone in the preheater system is a very important element in the cement manufacturing process. Each cyclone system is unique in its design and operating characteristics and is decided by the factors such as raw meal composition, the nominal and peak throughput, and the type and quality of the fuel burned in the calciner or smoke chamber. The cyclones are designed

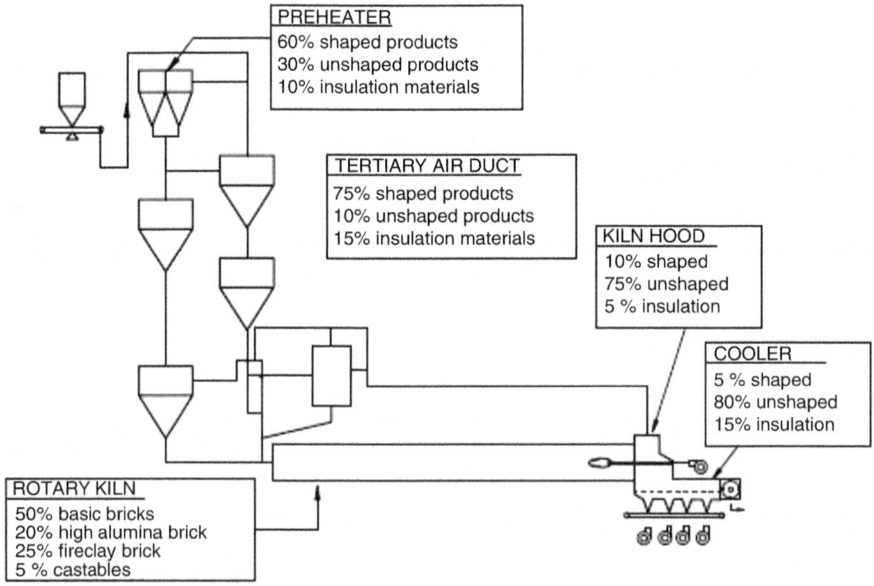

Fig. 11.4 Refractories used in different application areas, in a cement plant

always to meet the objective to achieve higher thermal efficiency and greater kiln capacity which place high demand on refractory lining and anchor system.

The criteria for a good refractory for lining in the cyclone and ducts are the following:

1. It must be able to withstand high temperature.
2. It must be able to withstand the chemical attack of alkali, chlorides, sulfur oxides, and transition metals.
3. The refractory must meet the construction requirements of increasing size of cyclones to form a stable lining.
4. The refractory surface must not support the buildups on its surface.
5. The refractory surface must be smooth to reduce the resistance to gas flow inside the cyclone.

Higher operating temperatures, particularly in the lower cyclones, demand use of refractories with high refractoriness and high strength. With higher temperature, corrosive vapors can penetrate deep into the cyclone refractory system, so alkali- and chloride-resistant refractories should be used [4].

Buildups, usually caused by chemical attack, in combination with the Venturi effect in the cyclone's complex design can also be caused by rapid temperature drops. Buildups reduce cyclone efficiency, and their removal ultimately requires a production stop. Regardless of the method used, removal is a risky job. In the most severe cases, explosives are used. Several mechanical alternatives exist, none of which is particularly satisfactory.

Chemical attack in the form of alkali penetration is unavoidable in cement production. The worst damage occurs in lower cyclone stages and riser ducts, kiln inlets, and even recliners. Alkali and acid vapors infiltrate the refractory linings and attack the binding phase at temperatures as low as 600–700 °C, thus endangering the lining. When these gases penetrate behind the refractories, the effects are even worse because Cl_2 and SO_2 combine with condensing steam to form acids that corrode anchors and steel shell. If proper attention is not paid, this can lead to lining collapse. The safest, most cost-effective method to deal with this is to reduce the amount of insulation used, in order to move the vapor's "dew point" outside the shell of the cyclone. If refractory and cyclone external shell temperatures remain above 100 °C, ambient moisture and chemical vapors will not condense and acid formation will be prevented.

The temperature varies from 300 to 900 °C from the topmost cyclone to the bottommost cyclone. The same quality of refractory can be used in all the cyclones with different thicknesses of working lining and insulation lining to achieve the skin temperature as per the design.

Some properties of the bricks and castable suitable for the lining of the cyclones are given in Table 11.3

The prevailing operating temperature and atmosphere inside the cyclones suggest the suitability of dense aluminosilicate refractories with 40 to 50% alumina, low porosity, and high alkali resistance which does not allow high buildup. As an insulation material, calcium silicate board will be the most ideal. Where buildup is

Table 11.3 Bricks and castable suitable for lining the preheater cyclone

Product	Al_2O_3%	SiO_2%	AP %	BD (gm/cc)	CCS (MPa)	RUL (°C)	Therm. cond. (W/mK)/500 °C	TSR (water) 1000°C/ Cycles
Brick A	35–40	58–59	17–18	2.10–2.15	40–50	1350	1.7–1.8	25
Brick B	30	65	13	2.15	65	1300	1.2	20
Castable-I	35	45	–	2.0	25	–	0.8	–
Castable-II	45	40	–	2.1	35	–	0.9	–

the major problem, aluminosilicate refractories containing SiC (silicon carbide) or zircon helps to reduce the buildup and alkali attack. SiC reacts at high temperature with atmospheric oxygen to form a glassy layer of SiO_2 on the refractory surface, which prevents buildup. Similarly, zircon at high temperature decomposes to form SiO_2, which forms a glassy coating over the refractory surface. Zircon is better because it has a lower thermal conductivity than SiC which helps saving anchors from overheating [5].

The ideal lining thickness for any area of the preheater is the minimum thickness required to attain the specified shell temperature. In case of monolithic lining, if it is too thin, the anchors may get overheated, which can accelerate corrosion, scale formation, fatigue, etc. and may lead to collapse of the lining. If the lining is too thick, then sharp temperature gradient inside the lining may cause crack and spalling of the lining.

In the preheater, the fine dust behaves like fluid, and very often the infiltration of red-hot dust behind the refractory lining causes failures to the lining. The failures are caused by, sometimes, shearing of anchors or the shrinkage of insulation due to sintering in contact with red-hot dusts or by bulging of the brick lining due to pushing of the lining from behind.

The refractory used can be both in shaped or unshaped form, but which one is more preferable is a controversial matter. Performance-wise, both are good and both have some merits and demerits. The advantages and disadvantages of brick and monolithic lining are shown in Table 11.4.

The installation of monolithic lining is faster and easier than brick lining. But the installation of monolithic refractory is more critical than brick lining. Brick is a finished product, but castable is a semifinished product and needs much attention and care during installation.

In the brick lining, the shape of the brick is important, especially when the diameter of the cyclone is larger. The brick shape and size must be designed to have an interlocking arrangement to avoid bulging out of the lining. In case of monolithic lining, the anchor design is very important to attain the stability of the lining.

The common mistake made during selection of refractory is to put the speed of installation ahead of the lining stability. If the preheater lining life is 2 to 5 years without repair, then installation time should never get the priority.

Table 11.4 Comparison of shaped and unshaped refractories

Brick lining	Monolithic lining
Very good product properties	Very good product properties
Difficult to introduce automation	Easy to introduce automation
Self-supporting lining. Only retainer plates are required	It needs anchors to be welded on the shell to hold the lining and it consumes time
Lining is difficult in case of complicated geometry. Special shapes are required and inventory goes up	Can be lined in the furnaces with any design and dimension. It can be installed in furnaces with even complex geometry. No special inventory is required
Installation is slower but heating up is quicker	Installation is faster but heating up is slower
Lining process is not influenced by the ambient temperature	Workability and the setting of the lining are influenced by the ambient temperature
Difficult to repair the lining in case of any damage	Easy to repair in case of any damage

11.6.2 Calciner

The thermal efficiency of the cyclone system improves a lot when the recycled hot air from the cooler is delivered to the calciner through the tertiary air duct. Fuel cost is also reduced by using the cheaper alternate fuel in the calciner. At higher operating temperature, chemical attacks are more severe in calciner compared to cyclone. The suitable refractory, for the calciner, must have high refractoriness, high thermal shock resistance, and high resistance to chemical attack by alkali, chlorides, and sulfur oxides. The refractory in the calciner also should have high abrasion resistance because large volume of dust-laden gas will be delivered in the calciner at high velocity through the TAD. The selection of the right kind of insulation is also very important to prevent energy loss. The use of alternate fuel like plastics, municipal wastes, etc. may pose serious problem like early wear in the central pipe.

Besides reacting with the refractory material itself, alkali vapor penetrates inside the refractory lining and attacks the metallic parts like anchors, and the stability of the lining may be greatly reduced.

The three main areas of the calciner are the cone, wall, and roof. Both shaped and unshaped refractories can be used to line all these areas of the calciner, but the refractories must have the desired properties as mentioned before to ensure satisfactory performance. It is better to line the roof with castable which has lower permeability compared to brick lining.

The typical properties of the refractories, suitable for the calciner, are shown in Table 11.5.

Table 11.5 Properties of refractories suitable for the calciner

Product	Al_2O_3%	SiO_2%	Fe_2O_3%	AP %	BD (gm/cc)	CCS (MPa)	RUL (°C)	Therm. cond. (W/mK)/400°C
Brick A	35–36	58–59	1.5	17–18	2.1–2.15	40–45	1330	1.50–1.60
Brick B	48–50	44–46	1.7	16–17	2.3–2.4	50–60	1400	1.50–1.60
Brick C	58–60	36–38	1.0	15–16	2.5–2.55	70–80	1450	2.3–2.40
Brick D	74–76	12–14	2.0	18–19	2.72–2.73	80–100	1480	1.85–1.90
Castable-I	35	SiC = 29–30	0.8–0.9		2.45	100–110	–	2.50

Table 11.6 Properties of refractories suitable for riser and tertiary air duct

Product	Al_2O_3%	SiO_2%	Fe_2O_3%	AP %	BD (gm/cc)	CCS (MPa)	RUL (°C)	Therm. cond. (W/mK)/400°C
Brick A	35–36	58–59	1.5	17–18	2.1–2.15	40–45	1330	1.50–1.60
Brick B	48–50	44–46	1.7	16–17	2.3–2.4	50–60	1400	1.50–1.60
Castable-I	35	SiC-29-30	1.0	–	2.45	100–110	–	2.50

11.6.3 Riser Duct

It recycles hot air from the lower stage cyclones to kiln improving the thermal efficiency of the process and reduces the fuel cost. Riser duct lining often faces the problem of buildup especially in the curved sections.

The closer the duct is to the hot end of the cyclone system, the greater is the alkali penetration and spalling of the refractory. The alkali vapor penetration can cause severe alkali bursting, where the refractory volume can increase by ten times from normal volume [6]. The typical properties of the refractories, suitable for riser duct lining, are shown in Table 11.6.

11.6.4 Smoke Chamber

It experiences the worst possible exposure in a cement plant, like high temperature (1300°*C*), alkali attack, buildup, anchor corrosion, etc. The refractory selected for lining inside this equipment must have low porosity and low permeability, and the best choices are dense bricks or low cement castables. The typical properties of the refractories, suitable for smoke chamber lining, are shown in Table 11.7

Table 11.7 Properties of refractories suitable for meal pipes

Product	Al_2O_3%	SiO_2%	Fe_2O_3%	AP %	BD (gm/cc)	CCS (MPa)	RUL (°C)	Therm. cond. (W/mK)/400°C
Brick A	35–36	58–59	1.5	17–18	2.1–2.15	40–45	1330	1.50–1.60
CastableX	43–45	CaO = 1.6	1.5	Water for casting 6.5%	2.2–2.25	65–70	–	1.45

11.6.5 Meal Pipes

Meal pipes connect the cyclones and have smaller diameters. The meal pipes are made in small sections. They are lined in the ground level and then are transported to the area of installation. The main property required here is to withstand the continuous abrasion of the meal material. It can be lined with both bricks and castables. The typical properties of the refractories, suitable for meal pipe lining, are shown in Table 11.7.

11.6.6 Insulation

The total heat loss by radiation in the preheater is less than 2% of the total heat input. Most preheater vessels and ducts are insulated to protect the steel shell. The use of refractory fiberboard as insulation, in the hotter sections of the preheater, may induce premature lining failure, if the fiberboards fail behind the dense lining. Fiberboard contains a certain amount of organic matter that, when exposed to moderately high temperature, burns out with emission of fumes and loses both its mechanical strength and insulating property. The loss of insulating property exposes both the steel shell and the anchor stem to higher temperatures resulting in relative movements and damage of the anchoring system and induces cracks on the dense lining. To avoid such problems, an insulating material with a minimum amount of loss on ignition, e.g., calcium silicate, diatomaceous earth or vermiculite boards, or insulating castables, that can be cast, gunned, sprayed, or pumped in place, should be selected.

11.6.7 Kiln Hood

The kiln hood for lining purposes can be divided in two halves. The upper half, consisting of the ceiling and sidewall, experiences high-temperature radiation, abrasion, exposure to alkali vapors and sulfur oxides from secondary air, and thermal

Table 11.8 Refractories for the upper half of the kiln hood

Product	Al_2O_3%	SiO_2%	Fe_2O_3%	AP %	BD (gm/cc)	CCS (MPa)	RUL (°C)	Thermal cond. (W/mK)/400 °C
Brick E	60–62	P_2O_5 = 1.5	1.5	17–18	2.40–2.45	85–90	1380	1.80
Brick F	75	19	1.5–1.7	17–18	2.7–2.75	70–80	1500	2.00
Castable-III	40	SiC-30	0.5	Water for casting 5%	2.50	90–100	–	2.65

Table 11.9 Refractories for the lower half of the kiln hood

Product	Al_2O_3%	SiO_2%	Fe_2O_3%	AP %	BD (gm/cc)	CCS (MPa)	RUL (°C)	Therm. cond. (W/mK)/400°C
Brick G	55–57	47–48	1.5–1.7	17–18	2.35	50–55	1400	1.60
Brick H	65	SiC = 10	2.0	17–18	2.60/2.65	80–90	1550	3.1
Castable-IV	50	42	1.0	Water for casting 5.5%	2.35	90–100	–	1.50

shock, when there is kiln shutdown. It is best to line that area by 50–75% alumina refractories or dense low cement castables. There is report of good result of silica sol-bonded castable in this area [7–9]. If the temperature of the secondary air is above 900 °C, then magnesia-spinel refractory can be a good choice also.

Another interesting alternative, found to gain more and more acceptance in cement plants, is the modular lining which offers the advantages of both bricks and castables. In this alternative, the castable is precast and prefired into blocks that have the insulation and the anchoring system preattached.

The lower half of the hood towards the cooler side can be lined with silicon carbide or zircon containing high alumina refractories. The buildup of the coating occurs in this area and the said refractories will repel the coating. The typical properties of the refractories, suitable for kiln hood lining, are shown in Tables 11.8 and 11.9.

11.6.8 Tertiary Air Duct and Cooler

TAD recycles the heat from the hot clinker to the kiln. Air from the grate cooler, laden with highly abrasive clinker dust, travels through the TAD at a very high velocity (25–30 meters/sec) at high temperature (about 1000 °C) and poses a very high abrasion on the refractory lining especially at bends and dampers. Therefore, the selected refractory to line the TAD must have very high abrasion resistance and alkali resistance.

11.6.9 Cooler

Clinker enters the cooler at a temperature of around 1200 °C and heats up the grate and surrounding refractories rapidly. High refractoriness, high abrasion resistance, and resistance to thermal shock are necessary, especially at the clinker downfall area and bull nose. The cooler walls directly above the grate are also subjected to extreme wear. The partition wall, designed to limit the inflow of cold air to the kiln system, must withstand the abrasive effect of clinker dust entering the tertiary air duct as well as the temperature fluctuations, resulting from its exposed position.

The cooler is effectively divided into two zones – the "hot" zone and the "cold" zone. The first zone is from the clinker downfall to the bypass duct and partition wall. The purpose is to exchange the heat from the hot clinker and route it back to the preheater via the tertiary air duct. The function of the second zone is to ensure a uniform reduction of temperature of the clinker body from 800 °C down to ambient. The refractories to be used here must have high abrasion resistance and high thermal shock resistance.

Both brick and monolithic linings can be used to line the kiln hood. To get good refractory life and the fastest possible installation at the clinker downfall, bull nose, front sidewalls and roof, and precast refractory blocks, which can be quickly bolted into place and just as quickly be removed, can be used.

11.6.10 Kiln

Kiln is the heart of the cement plant where the clinkerization takes place at high temperature. The kiln is a cylindrical horizontal tube with a very little inclination of 1–2 degrees. The kilns have a diameter of 3 to 7 meters and length of 50–100 m. The kiln rotates on its horizontal axis at a speed of 1.5–4 RPM.

The refractory lining of a modern cement rotary kiln is performing different specific functions at different zones of the kiln, which includes:

1. Protection of the kiln shell and tires from attaining an excessive temperature which can deform the shell
2. Protection of the kiln shell from the corrosive gases and liquids
3. Protection of the kiln shell from mechanical abrasion by the feed material and the clinker
4. Reduction of the thermal energy loss to keep it within tolerable limit
5. Increase of heat exchange between the hot gases and the feed material

Table 11.10 Length of different zones of the rotary kiln in terms of kiln diameter

	Zone							
Kiln type	Cooler type	Outlet zone	Lower transition zone	Burning zone	Upper transition zone	Preheating zone	Inlet zone	Inlet cone
Normal dry process	Grate	1 Mtr	1–1.5 × D	3 × D	3 × D	11–16 × D	0.5–2×D	1 Mtr
Suspension preheater	Grate	1 Mtr	1–1.5 × D	3–4 × D	2–3 × D	7–8 × D	0.5–2×D	1 Mtr
Pre-calciner	Grate	1 Mtr	1–1.5 × D	5–8 × D	2–4 × D	4 × D	0.5–2×D	1 Mtr

D diameter of the kiln

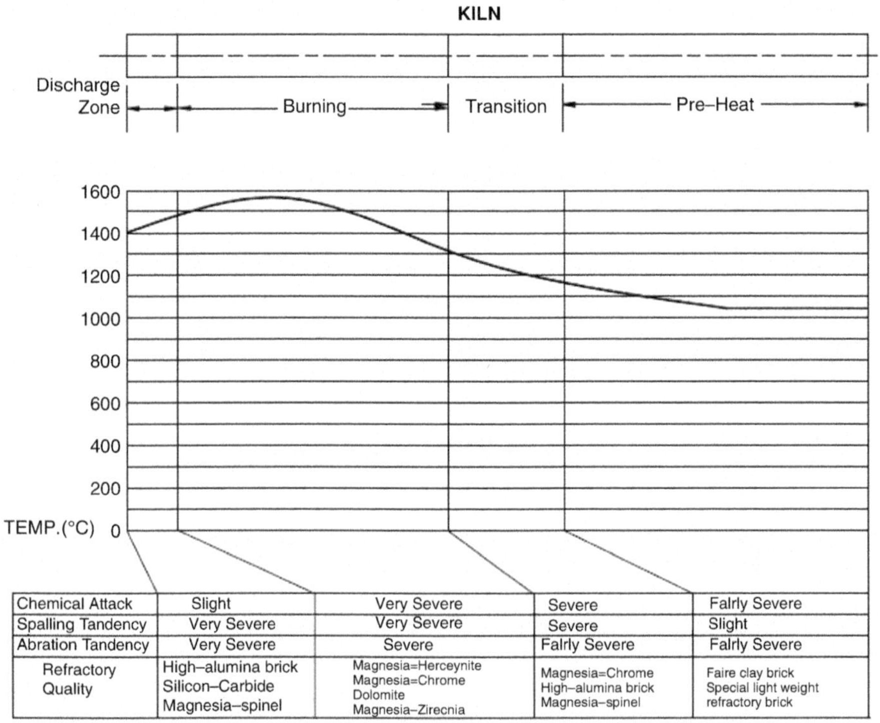

Fig. 11.5 Temperature profile at different zones inside the rotary kiln

The kiln can be divided into several zones depending on its temperature at different zones and also functions being performed. Table 11.10 shows the length norms of different zones in different types of kilns.

Refractory selection for different zones is done based on the operational environment. The temperature profile and the refractory practices followed for different zones in the kiln are shown in Fig. 11.5 [10, 11].

11.6.10.1 Inlet Cone

The most important considerations for the bricks here are that they must be able to withstand high abrasion, thermal shock, and alkali attack. Therefore, 40–50% alumina fireclay refractories are found to work best in this area. Both shaped and unshaped refractories can be used in this area. The typical properties of the refractories suitable for lining are given in Table 11.11.

11.6.10.2 Preheating Zone

The refractories in this zone must be able to withstand high mechanical abrasion by the moving calcined feed material and alkali attack. 50–60% alumina refractory bricks are the best lining material for this area. The typical properties of the aluminosilicate refractories suitable for lining different zones of the kiln are given in Table 11.11

11.6.10.3 Transition Zone

Here the refractory lining are exposed to:

1. High temperature and also cyclic temperature fluctuations, because there will be no coating or unstable coating
2. Radiant heat of the flame
3. Attack by alkali salts and alkali vapor
4. Attack by the liquid phase in the cement starts and attack by other corrosive gases if alternate fuels are used
5. Redox reaction cycles when there is variation in oxygen potential due to change in fuel

Magnesia-alumina-spinel refractories are the best material to line this area, especially when the furnace is with large diameter. The magnesia-alumina refractories have the property to withstand high thermal shock. It has low MOE to withstand high thermo-mechanical stresses, and it can withstand the chemical attack of alkali salts, corrosive gases at high temperature and cyclic oxidizing and reducing conditions.

With smaller diameter kiln, normal high alumina refractories and phosphate-bonded high alumina refractories with 70% alumina also work well. High alumina refractories because of its pyroplastic property can withstand the high deformation without getting crack. It has also have a good thermal shock resistance. High alumina refractories also take up good coating which protects it from all abuse.

Table 11.11 Properties of refractories for different zones of cement rotary kiln

	Property							
Kiln zone	Al_2O_3%	SiO_2%	Fe_2O_3%	AP %	BD (gm/cc)	CCS (MPa)	RUL (°C)	Therm. cond. (W/mK)/400°C
Inlet cone	35–45	53–57	1.5–1.7	15–18	2.15–2.30	40–50	1400	1.3
Preheating	50–60	35–38	1.3–1.6	17–18	2.32–2.45	50–60	1450	1.4–1.7
	35–50	SiC- 9-10	1.5	17–18	2.35–2.40	50–60	1500	2.5–3.0
Transition	75		2.0–2.5	17–18	2.70	60–70	1500	1.90
	70	P_2O_5–1.6	2.0	16–17	2.72	80–100	1500	1.90
Burning	70		2.0–3.0	18–20	2.65	50–60	1480	1.80
	70	P_2O_5–1.6	2.0	16–17	2.70	70–90	1500	1.85

11.6.10.4 Burning zone

Burning zone is the most important area in a cement rotary kiln where the clinkerization takes place. This zone experiences:

1. Highest temperature.
2. Reaction with the liquid phase formed in cement.
3. Continuous temperature fluctuation between when the lining is in contact with the clinkers at the bottom and when it is at the top, farthest away from the clinkers. The temperature fluctuation gets further aggravated, if the coating becomes unstable.
4. Reaction with the sulfur oxides and other gases, e.g., chlorine, generated from alternate fuels.
5. High abrasion from the clinkers.
6. Different kinds of thermo-mechanical stresses.

Fortunately, a thick coating of the cement clinker is formed on the refractory lining in the burning zone, which minimizes the damaging effects of the said factors on the lining and enhances its campaign life. The process people in the cement plant always pay attention to maintain a stable and thick coating on the burning zone refractory lining. The stability of the coating depends on many factors which will be discussed later.

Thermal load is one of the most crucial parameters to decide the lining quality. The thermal load (T_L) is defined as the heat consumption per hour per unit cross section of the kiln, and usually it is expressed in terms of GJ/m^{2}h:

$$T_L = \frac{F_i N_{HV} 10^{-6}}{\left(\pi D_i^2 / 4 \right)} \tag{11.1}$$

where F_i = fuel burning rate per hour (kg/hr) in the kiln, N_{HV} = net calorific value of the fuel (kJ/kg), and D_i = effective internal diameter of the kiln (meter) after refractory lining.

With increasing production per net kiln volume (ton/dm^2), the thermal load increases, and the lining material must be designed accordingly to be able to take up that thermal load.

It is found through experience that the normal thermal load of cement kiln remains in the range of 14–22 GJ/m^{2}h, whereas the recommended maximum thermal load in the burning zone is 24.3 GJ/m^{2}h. It is found by experience that up to a thermal load of 13 GJ/m^{2}h, high alumina bricks are working satisfactorily, when other operational parameters are within tolerable limit. Beyond this value of thermal loading, basic refractories, e.g., magnesia-spinel, work much better, and if the thermal loading exceeds 21 GJ/m^{2}h, special care must be taken in selecting the refractory. In such case, very pure grade of magnesia-alumina or magnesia-zirconia lining should be used.

For large diameter kilns where the thermal loading is high and alternate fuels are used, the basic refractories are preferred over high alumina products, but the high alumina refractories are also used very successfully in smaller diameter kilns. The raw material chemistry, fuel used, and operational discipline play a major role in deciding the refractory lining life in the burning zone. The selection of the refractory for the burning zone depends upon the kind of stresses the refractory lining is exposed to. The operating conditions that prevail in different cement plants are not same.

The typical properties of some basic refractories for the transition and the burning zone are shown in Table 11.12.

Before selection of the refractory, one should know the actual operating conditions in the cement plant, under consideration. Table 11.13 shows the suitable quality of basic bricks under different kinds of stresses [1].

11.6.10.5 Outlet Area (Retainer Ring and Nose Ring)

This is a very critical area in the kiln and highly stressed. The thrust of the whole kiln lining is borne by the retainer ring, and the refractory in contact with the retainer ring is to withstand it. Besides the thrust in this area, the bricks are to withstand high abrasion of the clinkers, and the bricks also experience high thermal shock due to temperature fluctuation.

Various designs are tried to minimize the thrust, e.g., to put two retainer rings or to make a conical design to transfer part of the axial thrust in radial direction, but none is found to be a foolproof solution. The stress generated on lining this zone, because of thermal expansion, was computer simulated and modeled. During initial

Table 11.12 Properties of some basic refractories for rotary kiln lining

Area of use	MgO%	Al_2O_3%	Cr_2O_3%	Fe_2O_3%	ZrO_2%	BD (gm/cc)	AP (%)	CCS (MPa)	Periclase combined with
Transition zone	84–90	6–12	–	0.5–0.7	–	2.88–2.92	15–17	50–80	Spinel
	85–90	4–10	–	4–8	–	3.0	15–17	80–90	Hercynite
	80–83	3–7	4–5	3–5	–	3.0	17–18	50–70	Chrome
Burning zone	84	7–12	–	0.5	–	2.88	17	50	Spinel
	85–91	3–4	–	3–4	–	2.95–3.0	15–16	80–90	Hercynite
	78–83	2–4	5–10	5–8	–	2.9–3.1	16–18	50–55	Chrome
	40–50	CaO = 40–55%		–	–	2.90	13–14	>35	Dolomite
	93	–	–	0.5	5.0	3.0	17	50	Zirconia

Table 11.13 Basic refractory type and the stress it can withstand

	Brick type				
Kind of stress	Magnesia-zirconia	Magnesia-spinel	Magnesia-hercynite	Magnesia-chrome	Dolomite
Thermo-mechanical	E	E	E	E	G
Thermal shock	E	E	E	G	F
Abrasion	M	M	M	M	M
Overheating	E	G	G	G	M
Salt corrosion	E	E	E	M	G
Salt infiltration	M	M	E	M	M
Silicate infiltration	M	M	G	M	G
Redox	E	E	G	M	G
Hydration	G	G	G	G	F
Coating adhesion	M	M	E	G	E

E excellent, *G* good, *M* moderate, *F* fair

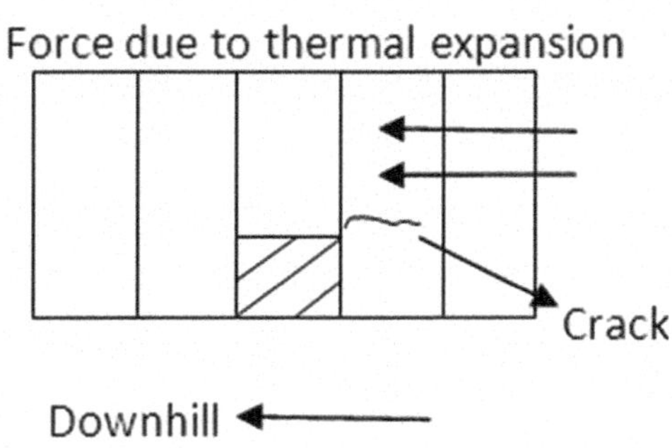

Fig. 11.6 The mechanism of crushing of the brick around the retainer ring

lining, the cardboards are used as expansion joints. It was found that in the initial heat up, the highest stress produced was confined to 1/3 of the brick height from the hot face [12], and the stress produced was lower than the crushing strength of the bricks. But a completely different situation was encountered after kiln shutdown and reheating. The expansion joint provided during initial lining is lost due to downhill movement of the lining.

Moreover, the open joints are all filled up with clinker dust and the residual expansion of the lining made it further worse. The brick just before the retainer ring in downhill direction is supported by steel block (hatched area in Fig 11.6) and 70 mm brick over it. In downhill direction the thrust of the total brick length of the kiln comes on this brick and very high shear force is generated which is much higher than the brick strength resulting in crushing of the brick.

To combat this problem, RHI has developed a design to increase the contact surface area by providing slanting surface as shown in Fig. 11.7. Because of higher surface area, the force per unit will be lower to save the bricks from getting crushed [13].

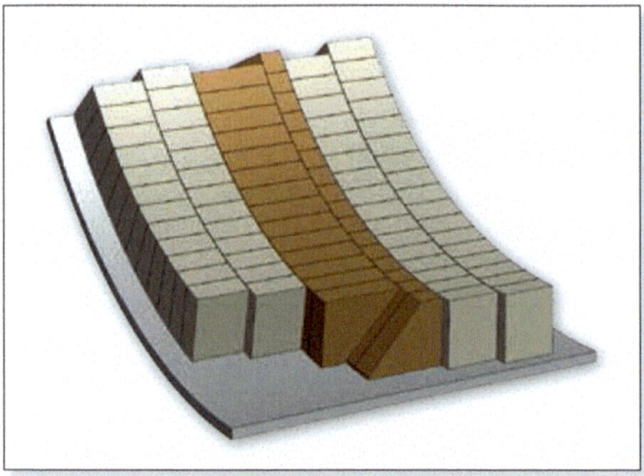

Fig. 11.7 Thrust lock design, of RHI, to reduce crushing force acting on the lining bricks

Table 11.14 Properties of refractories for outlet and nose ring

Product type	Al_2O_3%	Fe_2O_3%	CaO %	SiC %	BD (gm/cc)	AP %	CCS (MPa)	RUL (°C)	TC at 400 °C (W/mK)
Brick	82	2.0	–	–	2.75–2.77	18–19	85–90	1470	2.30
	40–45	0.5–1.0	–	15–20	2.50–2.60	12–14	70–120	1500	2.5
Castable	38–45	0.5–1.0	2.5	25–30	2.60–2.65	–	70–80	–	2.6
	95	0.1–0.2	2.5	–	2.74	–	100	–	2.5
	45	0.5–0.7	2.5	30	2.60	–	100	–	2.6

TC thermal conductivity

In the nose ring area, the main factors which make this area critical are high abrasion of the falling clinkers and high thermal shock as it is exposed to atmosphere.

The material to be used in the nose ring must have high volume stability, abrasion resistance, and thermal shock resistance. The properties of some ideal material used for lining this area are shown in Table 11.14. Silica sol-bonded or no cement castable is also tried in this area with great success [7].

Along with the castable quality, the design of anchors is also very important for the successful performance. It has been found that specially designed precast refractory blocks with very high content of stainless steel fiber work very well in this area.

11.6.10.6 Brick Shapes for the Kiln Lining

The brick shapes used for lining of the rotary kiln are side arch bricks. Normally, the kilns are lined with brick shapes with two different tapers. Two different systems of refractory sizes are very popular, one is VDZ and the other is ISO, as shown in Fig. 11.8.

ISO AND VDZ COMBINATION LININGS

International System

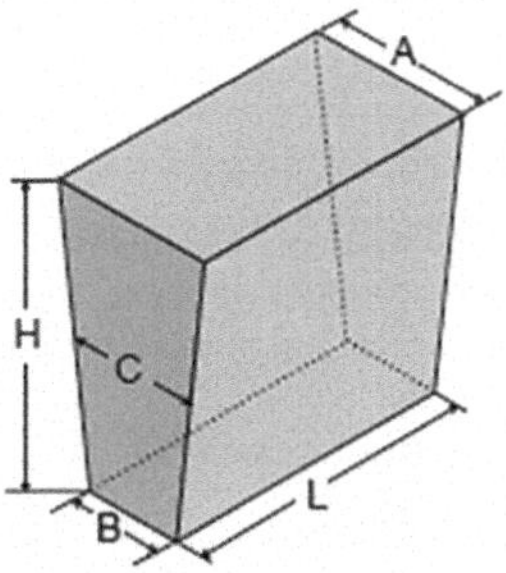

INTERNATIONAL SYSTEM			
		ISO	VDZ
BACK CHORD	(A)	103	VARIABLE
INSIDE CHORD	(B)	VARIABLE	VARIABLE
MIDDLE CHORD	(C)	VARIABLE	71.5
LENGTH	(L)	198	198
HEIGHT	(H)	160 180 200 220 250	160 180 200 220 250
BASIC LINING			X
NON-BASIC LINING		X	

Schematic diagram showing critical dimensions of ISO and VDZ shapes

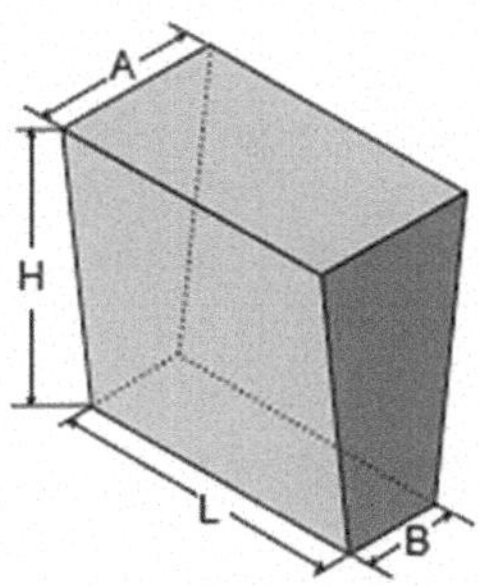

ISO brick outside chord (A) dimension is constant at 103mm (4.05")

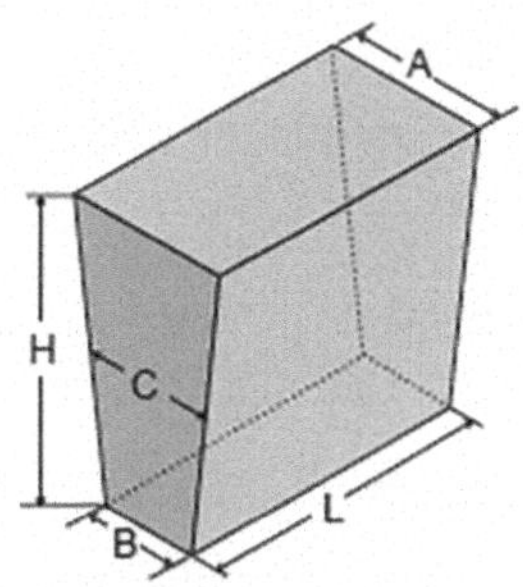

VDZ brick mean chord (C) dimension is constant at 71.5mm (2.81")

Fig. 11.8 Main shapes of the brick lining system of rotary kiln and their comparison

Two shape brick systems fit better in the rotary kiln lining because often the kiln shells are not perfectly circular. By changing the ratio of the two different brick sizes, the distorted area in the kiln can be lined better. Some practical thumb rules used for the lining are the following:

1. Basic bricks should be lined in the VDZ shapes because VDZ bricks are thinner compared to ISO sizes and the lining with VDZ shapes has more joints which can accommodate the higher thermal expansion of the basic refractory.
2. Kiln lining with larger diameter needs bricks with more taper and preferably ISO shapes should be used.

Table 11.15 Recommended lining thickness

Kiln diameter (meter)	Lining thickness (mm)
< 3.5	180
3.5–4.2	200
4.3–5.2	220
>5.2	250

3. At the kiln tires because of the higher shell ovality, VDZ brick shapes should be used because of their flexibility with greater number of radial joints.
4. Handling of ISO bricks is more stressful for lining crews because of their heavier weight compared to VDZ shapes.

There is no proven corelation between the brick lining thickness and its campaign life. Sometimes lining with 160 mm brick thickness gives better refractory life than 220 mm thick lining. Actually the radial stress, which crushes the brick, increases with lining thickness. The ideal thickness of the lining should be the minimum thickness, which will not allow the shell temperature to exceed the acceptable limit. The normal recommended lining thickness is given in Table 11.15.

It is to be borne in mind also that the key effect of the bricks in the lining becomes more prominent with higher length of the bricks, and in bigger diameter kiln, the brick lining with longer bricks is more stable.

Calculation of combination of two different brick sizes – The number of bricks of each size used in the lining can be calculated by using the mathematical formula given below.

If N_a and N_b are the number of pieces required of shape A and B to form a ring in a kiln with inside diameter D, and X_a and X_b are the a and b sides of shape A and Y_a and Y_b are the a and b sides of brick B, then

$$N_a = \frac{\pi\left(d - \frac{DY_b}{Y_a}\right)}{X_b - \frac{X_a Y_b}{Y_a}} \tag{11.2}$$

and

$$N_b = \frac{\pi\left(d - \frac{DX_b}{X_a}\right)}{Y_b - \frac{Y_a}{X_a}.X_b} \tag{11.3}$$

where d = Kiln ID – 2x lining thickness.

11.6.10.7 Burner Pipe

Refractory selection in the burner pipe must take into consideration the following factors:

1. High temperature
2. Thermal cycling
3. Alkali attack
4. Sulfur attack.
5. High abrasion of the clinker
6. Low thermal conductivity to protect the steel pipe inside

It is not possible to get the best of all these properties in one material, and therefore a balance is struck to optimize the different properties to get the best suitable material. Castable is the type of refractory used to line the burner pipe.

Since the main requirement of a burner pipe lining is to protect the metal pipe inside, an insulation cloth or paper must be wrapped on the pipe and then casting is done. The lining thickness is normally 100 mm. The front 1.5–2 meter of the burner pipe is the most vulnerable area where damage takes place in service. In order to reduce the cost, the zone lining can be done in the pipe. In zone lining, 2 meter from the tip is cast with the best castable, and beyond that, the lining can be made with normal castable with lower price.

The main reason for damage of the castable lining of the burner pipe is the differential expansion and contraction of the castable and the anchors [14]. To minimize this effect, the anchor design is very important. The anchor must be designed in such a way that it can accommodate the expansion of the castable lining material [15]. The anchors are to be covered with bitumen paint or wax, which, when burns off, creates a gap for the play of the castable. Instead of rigid anchors, the floating anchors, which can accommodate the movement of castable, enhance the performance. The stainless steel fiber reinforcement in the castable resists the abrasion and thermal shock better and improves the performance. Properties of some castable suitable for lining the burner pipe are given in Table 11.16.

Silica sol-based castable has been tried successfully in the burner pipe lining [9, 16]. The report also presents that specially developed magnesia-spinel castable shows very promising result in the lining of the burner pipe [17].

11.7 Kiln Refractory Performance

The most vulnerable area in the rotary kiln is the burning zone, where the refractory wear out is much more than that in other areas of the kiln, because of higher temperature, cyclic temperature fluctuation, and the presence of clinker liquid phase. Although the factors responsible for refractory performances are shown as discreet entities, they are all interdependent and overlapped with each other. Analysis of the refractory wear out and failures in the rotary kiln alone shows that about 35% of the

Table 11.16 Properties of castables for burner pipe lining

Type of castable	Al_2O_3%	Fe_2O_3%	SiC %	CaO%	BD (gm/cc)	CCS/110°C (MPa)	CCS/1000°C (MPa)	Th. cond (W/mK)/1000°C
Low cement	90–95	0.5	–	1.2	2.85–2.90	100	120	2.7
Low cement	40–45	0.5–1	25–30	1.5	2.50–2.55	100–110	100–110	2.3
No cement	45	0.5	25–30	0.2	2.6	100–110	100–110	2.3

Table 11.17 Frequency of occurrences of refractory problem in different areas

	Contribution of different factors (A to G given below) towards the refractory-related problems (%)							Total %
	Factors							
Area application	A	B	C	D	E	F	G	
Cyclone Calciner Inlet housing	–	–	–	5	-	3	4	12
Kiln								
Safety zone Calcining zone Upper trans zone	–	4	–	6	2	5	5	22
Burning zone	3	2	5	4	2	3	–	19
Lower trans zone Kiln outlet/hood Nose ring	7	3	10	6	2	9	–	37
Cooler sidewall Bull nose	–	3	–	–	–	7	–	10
Grand total								100

A, thermal load; D, salt infiltration; F, mechanical stress
B, thermal shock; E, redox condition; G, buildup and ring formation
C, liquid infiltration

problems are because of thermo-chemical reasons 28% are thermo-mechanical effect, 19% are due to thermal overloading, 3% are due to redox reactions, and another 15% are attributable to reasons other than those mentioned here.

Statistical analysis of the refractory-related problems and performances, on a large number of cement kilns worldwide, shows (Table 11.17) the frequency of problems occurring in different application areas of the kiln system [3]. The table also shows the effect of different factors, e.g., thermal load, thermal shock, clinker liquid-phase infiltration, salt infiltration, redox condition, mechanical stress, and ring formation on the refractory performances, in different areas of the total kiln system.

References

1. Sengupta P, Refractories for the Cement industry, Springer, ISBN 978-3-030-21339-8
2. Parkes PF (2008) Sourcing Refractories from a Cement manufacturer's point of view, Global Cement Magazine, March 2008
3. Kunneman H et al (2004) Modern Refractory solution for Cement rotary kiln- which brick suits which kiln section, Tehran International Conference on Refractories, 4–6 May, 2004
4. Yiakoumi A (2014) Preheater cyclone protection, International Review, October, 2014
5. Dominguez IR et al (2009) Alternative fuel in the Cement industry: attack and damage of Monolithic refractory in the Preheater zone. In: UNITECR Proceedings, 2009
6. Bovio A (2013) Riser duct castable, Seven Refractories, Italy, 88, ICR March, 2013
7. Von Der Heyde R et al (2012) Compaq sol- the success story continues with gunning mixes and new product development. RHI Bull 2:12–16
8. Rene Von Der Heyede et al (2013) New sol bonded product members – Gunning experiences and shotcasting application. RHI Bull
9. Shneider U et al (2015) No cement sol mixes achieve service life records at Solnhofer Portland Zementwerke. RHI Bull 2
10. Refractory lining for Cement kiln system, Holderbank Cement seminar, 2000
11. Refractory materials and coating formation- Holderbank Cement seminar, 2000
12. Marschall HU, Wiry A Computer aided thermo-mechanical stress modeling of different rotary kiln outlet design. RHI Refractories, Technical Center, Leoben
13. Wirsing H et al (2015) Magnesia brick containing Iron spinel-Troubleshooters for thermo-mechanically stressed kilns. In: UNITECR Proceedings, 2015, Vienna
14. Huixin L et al Simulation analysis on the thermal analysis field of the Cement burner. Wuhan University of Science and Technology, Wuhan
15. Gronwall F, Optimization of Burner kiln 7, Cementa Slite, SLU, Swedish University of Agricultural Science, Department of energy and technology
16. Rene Von der Heyde, The keys to improvement, World Cement, August 13
17. Wagner V, Malkamus P (2013) New type of basic castable for Cement industry. In: UNITECR Proceedings, 2013

Chapter 12
Refractories for Boiler and Waste Heat Recovery

12.1 Introduction

Boiler is an integrated part of many Chemical industries for the production of steam which is used as a reactant ingredient, e.g., in synthesis gas manufacturing.

Many different types of boilers are being used, but here we shall confine our focus on CFB (*c*irculating *f*luidized *b*ed) boiler which is the most efficient type of boiler (Fig. 12.1) [1]. The figure shows the different components of a CFBC boiler.

The function of circulation of the fluidized bed is shown in Fig. 12.2.

12.2 Boiler Components

The different components in the CFB boiler where refractories are used are as given below:

1. Combustion chamber with its wall and ceiling
2. Cyclone
3. Loop seal
4. Bottom Ash Hopper

12.3 Refractories for Different Components

Let us discuss the refractory used in different parts of the boiler in details. Once a CFB boiler is put into operation, the expected refractory life is minimum 10 years with minor repairs, considering the life of the boiler is 25 years. During this period

P. Sengupta, *Refractories for the Chemical Industries*,
https://doi.org/10.1007/978-3-030-61240-5_12

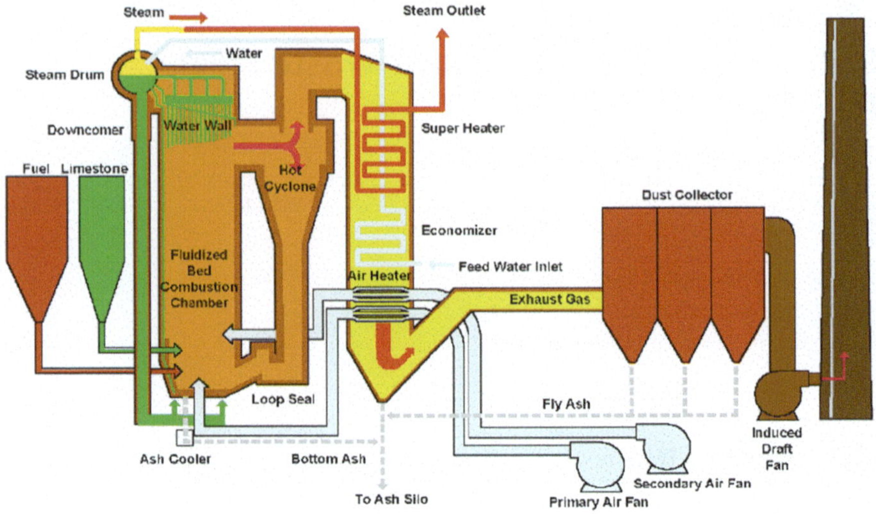

Fig. 12.1 CFB Boiler

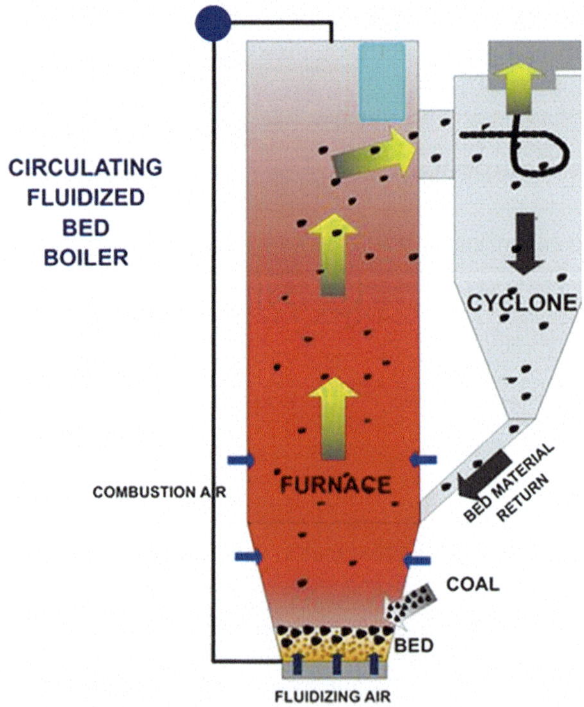

Fig. 12.2 Work function of CFB Boiler

the refractory life should not be a constraint for the boiler availability. The refractory suitable for the lining in boiler should be required to have the following properties:

1. High thermal stability
2. Volume stability at high temperature
3. Chemical stability at high temperature
4. High abrasion resistance
5. Low permeability to gases
6. Optimum thermal conductivity
7. Ease of installation

12.3.1 Combustor Wall

Lower part of the combustor is lined with 70 mm dense and highly abrasion-resistant low cement castable with 30 mm insulation at the back. Similarly the nozzle grate is lined with 100 mm insulation refractory followed by 100 mm dense castable lining. The exit area of the flue gas from combustor to cyclone is 100-mm-thick dense abrasion-resistant castable with 2% of stainless steel fiber as reinforcement or by plastics. The operating condition inside the combustion chamber is shown in Table 12.1, and the properties of the castables are given in Table 12.2.

Table 12.1 Operational parameters of a typical CFB combustor

Fluidized bed area up to bottom secondary nozzle	
Operating temperature	850 ± 50 °C
Operating pressure	200 mbar
Excess air coefficient	0.49–1.0
Gas velocity	Max 6 m/s
Particle loading	1000–1500 kg/Nm^3
Upper part area of the combustor	
Operating temperature	850±50 °C
Operating pressure	0–100 mbar
Excess air coefficient	0.73–1.2
Gas velocity	7 m/s
Particle loading	20 kg/Nm^3

Table 12.2 Castables used in the lining of combustor lower wall

	Castable type		
Properties	Low cement castable	Insulation castable	Plastics
Bulk density (gm/cc)	2.28	1.0 max	2.75
CCS (MPa)	110–130	2–2.5 at 800 °C	60 at 1200 °C
Abrasion resistance(ASTM C-704 test)	<10 cc	–	<10 cc
Al_2O_3 %	45–50	44	82
SiO_2	47	40	P_2O_5–4.5 max
Fe_2O_3 %	0.5 max	1.0 max	1.5

Table 12.3 Castable used for the combustor upper wall and ceiling

	Refractory type		
Properties	Dense castable	Insulation castable	Calcium silicate
Bulk density (g/cc)	2.72	1.0 max	0.22
CCS (MPa)	80	2–2.5	1.5
Abrasion resistance (ASTM C-704 test)	<10 cc	–	–
Al_2O_3 %	70	44	
SiO_2 %	–	40	45
Fe_2O_3 %	0.5	1.0	0.2
CaO %	–	–	45

12.3.2 Combustor Upper Wall and Ceiling

The lining configuration normally used for upper wall is 150–160 mm dense castable, 175 mm insulation castable, and 75 mm calcium silicate lining. For ceiling the lining configuration suggested is the same except the thickness of dense layer is 200 mm. The desired properties are given in Table 12.3.

Water wall or water panels are a very common choice for the upper wall and roof of the combustor. This is also called the membrane wall, and details of it are discussed in Chap. 7. In case of fluidized bed combustion, the ash particles and the bed material used have the effect of severe abrasion on the tubes carrying water. Unless the tubes are protected by a high abrasion-resistant refractory material, the tube wall gets eroded and finally gets leaked (Fig. 12.3) [2]. To protect the steel tubes, normally plastic refractories are used which are applied by tamping with a wooden mallet. The plastic refractory used must have good abrasion resistance and high thermal conductivity to make the cooling effective and also to utilize best the furnace heat.

The plastics applied on the wall have up to a maximum thickness of 40 mm. To hold the material over the tubes, small studs are welded on the tubes as shown in Fig. 12.4a [2, 3]. The specification of this material is shown in Table 12.4. This material uses chemical binder and does not have any hydraulic bond, and therefore its heat up can be much faster than the castable materials.

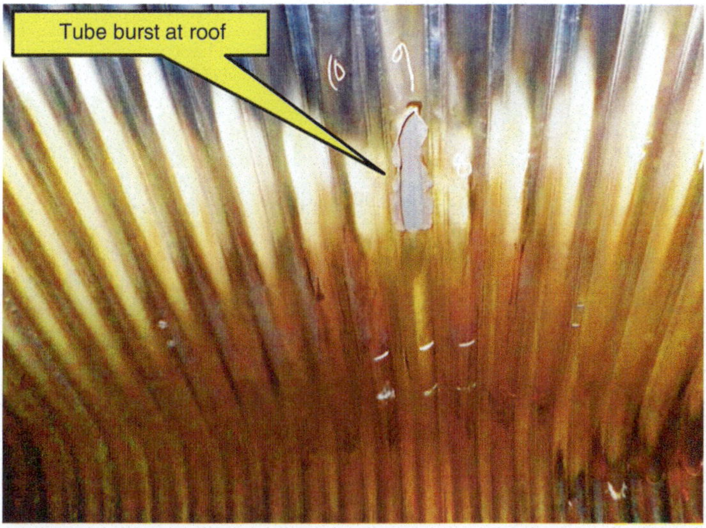

Fig. 12.3 Erosion of the steel tube of the water wall panel

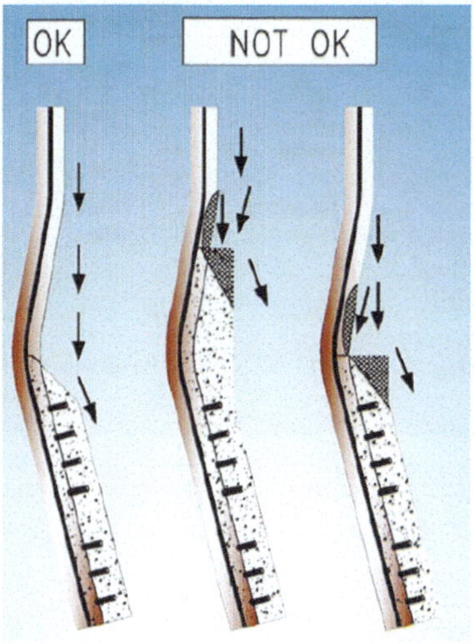

Fig. 12.4 (**a**) Studs welded on the tubes to hold the refractory mass, (**b**) correct way of application of the mass

The application procedure of the plastic mass over the tubes is also very important. When the tubes are bent in the way shown (Fig. 12.4b), the desired profile of the refractory masses to have minimum abrasion by movement of the dust laden gases is shown by arrow. There are references that low cement castables can also be used there, but for such low thickness of the lining, it cannot be used by casting or

Table 12.4 Properties of plastics and low cement castable for water wall panel

Products	Al_2O_3%	Grain size (mm)	B.D. (g/cc)	CCS (kg/cm^2)		Th. cond (W/mk)		Abrasion resistance (ASTM C704)
				110 °C	1000 °C	500	1000	
Alumina plastic	90	5	3.0	1000	1200	2.3	2.35	3 cc
Castable	90	4	3.0	800	1000	2.1	2.15	3 cc

Fig. 12.5 Guniting of refractories over the boiler tubes

trowelling, and it is to be applied by gunning only (Fig. 12.5). The properties of such castable are shown in Table 12.4.

12.3.3 Bottom Lining

In the bottom dense bricks of thickness of 150–160 mm are used with backup insulation brick thickness of 175 mm and calcium silicate slab of 75 mm thickness. The specification of the bricks is given in Table 12.4.

12.3.4 Cyclone

Solids, suspended in flue gas, entered from the combustor to cyclone are separated here from the bulk gas. The wall which is tangent to the cyclone and the target wall (Fig. 12.6), where the gas stream hits first, are especially vulnerable to high abrasion and wear [4] (Table 12.5). The operating condition of target area and the specification of the suitable refractory for target area are given in Tables 12.6 and 12.7.

The dense refractory layer is recommended as 150–160 mm with 175-mm-thick insulation layer and 75 mm of calcium silicate layer. The exfoliated vermiculite base brick is recommended as insulation brick layer.

12.3.4.1 Cylinder and Cone Wall of Cyclone

Cylinder and cone wall of cyclone are lined normally with low iron fireclay brick of same quality as it is recommended for the combustor bottom. Similarly the same insulation brick quality and calcium silicate with same thicknesses are recommended.

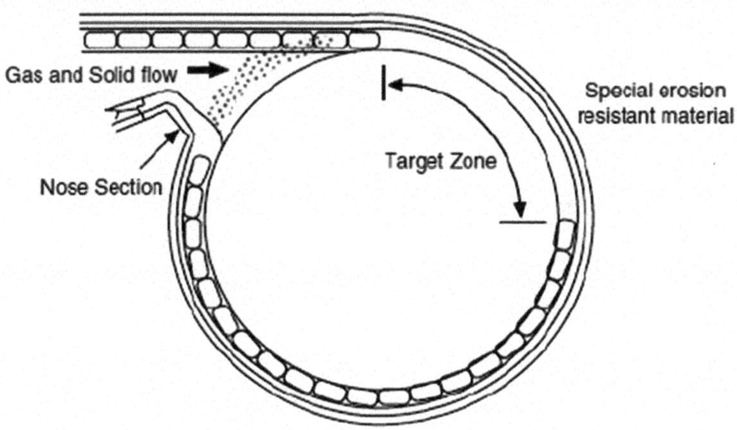

Fig. 12.6 Cyclone target area

Table 12.5 Specification of fireclay bricks for the bottom lining

Specification of fireclay bricks for the bottom lining		
Bulk density (g/cc)	2.22	0.75
CCS (MPa)	50–60	2.0
Abrasion resistance (ASTM C-704 test)	<14 cc	Base material – exfoliated vermiculite
Al_2O_3	35	13
Fe_2O_3 %	1.5	2.5–3.5
Na_2O %	0.2	
K_2O %	1.7	

Table 12.6 Operating conditions in target area in cyclone

Temperature	950 °C
Pressure	−25 to 40 mbar
Excess air coefficient	1.2
Gas velocity	6–28 m/s
Particle loading	1–15 kg/Nm^3

Table 12.7 Specification of refractories suitable for cyclone target area

Bulk density (g/cc)	2.5–2.6
CCS (MPa)	80–90
Abrasion resistance	<7 cc
Base material	Andalusite, corundum
Al_2O_3 %	58–60
SiO_2 %	36 min
Fe_2O_3 %	<1.0
P_2O_5 %	2.6 max

Table 12.8 Specifications of the castables for cyclone ceiling

	Dense castable	Insulating castable
BD (g/cc)	2.70	1.0
CCS (MPa)	80	7.0
Abrasion resistance	<10 cc	–
Al_2O_3 %	70	38
SiO_2 %	17	Alkali – 1.5
Fe_2O_3 %	0.5	2.5
Base raw material	Corundum	Insulating grog, perlite

For the ceiling of the cyclone, low cement dense abrasion-resistant castable is recommended with 200 mm thickness and backup insulation castable layer of thickness 175 mm and 75 mm calcium silicate. The specifications are given in Table 12.8. The insulating castable here should be designed looking into its application by gunning.

In case of bullnose construction, the same dense castable is used with additions of 2% stainless steel fiber by weight, which improves its abrasion resistance and toughness.

12.3.5 *Seal Pot*

It is an important area in the CFB boiler which serves as the nonmechanical seal between the cyclone, working under ambient pressure, and the lower section of the combustor, working under slight overpressure, and on the other hand between

Table 12.9 Specification of castables for seal pot

	Dense castable	Insulating castable A	Insulating castable B
BD (g/cc)	2.3	0.5	0.4
CCS (MPa)	80–90 at 800 °C	2.5	1.5
Abrasion resistance (ASTM C-704 test)	<8 cc	–	
Al_2O_3 %	50–52	9–10	14–16
SiO_2 %	42	48–50	65–67
Fe_2O_3 %	0.8	5–6	4–4.5
CaO %	4.0 max	12–13%	5–6
Base raw material	Sillimanite/ andalusite	Vermiculite	Perlite
Layer thickness (mm)	150–160	125	125

Table 12.10 Specification of castables for the FBHE chamber

	Dense castable	Insulating castable
BD (g/cc)	2.3	0.75
CCS (MPa)	90 at 800 °C	0.75
Abrasion resistance (ASTM C-704 test)	<8 cc	–
Al_2O_3 %	50–52	30–32
SiO_2 %	42	30
Fe_2O_3 %	0.8	8 max
CaO %	4.0 max	20–22

cyclone and FBHE. The operating conditions here are temperature, 950 °C; pressure, 250 mbar; gas velocity, 0.5–1.5 m/s; and heavy particle loading. The lining configuration and quality are shown in Table 12.9.

Part of the heat generated in the combustion can be transferred in the fluidized bed heat exchanger (FBHE). The part of the hot ash from the seal pot can be diverted to FBHE to extract its heat. The wall of the chamber housing the heat exchanger tubes is lined with 150 mm dense castable with 200–250 mm insulating castable at back. The specifications of the castables used here are shown in Table 12.10.

12.3.6 Ash Cooler

Bottom ash from combustor and the ash from FBHE are cooled in the ash cooler for further transport. The refractory to be used especially in the inlet and outlet chamber must have excellent thermal shock resistance and high wear resistance because the ash flow is discontinuous. The maximum temperature in this area is 850 °C and pressure is 250 mbar, and the gas velocity remains in the range 0.3–20 m/s. The

Table 12.11 Refractories for the ash cooler

	Application			
	Wall		Ceiling	
Properties	Dense castable	Insulating brick	Dense castable	Insulating castable
BD (g/cc)	2.3	0.5	2.70	1.0
CCS (MPa)	90 at 800 °C	2.0	80 at 850 °C	6–7 at 110 °C
Abrasion resistance (ASTM C-704 test)	<8.0 cc	Chemically bonded	<8.0 cc	–
Al_2O_3 %	50–52	9.0	70	38
SiO_2 %	40–42	48–51	17	36
Fe_2O_3 %	1 max	5–6	0.5	2.6
CaO %	4.0	MgO-24-27		
Base raw material	Sillimanite Andalusite	Exfoliated vermiculite	Corundum	Fireclay Perlite
Thickness (mm)	125	175	100	100

specification and thicknesses of the castables used in the wall and ceiling are given in Table 12.11.

12.3.7 Flue Gas Duct Between Cyclone and Backpass

Considering the stability of brick lining in circular cross section, the lining of circular duct is made with dense bricks. The intersection of the circular and the rectangular ducts and the wall and ceiling of rectangular duct is lined with castables. The temperature which exists here is 850 °C, gas velocity is 30 m/s, and particle loading in the gas is about 0.1 kg/Nm^3. The specification of the refractory brick and castable used for this duct and the thickness recommended are given in Table 12.12.

12.3.7.1 Ceiling of the Duct

The refractory lining configuration recommended is same as that of cyclone ceiling. Only in case of working lining the thickness can be reduced to 150–160 mm.

12.4 Waste Heat Utilization for the Power Generation

At present about 20–50% of energy used by the industry is discharged to atmosphere as waste heat. This waste heat can be used for making of steam which can be used for the purposes mentioned.

Table 12.12 Refractories used to line the ducts

Properties	Brick type Wall, bottom, and circular duct Dense brick	Insulating brick 1	Insulating brick 2
BD (g/cc)	2.2	0.7	0.23
CCS (MPa)	50–60	2.0	1.7
Abrasion resistance (ASTM C-704 test)	<12 cc	–	–
Al_2O_3 %	35–40	12	–
SiO_2 %	55	35	45
Fe_2O_3 %	2	3	0.2
CaO %	–	20–23	45
Base raw material	Chamotte	Vermiculite	Calcium Silicate
Thickness (mm)	160	175	75

The feasibility of utilization of waste heat needs characterization of the waste heat source. The important technical parameters which need to be analyzed are:

1. Temperature of the waste heat source.
2. Total heat of the stream available, which can be known from its flow rate.
3. Composition of the stream gas or liquid. It is corrosive or contains solid particle that may abrade the equipment.
4. Operating schedule and its availability whether continuous or intermittent.
5. The dew point of condensate and its nature.

12.5 Waste Heat Recovery

The waste heat temperature is the key factor to decide the usefulness of the waste heat. As per the thrmodynamic principle higher is the temperature and more is the difference with the sink temperature, better it is for running the engine that will produce the power. Figure 12.7 [5] gives an idea of wastages of energy from a petroleum refinery. The eight different units in the refinery have been chosen which consume 68% of the energy requirement of the refinery, and unit wise the energy getting wasted is shown.

In the financial analysis, the total cost of the plant, payback period, rate of return, and the depreciation of the plant are most important considerations for taking decisions to put up a plant.

Table 12.13 gives an idea of the temperature of waste gas from some relevant industries [6, 7]. If the volume of the gas is known with its specific heat, then total energy wasted through the emitted gases can be calculated.

For example, in cement industry, the waste heat is obtained from the cooler, and the preheater can generate almost 20% of the need of the plant.

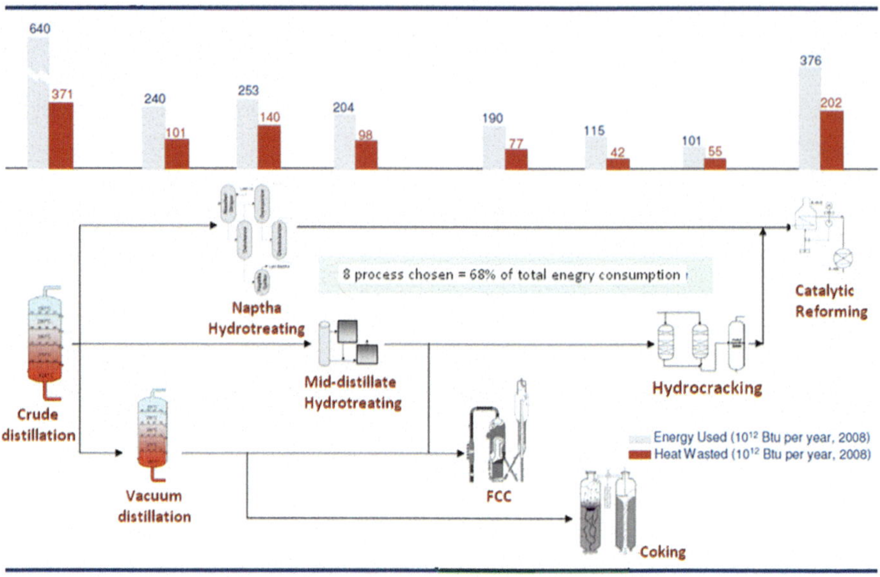

Fig. 12.7 Energy being wasted in different units of a petroleum refinery

Table 12.13 Temperature of flue gases emitted in different industries

Source industry	Temperature of waste stream (°C)	Characteristics	Exhaust gas flow
Glass industry	430–1430	Flue gas containing particulates	Constant
Cement industry (dry process)	450–620	-Do-	Constant
Lime calcination	200–600	-Do-	Constant
Carbon black		-Do-	Constant
Petrochemical	120–255	-Do-	Constant

The waste heat recovery system consists of the suspension preheater boiler to utilize the sensible heat which comes out from the preheater, and the air quenching cooler (AQC) boiler to utilize the sensible heat which comes out from the air after its heat exchange with hot clinker to cool it down fast. Generated steam from both the units goes to turbine generator to produce electrical power (Fig. 12.8) [8].

In the waste heat recovery system, the major use of refractories is in the lining of ducts to transport gases. The gases are being transported at high velocity, and the gases contain particulate matters. In most of the cases, the working lining is thin with thickness of 50–75 mm made of high abrasion-resistant low cement dense castable and the thicker backup lining with thickness of 75–150 mm to maintain the skin temperature of around 60 °C. Calcium silicate board or curved sections to match the contour of duct are used as insulation material.

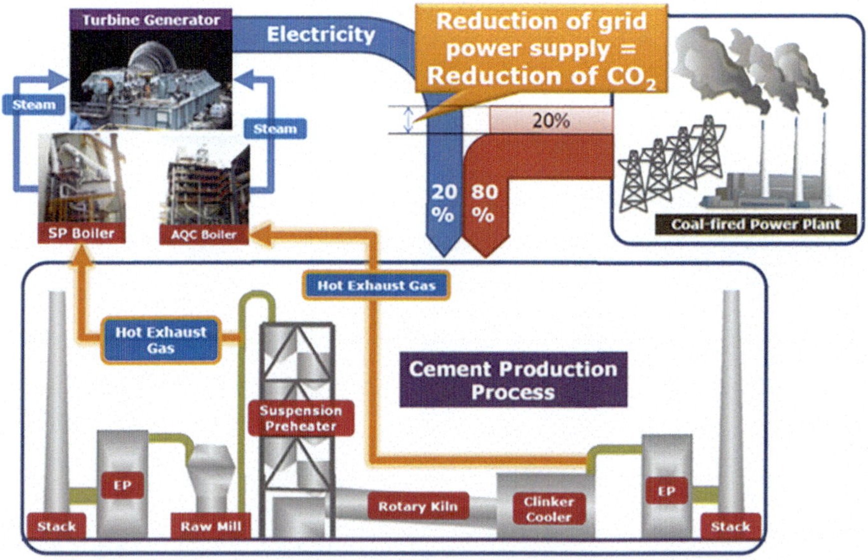

Fig. 12.8 Schematic drawing utilization of waste heat from cement plant to produce power

Table 12.14 Potential of power generation from the waste heat of a 6000TPD cement plant

Description	5-stage preheater	6-stage preheater
Total power input (MW)	213	207
Power utilized for clinkerization (MW)	111	111
Power lost in (MW)		
Preheater gases	44	40
Cooler gas	31	31
Radiation	21	19
Clinker heat	6	6
Total loss (MW)	102	96

Table 12.14 shows the waste energy and the possible utilization of that energy into the power [9].

References

1. Dr. Passi Makkonen. Control of wear in fluidized bed combustion. Foster Wheeler
2. Parthiban KK. Factors that affect the boiler tube erosion. Venus Energy Audit System, India
3. Kavidass S et al (1997) Overview and the status of the first 25 MW IR CFB boiler in India. In: 14th international ASME conference on Fluidized Bed Combustion May 11–14, 1997, Vancouver, BC, Canada
4. Kinkar AS et al (2015) Refractory failure investigation in CFBC Boiler. Int J Eng Res Gen Sci 3(1)

5. Executive summary– waste heat recovery opportunities in selective industries – Frost and Sullivan, 2010
6. Waste heat recovery – technology and opportunities in US industry – US Department of Energy, 2008
7. Industrial waste heat recovery – Benefits and recent advancements in technology and application, Cecilia Arzbaecher, Global Energy Partners, ACEEE, Summer study on energy efficiency in industries, 2007
8. Gupta SK, Kaul SK. Waste heat recovery power plants in cement industry. HOLTEC Consulting Pvt Ltd., India
9. Waste heat recovery for the Cement sector-International Finance Corporation, World Bank Group, June, 2014

Chapter 13
Refractory Performance and Mechanism of Damage

13.1 Introduction

Refractory performances in any industry depend largely on many other factors, besides the quality of refractories, as shown in Fig. 13.1 [1]. The diagram clearly shows that quality of refractory, quality of its storage and installation, mechanical features of the kiln or furnace, and operating condition, e.g., chemistry of the feed to the kiln, atmosphere inside the kiln, and the thermal parameters, play a very important role in deciding the refractory performance. The selection of the right kind of refractory also is very important, and selection sometimes is made by the user by providing a specification, and sometimes the quality is suggested by the refractory supplier, and in some other occasions, it is decided jointly.

According to Fig. 13.1, the factors, those that govern the performance of refractories, are mostly in control of the user of refractories, rather than in control of supplier of refractories. The best refractories also cannot provide desired performance, if the installation is improper or the operating conditions prevailed are not friendly to refractories. The storage quality of the refractories is particularly important for basic bricks and monolithic refractories. The influence of the chemistry of input material and different thermo-mechanical parameters of the kiln operation, on the refractory performance, are shown in Fig. 13.2 [2, 3].

Experience shows that in overwhelming majority of the cases, there is never one, definite cause, but instead a "chain of events" that lead to the "failure." In other words, there were multiple variables that impacted the refractory material in service [4].

The refractory material is only one component in the overall design of process equipment and industrial furnaces. It is sensitive to (and influenced by) time, tem-

P. Sengupta, *Refractories for the Chemical Industries*,
https://doi.org/10.1007/978-3-030-61240-5_13

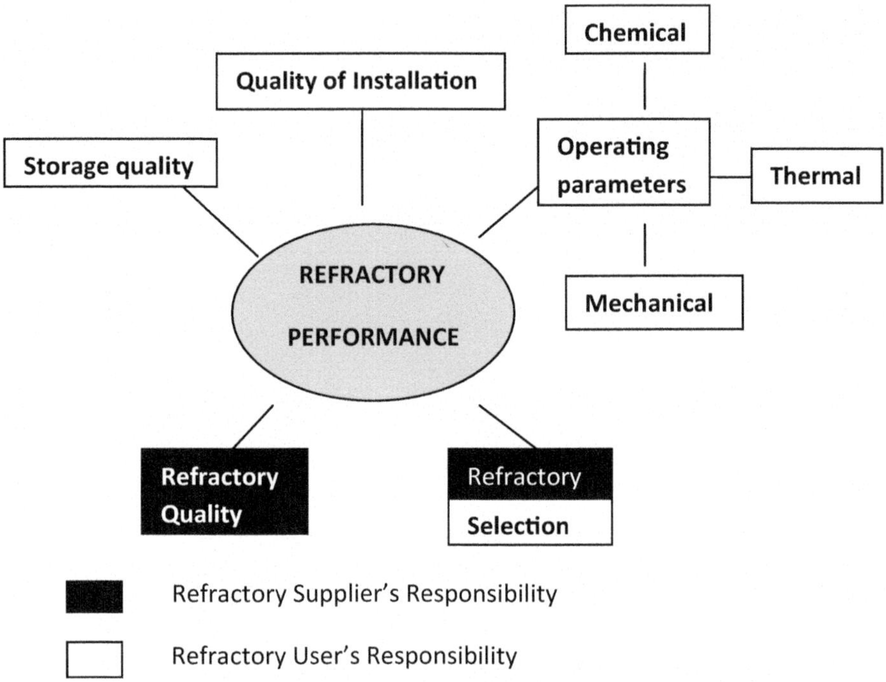

Fig. 13.1 Diagram showing responsibility distribution between refractory supplier and user

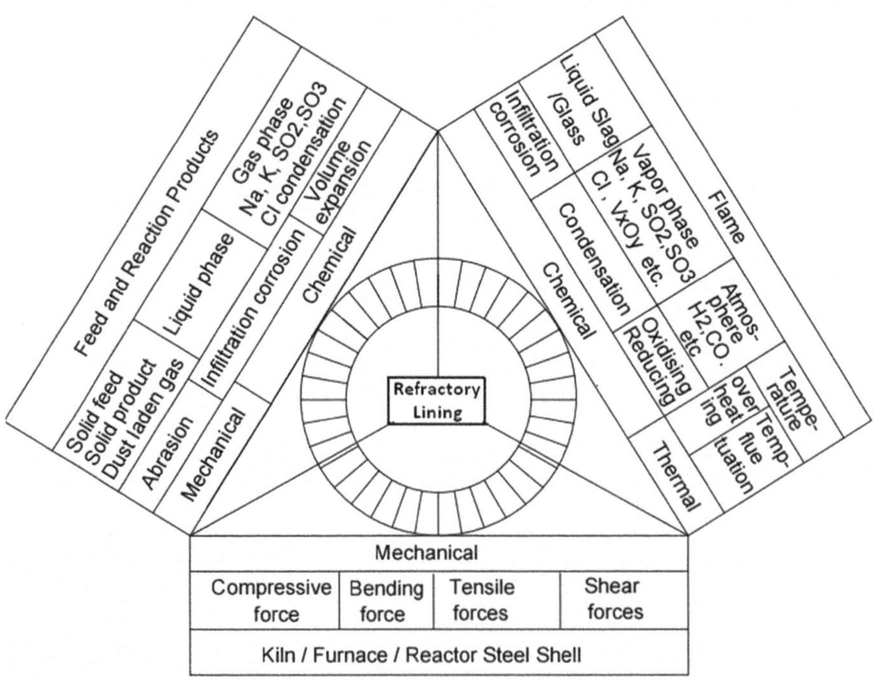

Fig. 13.2 Effect of different factors on the performance of rotary kiln refractory lining

perature, and operational factors. The experiences say the culprit will likely be one (or a combination) of the following:

1. Improper combustion (fuel, ratios, atmospheres, flow, etc.)
2. Flame impingement (flame geometry, improper set-up, etc.)
3. Improper furnace controls (PLC's controls, circuitry, thermocouples, dampers, etc.)
4. Operational change (cycling, throughput, load, etc.)
5. Operational procedures
6. Maintenance practices
7. Shell and lining movement due to a difference in expansion rates
8. Mechanical stresses
9. Environmental factors
10. Catalysts

From a refractory perspective, all these factors have an impact on refractory and can cause:

- Excessive temperatures (lead to shrinkage, etc.)
- Hot spots
- Excessive cracking due to spalling, thermal shock (a result of temperature fluctuations), mechanical shock or compressive forces, linear change, bending, etc.
- Structural issues
- Erosion and abrasion of the refractory lining
- Anchor failure
- Failure of the steel shell which supports the refractory
- Mechanical damage

All of the above factors can cause the refractory material to end its campaign life sooner than originally expected. As variables do change, the selection of the refractory used may need to be reconsidered. If there are new variables that the product must face, then an upgradation or change in refractory lining may be warranted.

The performance of refractories are industry specific; in some furnaces or kilns, it lasts for several years (e.g., Lime calcinations Kiln), but in another industry, it may last only for a year (burning zone in cement rotary kiln). But the basic factors to decide the refractory life are same, i.e., the degree of temperature, the extent of cyclic temperature fluctuation, the volume and properties (physical and chemical) of the slag or the gases in contact with refractory, chemistry of environment (oxidizing, reducing, alkali vapor, etc.) inside the furnace, and steadiness of operational conditions. The operational conditions that refractories are exposed to in furnaces, kilns, reactors, ducts, etc. in different types of chemical industries mentioned before confirm this. Under different conditions, how different types of refractories behave and the factors influence its performances will be discussed next.

13.2 Thermo-chemical Effects

Primarily furnace or kiln temperature is the most important operational parameter to decide the refractory performance, and higher operational temperature initiates more problems with refractory lining.

Rate of chemical corrosion increases exponentially with rise in temperature, and similarly the thermo-mechanical stresses also increase with temperature. Therefore to get output of desired quality, optimization of operational temperature is very important, and, for that, control of composition of feed material is most essential. For example, in cement industry, some very proven tool used to control the feed chemistry, vis-à-vis the firing temperature of the rotary kiln, is the control of different cement moduli, which are given in tabular form (Table 13.1).

At high temperature, especially, the refractories in the kiln or furnace react with different input materials in the system through the feed and fuel. The reaction takes place in solid-solid, solid-liquid, and solid-gas phases which ultimately causes the refractory wear and reduces the campaign life of refractory lining. The industrial furnaces or kilns are quite large, and the conditions prevailing inside the furnace are not uniform everywhere inside. The mechanism of degradation of refractory varies with the type of refractory and its place of use inside the furnace. The reactions responsible for degradation of refractories are mainly due to solid-liquid and solid-gas phase reactions.

Different chemical ingredients enter into the kiln either through the input raw material or through fuel and take part in different chemical reactions with the refractory lining at high temperature. The mechanism of reaction of refractory and the

Table 13.1 Desired values of different moduli for optimum control of the process

Moduli	Definition	Optimum value	Significance
Lime standard	Content of CaO available to combine with SiO_2, Al_2O_3, and Fe_2O_3	85–95	A value near 100 indicates hard burning material and increases thermal load
Hydraulic modulus (HM)	$CaO/(SiO_2 + Al_2O_3 + Fe_2O_3)$	1.7–2.3	Higher value of HM indicates requirement of more energy or higher thermal load
Silica ratio (SR)	$SiO_2/(Al_2O_3 + Fe_2O_3)$	1.9–3.2	Characterizes the solid phase to liquid phase ratio in the clinker
Alumina ratio (AR)	Al_2O_3/Fe_2O_3	1.5–2.5	Increase in AR indicates lower content of liquid phase and hence less coating which ultimately exerts more thermal load on the lining
Alkali sulfate ratio (ASR)	Mole ratio of alkali to sulfates $ASR = \frac{(Na_2O)+(K_2O)-(Cl_2)}{(SO_3)}$	0.8–1.2	If ASR is <1, then SO_3 is in excess to react; if ASR is >1, free alkali is available to react with refractory. When ASR = 1, no excess of either SO_3 or alkali is there

liquid phases formed in the product, can be divided into three different stages, for a better understanding of the reaction, namely:

1. Penetration of the liquid into the brick pores
2. Wetting of the refractory by the liquid phase
3. Chemical reaction with the brick matrix and grains and dissolution, in the liquid phase

The depth of slag penetration, l in the refractory, can be quantified by the expression [5, 6]:

$$l^2 = \frac{r\cos\theta}{2} . \frac{\nu t}{\eta} \tag{13.1}$$

r = pore radius, ν = clinker liquid surface tension, θ = contact angle of clinker liquids and the refractory, η = viscosity of the liquid phase, t = time.

The penetration of the liquid phase can be suppressed by increasing the viscosity or the contact angle and the decrease of the surface tension of the liquid phase. The values of the variables, i.e., η, θ, and γ, change as the reaction progresses.

The reaction products of the abovementioned reactions can be solid, liquid, or gases. The reaction products may stick to the solid surface to form a barrier for further attack of the liquid or gas. If the reaction products are liquid and gas and get removed from the surface of reaction easily, the solid surface is easily available for further reaction. The reaction may progress very fast when the reaction products are soluble in the liquid phase. The corrosion rate is the function of many factors including temperature solid-liquid interface composition liquid density, viscosity, diffusivity and degree of turbulence.

Aluminosilicate Refractories Aluminosilicate refractories contain Al_2O_3 from 30 to 85% and are used widely in all type of kilns and furnaces mentioned earlier.

The damage of this kind of refractory usually happens due to reaction with slag, glass, or the liquid phases developed in the product (e.g., cement clinker), with alkali vapors and with reducing gas (e.g., CO or H_2).

Considering the slag, glass or cement raw mix contains mainly four oxides, i.e., CaO(C), SiO_2(S), Al_2O_3(A), and Fe_2O_3(F); the liquid phase starts appearing at 1338 °C, i.e., the eutectic point of the C-S-A-F system. At this eutectic temperature, the liquid which is formed has a chemical analysis of CaO – 55%, SiO_2 – 6%, Al_2O_3 – 23%, and Fe_2O_3 – 6%. This liquid can take more SiO_2 in solution, and therefore it can attack the refractory containing SiO_2 and can corrode it, taking it into solution. The impurities in the raw mix like MgO, Na_2O, K_2O, and SO_3 can further bring down the liquidus to 1280 °C and make it more corrosive.

The liquid phase (23–30%) in the cement clinker in burning zone or the slag in the coal gasifier or the liquid glass penetrates through the brick pores and causes densification of the refractory, which increases its modulus of elasticity and makes the refractory brittle.

Moreover, the densification changes the physical properties, e.g., thermal expansion coefficient (α) of the densified portion of the refractory differs from the rest of the brick, and during heating and cooling of the bricks, crack develops because of dissimilar α value.

As shown in the Eq. 13.1, the viscosity of the liquid melt is a very important parameter to determine the extent of infiltration of the refractory by the liquid phase, and the lower is the viscosity, the more will be the infiltration. The viscosity decreases rapidly with temperature rise. It is found that increase of temperature by 93 °C reduces viscosity of the clinker liquid phase by 70% [3]. Presence of MgO, alkali sulfates, fluorides, and chlorides in the clinker also helps in reducing the viscosity. For example, a regular clinker liquid phase has a viscosity of 0.16 N.s/m^2 which is reduced to 0.05 N.s/m^2 in presence of 2% SO_3 in the clinker [7]. Free alkali and phosphorus increases the viscosity of the melt but is offset by presence of MgO and SO_3.

Another important property of the liquid is its surface tension which indicates its ability to wet the refractory surface. A liquid phase with higher surface tension has lesser ability to wet the refractory surface and to corrode it. The wetting angle of glass on refractory is governed by the equation:

$$\cos\theta = \frac{\sigma_{sv} - \sigma_{sl}}{\sigma} \tag{13.2}$$

θ = the wetting angle
σ = the surface tension of the glass
σ_{sv} = the surface energy of the refractory
σ_{sl} = the surface energy between the refractory and the glass melt

It has been found that the corrosion of the refractory is lessened if a reducing atmosphere can be maintained at the glass line in the glass tank furnace. It has been demonstrated that the surface tension of a glass melt at 1000 °C has been increased from 76.2 in air to 102.5 in an oxygen-deficient atmosphere [8, 9].

13.2.1 Interaction with Alkalis

Alkali compounds remain as impurities in all kind of naturally occurring raw materials used for different chemical process industries like Portland cement, lime calcinations, gasification, etc. It is also used as a raw material for certain process industries like glass. It enters into the reactor or furnace also through solid fuels like coal, biofuels, and alternate fuels used for cement manufacturing. The alkali gets circulated inside the kiln, and gradually its concentration increases. In case of cement rotary kiln, the alkali gets vaporized from the burning zone and is carried to the calcination zone, where it gets condensed and again is carried to the burning zone with the feed and gets evaporated, and this cycle continues (Fig. 13.3).

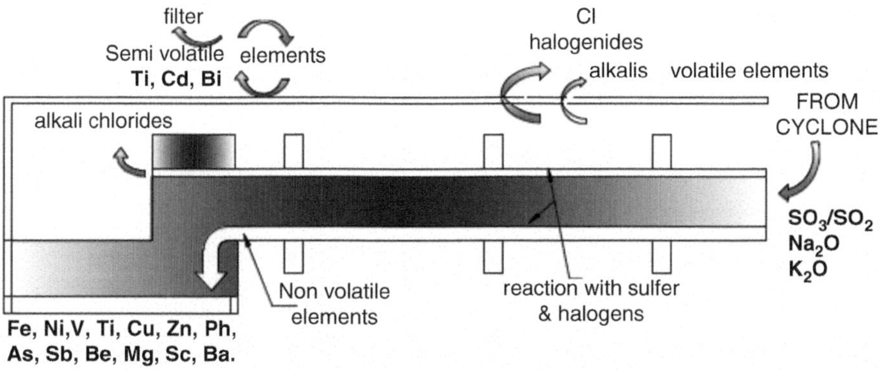

Fig. 13.3 The alkali cycle in cement rotary kiln

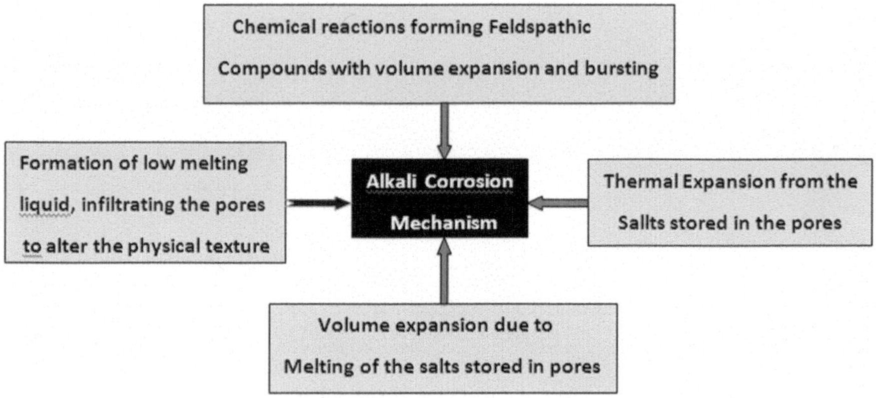

Fig. 13.4 Modes of destruction of refractories by alkali

The mechanism of destruction of refractories on interaction with alkalis can be shown schematically in Fig. 13.4. The alkali in the kiln combines with chlorine and oxides of sulfur to form alkali chlorides and sulfates, and their mixtures form low melting liquid phases reacting with refractories at high temperature, as shown in Table 13.2.

These liquids get infiltrated inside the pores of the refractory and form a densified layer with altered thermal properties and during heating and cooling, crack forms at the junction with unaltered brick causing gradual destruction of the lining.

The alkali salts in the vapor phase get infiltrated in the brick pores and get condensed inside the pores of the bricks in colder zone. Thermal expansions (α_{lin}) of these salts (Table 13.3) are much higher than the refractory body, and during heating and cooling, they exert pressure on refractories causing cracks [10–15]. These salts and the liquids get infiltrated in the brick pores and alter it physically to reduce its MOE and to change its coefficient of thermal expansion to make the refractory more brittle and prone to thermal spalling.

Table 13.2 Melting behavior of mixture of alkali salts

Refractory oxide	Alkali compound	Melting temperature (°C)
Al_2O_3	Na_2O	1410
	K_2O	1450
SiO_2	Na_2O	789
	K_2O	742
$Al_2O_3 + SiO_2$	Na_2O	732
	K_2O	695
$CaO + SiO_2$	Na_2O	725
	K_2O	720
TiO_2	Na_2O	986
	K_2O	950
	$K_2SO_4 + K_2O$	804
$MgO + SiO_2$	Na_2O	713
	K_2O	685

Table 13.3 Thermal expansion coefficients of different alkali compounds and their mixtures

Solid salt	α_{lin} measured 10^{-6} 1/K (20/600 °C)
KCl	52
K_2SO_4	90
K_2CO_3	58
Ca SO_4	16
(K_2SO_4/K_2CO_3)	58
$(K_2SO_4/K_2CO_3/KCl)$	50
$(K_2SO_4/K_2CO_3/KCl/CaSO_4)$	34

There is a difference in the density of the solid and molten alkali salts, accompanied by a volume change (Table 13.4). As a result when the refractory lining, containing the solid salts deposited in the pores, is heated and cooled, the melting and solidification of the salts exert a pressure on the refractory body, and repetition of this phenomena causes crack to the body.

The alkali salts react with alumina, silica, and mullite at high temperature with formation of different feldspathic compounds associated with large volume expansion which ultimately destruct refractory (Table 13.5). If alkalis are very high and are not balanced by sulfur, it will be very difficult for them to exit the kiln. The alkalis will therefore continue to recirculate within the kiln/preheater system and increase the probability of kiln rings and preheater buildups. Clinker quality may suffer because free alkalis can enter into solid solution within the clinker minerals affecting their reactivity.

The abovementioned mechanism of alkali vaporization and destruction of refractory is also applicable for the other furnaces, where substantial quantity of alkali is

Table 13.4 Densities of alkali salts in solid and liquid state

Solid salt	Density of solid (g/cm^3)	Density of melt (g/cm^3)	Volume increase (%)
KCl	1.99	1.52	31
K_2SO_4	2.66	1.89	41
K_2CO_3	2.43	1.96	24

Table 13.5 Densities of products formed due to alkali reaction on aluminosilicate refractories

Refractory oxide	Density (g/cc)	New compound formed after reaction with alkali	Density (g/cc)	Volume change (%)
Al_2O_3	3.99	$(N,K)_{1...6}A_{1...11}$	2.63–2.42	+17 to +52
SiO_2	2.65	$(N,K)_{1...3}S_{1...4}$	2.26–2.96	−10 to +17
3 Al_2O_3.2 SiO_2	3.17	$(N,K)_{1...3}AS_{1...6}$	2.40–2.62	+21 to +32
		$N_3CA_3S_6.(SO_4)$		

Table 13.6 Dust emission from glass furnaces in mg/Nm^3

	Oil-fired container glass	Gas-fired container glass	Gas-fired flat glass
Na_2SO_4	140	180	100
$NaHSO_4$	140	–	–
NaCl	–	–	20
K_2SO_4	5	5	100

Based on 8 Vol % O_2 in stack gas

present in the system, for example, glass furnace and regenerator. The vaporization of alkali from the glass bath is enhanced by water vapor in the atmosphere:

$$Na_2O(\text{glass}) + H_2O(\text{gas}) = 2NaOH(\text{gas}) \tag{13.3}$$

$$Na_2O + SO_2 + ½\,O_2 = Na_2SO_4 \tag{13.4}$$

The emission of different compounds from the glass industry is shown in Table 13.6. The size of the sodium sulfate particles varies from 0.05 to 20 μm. The reduction of particulate emission is possible:

- By changing melting behavior, it is found that reduction of melting temperature by 30 °C reduces emission by 30%.
- By batch pretreatment, like agglomeration of the batch.
- By filtration process separating large amount of dust.

The dust and chemicals emitted from the furnace get deposited in the checker brickwork in regenerator, and they reduce the thermal efficiency of the regenerator almost 4% every year [16].

The thermodynamic calculations show that alpha and beta alumina are more stable compared to silica at high temperature in the presence of alkali vapor [17].

Table 13.7 Melting points of the reaction products of alkali and basic refractory

Refractory oxide (m.p.)	Alkali compound	Temperature of melting
MgO (2840 °C)	Na_2O	No miscibility
	K_2O	No miscibility
	Na_2SO_4	1067 °C
	K_2CO_3	895 °C
CaO (2580 °C)	Na_2O	No miscibility
	K_2O	No miscibility
	NaCl + KCl	645 °C
	$CaO + CaSO_4$	1365 °C
Cr_2O_3 (2200 °C)	K_2O	669 °C
	$KCl + K_2O$	366 °C
$MgO + SiO_2$ (1543 °C)	Na_2O	713 °C
	K_2O	685 °C
$CaO + SiO_2$ (1436 °C)	Na_2O	725 °C
	K_2O	720 °C

Table 13.8 Expansion due to reaction of alkali with different minerals in basic refractory

Refractory oxide	Density (g/cc)	Reaction product formed	Density (g/cc)	Volume change
Cr_2O_3	5.25	Na_2CrO_4	4.36	+20%
$MgO.Al_2O_3$	3.55–3.70	$NM_{0.8...4.0}A_{5...15}$	3.28–3.33	+ 7 to +13%
$2MgO.SiO_2$	3.22	$(N,K)_{1...2}M1..._5$	2.56–3.28	−2 to +23%
$CaO.SiO_2$	2.92	$(N,K)_{1...2}C_{1...23}S_{1...12}$	2.72–3.36	−13 to +7%

Basic Refractory The major ingredients of basic refractories react with alkalis to form low melting liquid phases. These liquids penetrate inside the brick pores and densify the brick hot face. Table 13.7 shows the composition and temperature of melting of the liquid phases.

The eutectic of $MgO\text{-}K_2SO_4$ is 1067 °C, and eutectic temperature of $MgO\text{-}K_2CO_3$ is 895 °C. Therefore the MgO-based refractories are not really alkali resistant. Ingredients of basic refractories react with alkalis, and some are associated with considerable volume changes (Table 13.8) causing destruction of the refractory [13, 18].

13.2.2 *Interaction with Sulfur and Chlorides*

The major source of sulfur in manufacturing of cement clinker, glass, syngas or calcined lime is petroleum coke and coal and its content varies with origin. The lower price and higher calorific value has made petcoke a very attractive fuel for cement manufacturing. The obvious effect of using petcoke as fuel is the disturbance of the alkali/sulfur ratio, and the raw meal formulation needs to be changed to maintain the desired ratio in cement manufacturing.

Most cement plants burning petcoke have severe sulfur builtup problems revealing the sulfur is not being adequately removed from the system. The most obvious effects of the high concentration of sulfur are, in the amount and hardness of the coating, sulfur ring formation and tower builtups. All these contribute to a reduction in gas flow, which, without adjustment, causes reduced oxygen levels and a worsening effect. It is estimated that a minimum of 15% flow is required to the bypass to reduce the sulfur compared to only about 2% to reduce chlorides to a "refractory" acceptable level. The removal of the buildup by mechanical means damages the refractory lining. The prevention of buildup requires control in alkali/sulfur ratio and oxygen level.

Minimum availability of chlorides helps to prevent buildup [19]. The reducing condition in the kiln increases the volatility of sulfur, and excess oxygen decreases it. All these facts support the statement made, in the beginning of this chapter that how important the effect of operational parameters on the refractory performance.

The fluctuation in quality and quantity of fuel in the process industry can have a major impact on the refractories. The variation can be direct like the thermal input can fluctuate the temperature which may cause thermal spalling of the refractory.

The indirect effect is found in cement rotary kiln burning zone, where the stability of coating is affected by both qualitative and quantitative changes of fuel, and it is a known fact that stability of coating in burning and transition zone of cement rotary kiln has a very strong effect on refractory performance. Every instance the coating is dropped from an area of brick; the brick is subjected to an instant thermal shock. Also when the coating falls, the brick is directly exposed to the process heat and environment. This allows the volatiles (sulfur/alkali) to condense on the refractory surface and penetrate deeper into the brick. Due to the higher temperature at the brick hot face and the additional elements available to react with, a liquid phase will result on the brick surface and destroy the brick properties in all affected areas. Upon repeated exposure the liquid phase will penetrate deeper into the brick. It is also possible that if a liquid phase is formed and then quenched, a slick glass layer will form making the surface unable to adhere to further coating. There is no direct evidence that sulfur oxides chemically attack alumina or silica at high temperatures and these oxides are stable in the presence of sulfur oxides.

Free SO_2 and SO_3 in the presence of O_2 when exposed to basic refractories at different temperatures for different time periods, it is found to react with grains and matrix to form $CaSO_4$, $MgSO_4$, and $CaMg_3(SO_4)_4$. Different spinel phases react with sulfur oxides at temperature above 700 °C:

$$2SO_2 + O_2 = 2SO_3 \quad (13.5)$$

$$MgO + SO_3 = MgSO_4 \quad (13.6)$$

$$MgCr_2O_4 + SO_3 = MgSO_4 + Cr_2O_3 \quad (13.7)$$

$$MgCr_2O_4 + 4SO_3 = MgSO_4 + Cr_2\left(SO_4\right)_3 \quad (13.8)$$

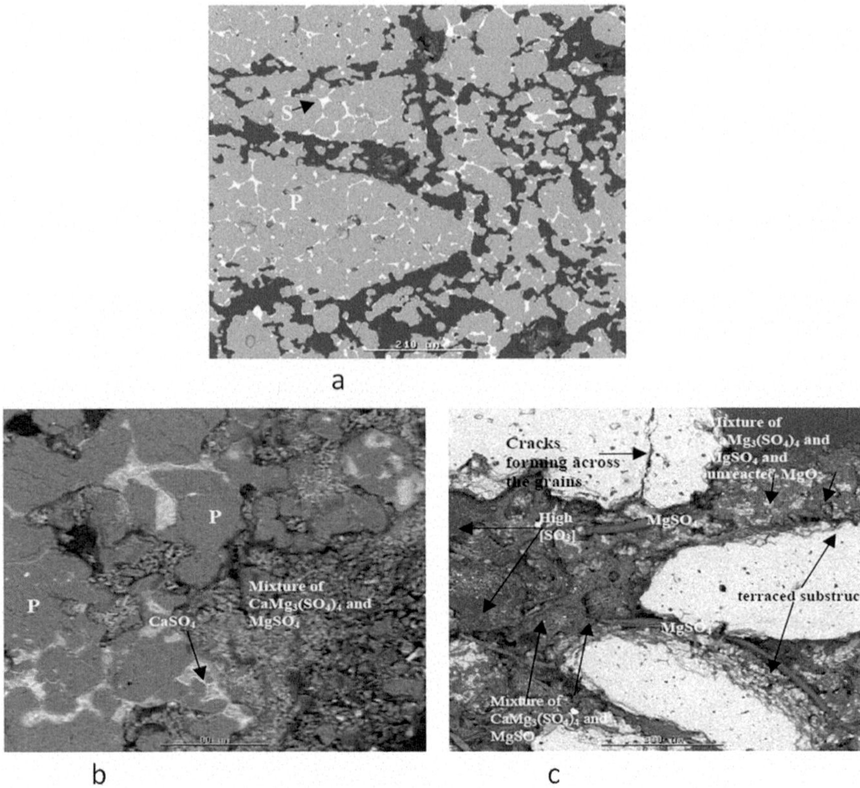

Fig. 13.5 SEM Photographs of magnesite bricks, (**a**) original brick, (**b**) after reaction with SO_2/O_2 at 700 °C for 1 week, (**c**) after reaction with SO_2/O_2 at 1350 °C for 1 week. *P* periclase, *S* silicates

$$MgFe_2O_4 + SO_3 = MgSO_4 + Fe_2O_3 \tag{13.9}$$

$$MgFe_2O_4 + 4SO_3 = MgSO_4 + Fe_2\left(SO_4\right)_3 \tag{13.10}$$

$$MgAl_2O_4 + SO_3 = MgSO_4 + Al_2O_3 \tag{13.11}$$

$$MgAl_2O_4 + 4SO_3 = MgSO_4 + Al_2\left(SO_4\right)_3 \tag{13.12}$$

The brick's microstructure, before and after reaction, was studied, and Fig. 13.5 shows the SEM photographs [20] of the brick before and after reaction. When both excess SO_3 and chlorides are available, the K_2O present reacts preferentially with chlorides, and the SO_3 reacts with CaO partly available as free lime and the Ca-silicates to form $CaSO_4$. That is why the basic bricks with lower content of CaO are an advantage [18].

The free SO_2/SO_3 attacks the CaO containing silicate matrix of basic refractories in the burning zone. By this, the highly refractory silicate, belite (C_2S), decomposes to low melting silicates, e.g., merwinite (C_3MS_2) and monticellite (CMS), as CaO is released and MgO from Magnesite refractory is absorbed. The released CaO reacts with SO_3 to form the sulfate salt $CaSO_4$ anhydrite ($CaSO_4$ = + Volume of 27.5%), which densifies the brick structure in deeper brick zones. This type of reaction leads to a corrosion of the brick structure, lowering of refractoriness, as well as structural flexibility within the affected brick zones.

The wear mechanism can be shown by the following equations:

$$2C_2S + MgO + SO_3 \rightarrow CaSO_4 + C_3MS_2, \quad \text{Melting point}: C_3MS_2: 1575^\circ C \tag{13.13}$$

$$C_3MS_2 + MgO + SO_3 \rightarrow CaSO_4 + 2CMS, \quad \text{Melting point}: CMS: 1500^\circ C \tag{13.14}$$

$$CMS + MgO + SO_3 \rightarrow CaSO_4 + M_2S, \quad \text{Melting point}: M_2S: 1890^\circ C \tag{13.15}$$

Chloride contained in the raw material input as well as in the kiln gas and water vapor (in the kiln gas) increases the volatility of the alkalies. If chloride is very high in the system, it will first combine with all of the alkalis present forming alkali chlorides which will recirculate in the kiln and increase the probability of buildups in the preheater. Any remaining chloride will then combine with CaO to form $CaCl_2$ which has a very low melting point (770–780 C. This will make the hot meal extremely "sticky" at this temperature and increase the chance of builtups higher up the preheater.

13.2.3 Interaction with Vanadium Oxides

Vanadium as an impurity remains in fuel oil and petroleum coke. Vanadium reacts with oxygen to form vanadium oxides, V_2O_5, which melts at 690 °C. When the availability of oxygen is less, V_2O_3 and V_2O_4 are also formed, having melting points of 1970 and 1634 °C. The vanadium oxide V_2O_3 reacts with Na_2O to form $Na_2O.3V_2O_5$ and $Na_2O.6V_2O_5$. Melting temperature of both these materials decreases as the partial pressure of oxygen increases [21].

Presence of small amount (0.1% by weight) of V_2O_5 can cause grayish green coloration to aluminosilicate refractories. The combustion products, V_2O_5 and $NaVO_3$, react with aluminosilicate refractories, including insulation. Fibrous refractory insulations are more prone to this reaction because of their larger surface area than dense refractory products.

Vanadium pentoxide forms a low-temperature eutectic at 650 °C with Al_2O_3. In comparison, the eutectic temperature of V_2O_5 and silica is 649 °C. Mullite exposed to sodium vandate may decompose according to the following reactions:

$$3Al_2O_3.2SiO_2 + 2NaVO_3 \rightarrow Na_2O.Al_2O_3.2SiO_2 + 2Al_2O_3 + V_2O_5 \quad (13.16)$$

$$11/3(3Al_2O_3.2SiO_2) + NaVO_3 \rightarrow Na_2O.11Al_2O_3 + 22/3SiO_2 + V_2O_5 \quad (13.17)$$

$$3(Al_2O_3.SiO_2) + 2NaVO_3 \rightarrow Na_2O.Al_2O_3.6SiO_2 + 8Al_2O_3 + V_2O_5 \quad (13.18)$$

Similarly, sillimanite and corundum may react with the combustion product $NaVO_3$ as follows:

$$2Al_2O_3.SiO_2 + 2NaVO_3 \rightarrow Na_2O.Al_2O_3.2SiO_2 + Al_2O_3 + 2V_2O_5 \quad (13.19)$$

$$11Al_2O_3.SiO_2 + 24NaVO_3 \rightarrow Na_2O.11Al_2O_3 + 11Na_2O.SiO_2 + 12V_2O_5 \quad (13.20)$$

$$11Al_2O_3 + NaVO_3 \rightarrow Na_2O.11Al_2O_3 + V_2O_5 \quad (13.21)$$

Vanadium pentoxide in the combustion ash will reduce the amount of free Na_2O available to react directly with the refractory because the V_2O_5 reacts with Na_2O to form sodium vanadate. However, sodium vandate acting both as a flux and a mineralizer may react with the Al_2O_3-SiO_2 compositions at much lower temperatures and generate more liquid phase than would either Na_2O or V_2O_5 alone. Indeed, in the presence of both Na_2O and V_2O_5, nepheline is formed at 800 °C, whereas it is not formed even at 900 °C when Na_2O alone is present. Albite, which also is not formed in the presence of Na_2O alone, is formed if $NaVO_3$ is present with Na_2O because of the powerful mineralizer action of $NaVO_3$.

The binary system CaO-V_2O_5 has a series of eutectics and intermediate compounds (Fig. 13.6).

The castable refractories contain the hydrates of the different calcium aluminate phases like CA, CA2, etc. which may react with sodium vandate at high temperatures:

$$2CaO.Al_2O_3 + NaVO_3 \rightarrow CaO.V_2O_5 + Na_2O.Al_2O_3 + Al_2O_3 \quad (13.22)$$

$$2(CaO.2Al_2O_3) + 2NaVO_3 \rightarrow 2CaO.V_2O_5 + Na_2O.Al_2O_3 + Al_2O_3 \quad (13.23)$$

The presence of V_2O_5 increases the amount of the liquid phase, in the pseudoternary system CaO-MgO-SiO_2, which becomes glass upon quenching. Dicalcium silicate and C3MS2 will be converted by V_2O_5 to liquids at 1450 °C [36]. The liquid phase between grains will lead to grain growth, reduction in hot strength, and consequent failure of the refractory. For magnesite bricks containing forsterite with low

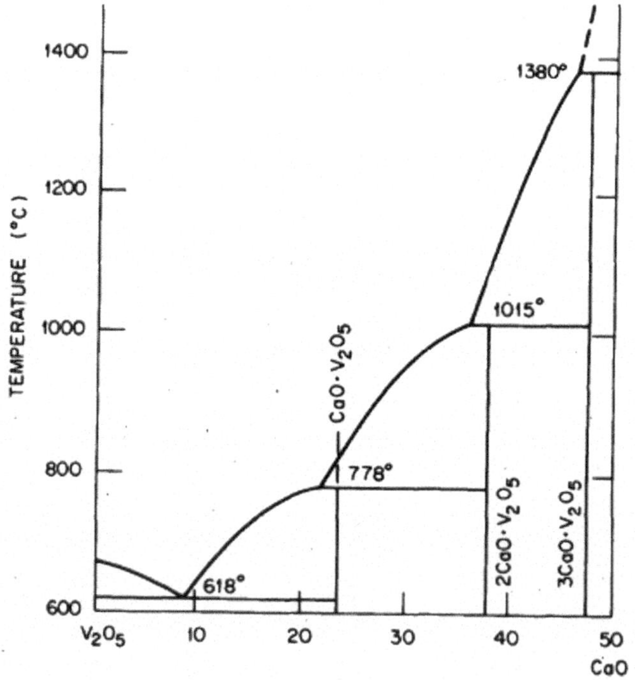

Fig. 13.6 V_2O_5-CaO phase diagram

lime content, the forsterite will act as a stable second phase that will limit the liquid penetration and the grain growth and will promote bonding of the grains. The corrosion, when both V_2O_5 and Na_2O present together, will be more prominent in basic bricks. Sodium vandate may react with the Di calcium silicate ($2CaO.SiO_2$), Monticellite ($CaO.MgO.SiO_2$), and $2CaO.Al_2O_3.SiO_2$ according to the reactions:

$$2NaVO_3 + 2CaO.SiO_2 \rightarrow 2CaO.V_2O_5 + Na_2O.SiO_2 \quad (13.24)$$

$$4NaVO_3 + CaO.MgO.SiO_2 \rightarrow 2CaO.V_2O_5 + 2MgO.V_2O_5 + 2Na_2O.SiO_2 \quad (13.25)$$

$$2NaVO_3 + 2CaO.Al_2O_3.SiO_2 \rightarrow 2CaO.V_2O_5 + Na_2O.SiO_2 + Al_2O_3 \quad (13.26)$$

13.2.4 Redox Reactions

At a given temperature and standard pressure (ambient = 1 atm), the stability of a pure metal and its oxide depends on the partial pressure of oxygen or CO_2/CO ratio in the surrounding atmosphere in a furnace. The main reaction involved is

$$2CO(g) + O_2(g) = CO_2(g) \quad (13.27)$$

Reducing and oxidizing or redox operating conditions are mostly linked to a not optimum combustion of the utilized secondary fuels. These conditions are mainly limited to kiln sections where coarse fuel material is burning directly on the lining. As a consequence, local overheating of bricks occurs, and depending on the oxygen, partial pressure elemental carbon condensation in deeper horizons of the lining may take place. Fundamental signs of thermal overload are wavy, concave surfaces or surfaces, which appear to have melted and solidified. Significant low oxygen partial pressure activates the Boudouard reaction:

$$CO_2 + C \rightleftarrows 2CO \quad (13.28)$$

Due to presence of CO in a reducing atmosphere on the hot side, elemental carbon may deposit in lower horizons of the lining in a temperature range below 600 °C. Carbon deposits in the form of soot on the internal shell are also observed. Damage to the brickwork by extensive spalling may occur with this so-called carbon disintegration.

In practice, the stability of a metal or oxide is studied by considering three chemical reactions of oxidation: metal into metal oxide, carbon into carbon monoxide, and carbon monoxide to carbon dioxide (Chap. 1, Sect. 1.2.10). The oxygen partial pressure and CO_2/CO ratio (at equilibrium) can then be determined by using thermodynamic calculations. As with any chemical reaction, prediction based on purely thermodynamic ground, the oxide with the more negative ΔG will be formed, and the one with less ΔG will be reduced. The Fe_2O_3, Mn_2O_3, or Cr_2O_3 in different types of refractories can be reduced by CO_2/CO gas mixture, and for each oxide, there is specific temperature and the ratio of CO_2 and CO in the mixture when the reduction reaction takes place. Magnesioferrite ($MgFe^{3+}{}_2O_4$), contained in the magnesia brick, is reduced to magnesiowusite ((Mg, Fe^{2+}) O), and this involves a significant reduction of volume (>20%). If redox burning cycles get repeated, there are frequent changes between magnesioferrite and magnesiowusite, resulting in the structural weakening of the affected brick lining. This can lead to a premature wear as a result of spallings of the redox-subjected brick lining.

If infiltrated sulfate salts are present in the structures of the refractory lining, and particularly under the influence of sulfur, sulfide compounds can be formed from these, under strongly reducing conditions. Potassium sulfide K_2S, oldhamite (CaS), K_2S_3, and $KFeS_2$ are found in analysis of used brick samples. When oxidizing conditions predominate again, an oxidation of the sulfides takes place, accompanied by a significant volume increase. This leads to expansion of the brick structure, resulting in the destruction of the brick.

Some of the reactions, which occur in burning zone under reducing condition, are as follows [143]:

$$MgO.Fe_2O_3 + CO = MgO + 2FeO + CO_2 \quad (13.29)$$

$$K_2SO_4 + 4CO = K_2S + 4CO_2 \tag{13.30}$$

$$K_2SO_4 + 8CO + 2SO_2 = K_2S_3 + 8CO_2 \tag{13.31}$$

$$MgO.Fe_2O_3 + K_2SO_4 + 13CO + 3SO_2 = 2KFeS_2 + MgO + 13CO_2 \tag{13.32}$$

Under oxidizing condition, some of the reactions occurring are as follows:

$$MgO.2FeO + \%oO_2 = MgO.Fe_2O_3 \tag{13.33}$$

$$2KFeS_2 + 8O_2 = K_2SO_4 + Fe_2\left(SO_4\right)_3 \tag{13.34}$$

Bricks, containing high Fe_2O_3, are particularly affected, because of transformation of Fe^{2+} to Fe^{3+} and vice versa. For example, magnesioferrite ($MgFe_2O_4$) under reducing condition gets converted to MgO.FeO, which under oxidizing condition reverts back to $MgFe_2O_4$ as per the reaction given below:

$$MgFe_2O_4 + CO \rightarrow 3\left(Mg,Fe\right)O + CO_2 \tag{13.35}$$

$$3\left(Mg,Fe\right)O + 1/2O_2 \rightarrow MgFe_2O_4 \tag{13.36}$$

The repeated redox burning cycle cause frequent changes between magnesiowustite (MgO.FeO) and magnesioferrite ($MgO.Fe_2O_3$), resulting in weakening of the brick structure and destruction of the lining.

13.2.5 Hydration

The storage of basic refractories is very important. Basic bricks are prone to hydration. MgO in basic refractories reacts with moisture in atmosphere to form magnesium hydroxide:

$$MgO + H_2O = Mg\left(OH\right)_2 \tag{13.37}$$

This reaction is associated with volume increase, which results in crack formation in the brick as shown in Fig. 13.7.

The hydration reaction of basic bricks is influenced by the following factors [22–25]:

1. Temperature of storage – The higher is the temperature, the higher is the speed of the hydration reaction. It has been seen that even rise of 10 °C in the ambient temperature increases the hydration reaction speed by 60%. Therefore storage

Fig. 13.7 Hydration of basic brick

should be done at a place where temperature remains low. Impact of temperature on the rate of hydration [23] is shown in Fig. 13.8.

2. State of occurrence of water – The hydration reaction proceeds faster, when the water, as reactant, remains in vapor phase rather than in the liquid phase. It means the atmosphere with high humidity is not good for storage of basic bricks.
3. The nature of the base raw material of the magnesite refractory – The higher is the processing temperature of the periclase, used in the brick, the higher will be its hydration resistance.
4. The composition of the impurity phases – In the basic bricks, besides periclase, other impurities are present, like forsterite (Mg_2SiO_4), merwinite ($Ca_3MgSi_2O_8$), dicalcium silicate ($CaO.2SiO_2$), etc. It has been found that when the matrix or the periclase grains contain dicalcium silicate or tricalcium silicate as impurity, the rate of hydration is much faster.

The basic bricks, if found, have a whitish coating on the surface; it indicates the hydrated layer on the surface. To check the penetration of hydrated layer, the brick should be broken to see if there is whitish coating inside also. If there is any indication of hydration, the bricks should be checked for its suitability for use, by the method described below.

A portion of the brick having a weight of about 2–3 kg is taken and dried at 110 °C for 4 h. The weight is noted. The brick sample is then heated at 1050 °C for 12 h. The brick sample is cooled and weighed. The difference between the two weights is the indication of hydration. If the weight loss is below 0.2%, the brick is

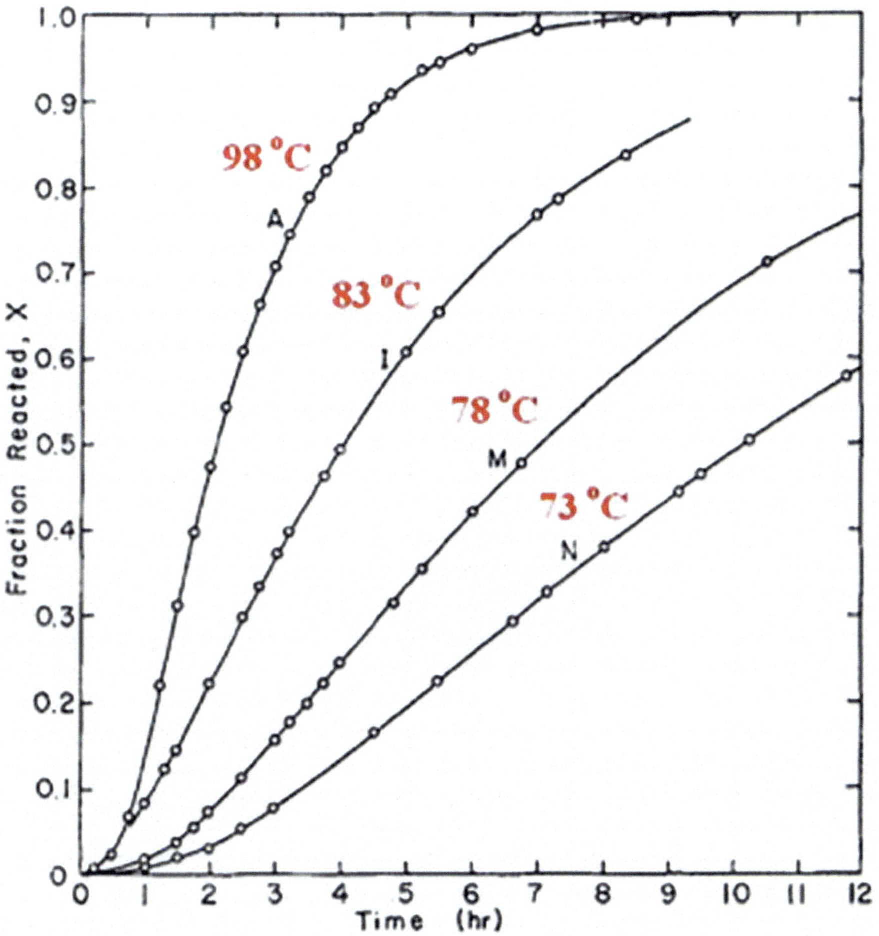

Fig. 13.8 Effect of the rate of hydration reaction with temperature

in good condition. If the weight loss is up to 0.6%, it can be used, but the preheating should be done immediately after the installation, and during preheating, it must be maintained at 1200 °C for 24 h. The bricks having weight loss more than 0.6% should not be used [22].

Case History – In a steel plant in India, a 500 TPD vertical shaft kiln was constructed for lime calcinations. The installation of complete refractory was done. The burning zone was lined with basic refractories. But after refractory installation, the kiln could not be taken into operation for a long time due to some logistic problems and a rainy season passed by, and it had to wait more than 6 months to start. When the kiln was lighted up, the whole lining got collapsed. On investigation it was found that the basic bricks got hydrated inside the kiln because of long storage and high atmospheric humidity and because the bricks were in close confined space inside the kiln.

13.2.6 Ruby Formation

In the secondary reformer, over the alumina support dome covered by alumina guard tiles, alumina lump and catalysts are there. Ruby formation (purple coloration), melting, and lump formation of the catalyst supports are found sometimes below 30–50 mm of the target bricks, to cause problems in the operation of reformer (Fig. 13.9). As the effective cross flow area of the catalyst bed is diminished, the gas velocity increases in the free area. The increased pressure drop has forced the gas to pass in small stream but with high velocity between the refractory and steel shell and causes hot spot. If prompt action is not taken, this may cause the rupture of the steel shell.

When investigated the phenomenon of ruby formation, its mechanism was revealed as below. The purple coloration was because of solid solution of minor quantity of Cr_2O_3 evaporated from stainless steel of the reactor body in the α-Al_2O_3 from the alumina lining of the reactor. It is found that at high temperature, Al_2O_3 reacts with water vapor to form AlOOH according to the reaction:

$$Al_2O_3(s) + H_2O(g) = 2AlOOH(g) \tag{13.38}$$

In the temperature range exists in secondary reformer, an increase of temperature by 100 °C, increases the formation of AlOOH by tenfold. It has been observed that in a reformer, in operation for some time, the surface of the alumina lining, the target tiles, and the alumina lumps are having a soapy feeling instead of its characteristic surface roughness, and that is the indication of the deposition of the AlOOH on the surface. The effect of temperature on the hydration reaction and Ruby formation is given in Table 13.9.

Apart from the temperature, the other factors that influence this hydration reaction are:

(a) Impurities and porosity of the refractory

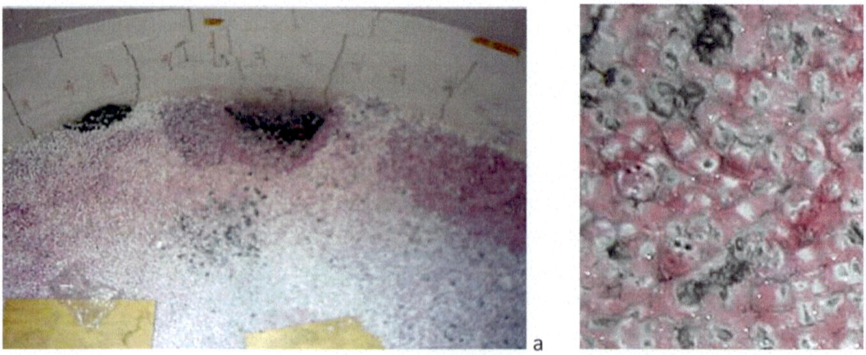

a b

Fig. 13.9 (**a**) Ruby formation in catalyst bed, (**b**) the close-up view of ruby formed catalyst pallets

Table 13.9 Effect of temperature on ruby formation

Temperature of alumina surface (°C)	Ruby formed in catalyst bed (kg/year)
1100	0.1
1200	1.0
1300	10.0

Fig. 13.10 Agglomerated catalyst pallets

(b) Reactivity of the refractory material
(c) Reactivity of the gas mixture

This AlOOH is reconverted to α-alumina as soon as the temperature drops. Because of strong endothermic reforming reaction, there is a steep temperature gradient on the upper part of the catalyst bed which helps in deposition of the α-alumina as mentioned before [26].

The major part of this alumina thus deposited grows in crystal size and binds the pallets together (Fig. 13.10) forming a compact layer which hinders the gas flow through the bed. Silica and calcium aluminate also act as a glue to bind the pallets.

13.3 Thermal Load

13.3.1 Overheating

Overheating is one of the important factors which cause severe damage to refractory lining. The overheating can be caused by thermal imbalance in the kiln or furnace or reactor when the input rate of feed and withdrawal rate of product get disturbed and the thermal input remains the same. The overheated refractory lining in cement rotary kiln can be recognized by its looks as if it is washed out in plane or concave pattern.

The ability of a refractory to withstand certain temperature is judged by its refractoriness or PCE value (refer to Sect. 1.1.2.6). Although this test gives the idea of its softening behavior, this test does not simulate the actual operating condition to which the refractory is exposed during operation. The laboratory test sample is pure and not infiltrated; there is no mechanical stress and the heat up is slow. The risk factor of a refractory lining to get damaged through overheating in a rotary kiln can be assessed by risk factor which is defined as

$$\text{Risk factor} = \frac{\text{Cross sectional thermal load}}{\text{Tangential speed of the kiln} \times \text{Feed rate}}$$

The risk increases with higher thermal load, decreasing kiln speed, and decreasing feed rate. It is easy to control the fuel burning rate and tangential speed of the kiln, but it is beyond the control of the kiln operator on how the feed material will flow along the length of the kiln. Ring formation, variations in fuel chemistry, excessive coating, massive coating losses, variation in fuel properties, variation in kiln feed properties, kiln slowdown, and burner pipe deformation are just a few factors that can cause fast and localized brick overheating.

In case of lime kiln also, the reduction in feed rate can suddenly cause overheating because the decomposition of $CaCO_3$ is an endothermic process.

13.3.2 *Flame*

In a furnace or kiln, the heat exchange between the burner flame and the material in the inside takes place by three principal mechanisms: radiation, conduction, and convection. The objective is to maximize the transfer of heat generated by the flame to the incoming material in the burning zone. The most important mechanism of heat transfer from flame to the clinker is radiation, and in the burning zone, about 95% of the heat transfer is through radiation. Radiation between two materials takes place when the materials are not in contact with each other. The flame, refractory, and coating radiate heat to the feed in the kiln.

The mathematical expression for heat transfer by radiation is

$$Q = \sigma \varepsilon A \left(T_{\mathrm{F}}^{4} - T_{\mathrm{P}}^{4} \right) \tag{13.39}$$

where

Q = rate of heat transfer in J/s
σ = Stefan-Boltzmann constant in J/(m^2 * s * °K)
ε = emissivity (with a value, 0.0–1.0)
A = area available for heat transfer in m^2
T_{F} = surface temperature of the flame in °K

T_P = surface temperature of the product in °K

The heat is transferred from the flame to the bed of material mainly by radiation. In the formula for radiation, the heat transferred is proportional to the fourth power of the temperature of the flame. Therefore the flame temperature got a very strong influence on the heat transferred. For example, a temperature increase by 10% gives a radiation, increased by $(1.1)^4 = 1.464$, i.e., 46%. This gives a clear idea about the strong influence of temperature increase on radiation [27].

Flame temperature can be approximated by the mathematical formula:

$$T = H_v / 1.11AS \tag{13.40}$$

where

T = flame temperature
A = combustion air required, kg/kg of fuel
H_v = heating value of the fuel, Kcal/kg
S = sp. heat of combustion gas

The mathematical relation shows that the radiation depends upon the emissivity of the flame which is different for different fuels. The emissivity of the flame from gas, oil, and coal is given in Table 13.10 [27].

The burner design and the flame quality have profound effect on the performance of refractory. The effect is more prominent in case of larger diameter kilns using multichannel burners. For better and uniform burning, the desired flame must be hot and convergent.

When the flame is short and divergent, it cuts the refractory lining like the knife cuts butter. An index, called swirl number, can be used to measure the damaging effect of the flame. If the swirl number exceeds 0.8, there is a chance that even the best quality of refractory lining will get damaged. The flame impingement occurs because of high quantity of swirled air, high velocity of the flame, and the distorted burner tip. The ideal flame must be stable and short. When the primary air quantity, speed, and swirl increase, the flame becomes short. Increase in temperature of secondary air, increase in fineness of solid fuel, lowering of ash content, and its volatile matter make the flame short.

Impingement of partially burned fuel upon the feed bed or kiln wall must always be avoided when a kiln is fired by coal or oil. The impingement of the flame, on the kiln wall, causes damages to refractory lining, reduces the fuel efficiency, and affects the kiln stability.

Table 13.10 Emissivity of flame from burning of different fuels

Fuel	Flame emissivity
Gas	0.3
Oil	0.5
Coal	0.85

13.4 Thermo-mechanical Load

Refractory lining in any kiln or furnace or reactor is not an independent entity, and the total structure can be considered as a composite structure, consisting of metal and ceramics. The movements of refractory and steel shell are not really independent. Movement of one is restricted by the movement of other components. This fact is very important to develop the thermo-mechanical stress both during heat up and cool down and during the operation of the kiln or furnace or the reactor. Let us discuss the case of rotary kiln used for cement manufacturing and lime calcinations. It has a specialty over other furnaces or reactors that it is not static and is dynamic in nature and it rotates on its axis very slowly (1.5–4 RPM) while is supported on two or three pairs of rollers.

Kiln shells are made with structural rolled steel plate, such as ASTM A36. The properties for this type of steel are:

Carbon – 0.25%
Manganese – 0.80% to 1.20%
Phosphorus – 0.04% Max
Sulfur – 0.05% Max
Silicon – 0.40%

The mechanical properties of this type of steel at room temperature are:

Tensile strength – 345 to 550 MPa
Yield strength – 250 MPa Min
Elongation – 20% Min
Linear thermal expansion coefficient – 11.7×10^{-6}/°C
Elastic modulus – 207 GPa
Poisson ratio – 0.3 in the elastic range, 0.5 in the plastic range

Steel loses its mechanical strength at elevated temperature. At 430 °C the ultimate strength of the steel drops by 33%, from 517 to 345 MPa. Some investigators report a 50% strength loss for the same temperature range. This loss in strength may cause deformation of the kiln shell depending upon its design and construction. Cement rotary kiln or lime calcination kiln, a huge steel tube with refractory lining, and the process material inside, weighing about 1000 Ton, rest on two or three supports. It rotates slowly at high temperature, and the load acts on the middle of two supports. Under such condition the kiln tends to deform. Thermo-mechanical stress generated in the rotary kiln, during the operation of the kiln, is because of various factors, e.g., the ovality of the kiln shell, the restricted expansion of the refractory, etc. Thermal expansion of the refractory lining, the rotary kiln shell, and the tire is a fundamental issue, related to the stress development. The developed thermo-mechanical stresses have direct effect on the refractory performances, and it appear in different forms like falling of bricks from the ring, crushing, slabbing of refractories, etc. The gap between the tire and the chair pads has to be sufficient for the thermal expansion of the steel shell. Too tight riding tire can inhibit thermal expansion of the steel shell, leading to the failure of the lining or even the tire. On the other hand, the gap cannot be too large as it affects ovality negatively.

13.4.1 Ovality

Steel is the only viable material for the construction of the kiln but presents the challenge to contain the maximum temperature of the feed inside the kiln which is over 1400 °C while the gas temperature is at 1900 °C in cement rotary kiln. In case of lime calcinations, the operational temperature is much lower, and naturally the problem of kiln deformation is also less.

Brick lining is tightly fitted to the steel casing of the kiln. During the usage of the kiln (starts, stops and rotation), the steel casing and therefore the lining are subjected to radial and longitudinal bending, vibrations, and torsion. Additional stresses can typically arise from misalignment of the kiln or other abnormalities. This results into different stress-controlled loads in the lining. Here, stress-controlled loads define external loads such as gravity load, pressure load, or any type of mechanical load.

Radial bending of the kiln, known as the ovality of the kiln, traditionally belongs to one of the most important load generators in the refractory lining and directly affects its campaign life. Ovality is an elastic distortion of the kiln shell that arises due to the gravity force. The weight of the casing, the lining, and the coating over kiln's hollow shape make the shell somewhat oval rather than circular (Fig. 13.11). The difference (δ_v and δ_h), between deviations of vertical and horizontal diameter of the kiln from the actual shell diameter, generally, amounts to 0.3%, and it can go up to 0.6–0.7% for an old and large diameter kiln.

Due to the ovality of the steel shell, the lining will experience load oscillations during rotation of the kiln. This may lead to the formation of cracks and eventually to the slabbing of the refractory lining (Figs. 13.12 and 13.14).

Additionally, bricks in the lining are forced to shift their relative position to each other due to the ovality. By that, opening up of the joints leads to worsen the integrity of the brick lining and may cause unhealthy stress concentrations.

Ovality is commonly presented in percent, as relative deformation to the nominal diameter. If the deformation is known, then relative ovality (ω_r) is found by equation:

$$\omega_r = \frac{100(\delta_v + \delta_h)}{D_o}\% \qquad (13.41)$$

δ_h and δ_v are the horizontal and vertical deviations from the circular line, and D_o is the nominal diameter of the kiln shell. A rule of thumb based on the experience has been established, suggesting that the ovality of the steel casing should not exceed 10% of the nominal inner diameter. For example, a kiln with a nominal inner

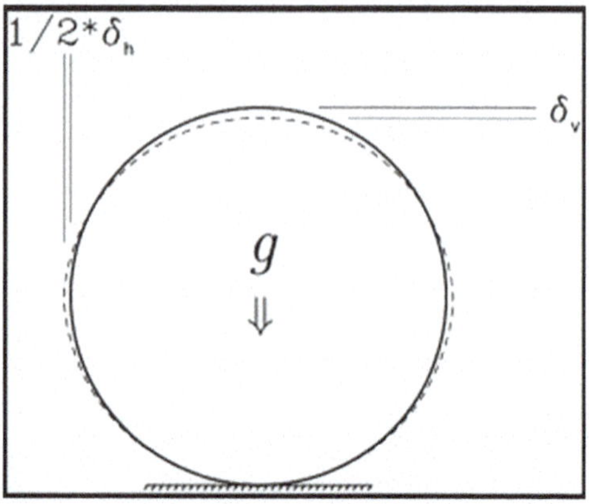

Fig. 13.11 The deformation of the kiln shell

Fig. 13.12 Twenty percent of the brick thickness fell down within 3 months of installation of refractory lining, due to ovality in the kiln shell

diameter of 6 m should not have relative ovality of more than 0.6%. The ovality of the kiln shell increases with diameter of the kiln (Figs. 13.11 and 13.13).

Let us imagine the condition of a brick lining inside a rotary kiln when the kiln is in rotation. As shown in Fig. 13.14a, the pieces of bricks in the bottom position experience brick to brick compression, and the same bricks, during rotation, when going to top position, experience a brick to brick tension. The compression and relaxation in the lining is shown by arrowheads in Fig. 13.14a.

Magnitude of compressive stresses in the lining, induced by the ovality of the kiln, can be expressed by the following equation [28]:

$$\sigma_{\mathrm{L}} = \frac{3\left(\delta_{\mathrm{h}} + \delta_{\mathrm{v}}\right)}{D_0^{\,2}} . E_{\mathrm{r}} . t_{\mathrm{r}} \tag{13.42}$$

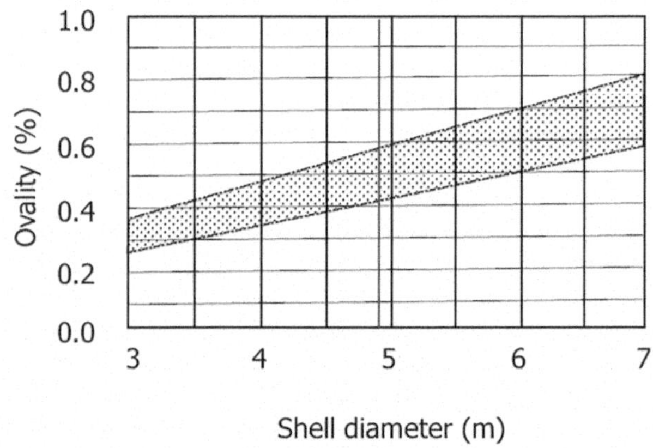

Fig. 13.13 Relation between kiln diameter and ovality

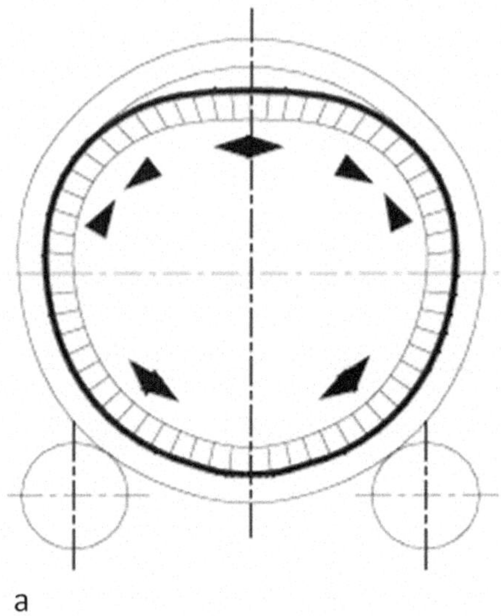

Fig. 13.14 (**a**) Compressive and tensile stress are shown by arrow heads. (**b**) Ovality vs compressive stress (6 & 7 are diameter, 200 & 250 are lining thickness)

where σ_L = compressive stress induced on the inner wall of the lining due to ovality of kiln shell, E_r = Young's modulus of the refractory used for the lining, and t_r = the thickness of the refractory lining. According to this expression, the stress increases with increased ovality, Young's modulus, and lining thickness. The compressive stress increases sharply with increasing ovality and with increased lining thickness (Fig. 13.14b).

A rotary kiln, having three rotations per minute, will produce 13,000 cycles of alternating compression and relaxation per day per brick. Such alternating compression

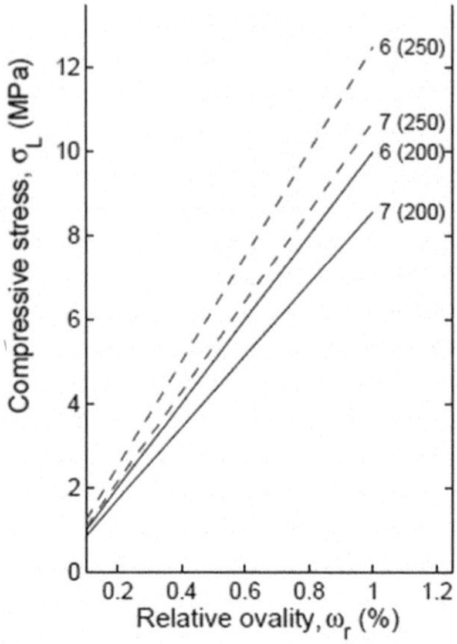

b

Fig. 13.14 (continued)

and tension among the bricks can lead to the brick falling out or chipping out and can have a fatigue effect. Typical mechanisms of lining degradation are shown in Fig. 13.15.

A perfect lining situation, that is, when the shell is not deformed and bricks correctly adjoin each other, is shown in Fig. 13.15a. The distribution of forces in the contact areas between the bricks is shown. Figure 13.15b shows a situation, when, during rotation of the kiln, the radius of curvature increases excessively, the load on the surface of the bricks decreases, and even the gaps between them become wider. When the shell radius becomes large enough, the bricks may move downwards or even fall out totally.

When a brick falls out, the adjacent bricks become loose, so subsequently it might lead to falling of other bricks, thus exposing the steel shell to high temperatures. If a loosened brick does not fall out but stays in place (Fig. 13.15b), then during rotation of the kiln at the same circumferential position of the shell with a reduced radius of curvature, the bricks will experience a compressive stress (Fig. 13.15c) [29].

Such a situation might cause the breaking of the downwardly moved brick, in the cross section, compressed by the edges of adjacent bricks and the chipping out of a large portion of this brick (Figs. 13.12 and 13.16).

There is a linear relationship between the ovality and the wear out speed of the lining [30]; the higher the ovality, the more is the speed of wear out (Fig. 13.17).

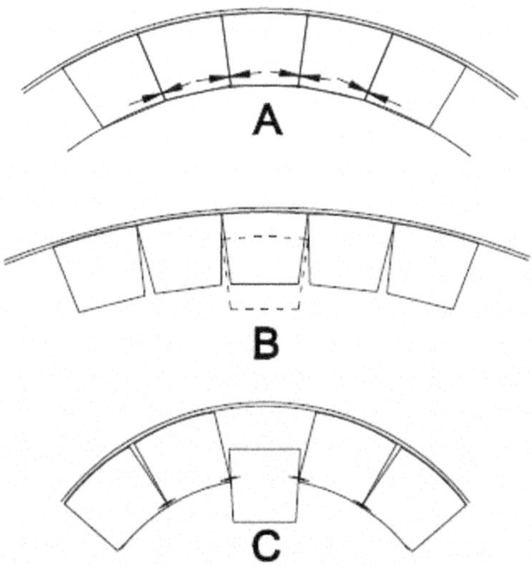

Fig. 13.15 Mechanism of destruction of the lining under the stress due to ovality

Fig. 13.16 Chipping out of the lining bricks due to kiln shell ovality

13.4.2 Strain-Controlled Load

Refractory lining expands at elevated temperature during service. Part of the expansion of the refractory lining is compensated by the expansion of the steel shell. Additionally, some part of the expansion is compensated by the natural joints and mortar joints between the bricks. However, a part of the expansion leads to increased mechanical stresses in the lining. The advantage of the expansion is the increased stability of the lining because the individual bricks in the lining remain tight.

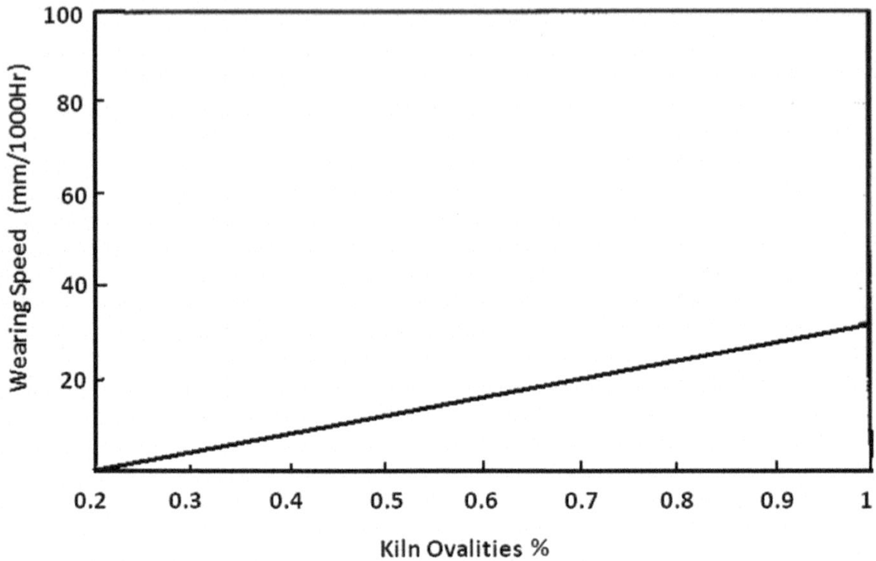

Fig. 13.17 Relation between kiln ovality and the wearing speed of refractories

However, it is important that the expansion is within the technical limitation of the material to withstand the stress. Some additional flexibility to a brick lining can be provided by installing the lining with mortar or steel shims between the radial joints. The steel shims and mortar shrink during heating up and create additional expansion space to the lining by lowering the stresses. In cold state, the lining becomes loose than normal, after erosion of steel shim or mortar, which might be problematic during cooling and reheating of the kiln. Typical heating and cooling procedure requires 24–36 h. In addition to the thermal stress due to expansion, the hot face (face of the lining exposed to the flame) of the lining also experiences thermal load oscillations during rotation. This can be due to different factors, e.g., relatively fixed flame direction and gas flow and the cooling effect on the lining when it passes below the bed of material. The thermal oscillation leads to local stresses which risk spalling of the lining.

When the furnace or kiln is in operation and reaches an equilibrium regarding the heat flow, a temperature gradient is established across the lining. If the lining is too thin or the lining material has very high thermal conductivity, the shell may attain a high temperature. The thermal expansion of steel is much more than refractory material. If the temperature of the steel shell goes high and it expands more than the refractory bricks in contact, it will exert a tensile stress on the refractory lining and may cause loosening of the bricks. On the other hand, if the lining is too thick or the lining material has low thermal conductivity, the steel shell cannot attain a high temperature and cannot expand much, but the refractory lining at high temperature will try to expand, and the lining experiences a high compressive stress. If the kiln is in open area without any shed over it and it starts raining, the kiln should never be stopped. If the kiln stops rotating, the upper half will get cooled faster than the lower half, and because of uneven expansion, the severe deformation of the kiln may happen. Similarly during preheating if the rain water falls on the kiln, the heating must be stopped, and the kiln must be kept in rotation till it becomes cool. The cold kiln should not be run for long because the refractory lining will get loosened.

Expansion joint for the refractory lining is especially important for long ducts where the overall thermal expansion for steel as well as the refractory lining is very high. Different types of expansion joints are used in the lining depending upon the requirement. This may be:

Single expansion joint – To absorb small amount of lateral, axial, and angular movements.

Universal expansion joint – To take care of axial movements and large amount of lateral deflection and angular rotation.

Hinged expansion joint – Contains hinges or pivots which allow the units to bend in a single plane.

Rectangular expansion joint – To take care of axial movement, lateral deflection, and angular rotation.

Externally pressurized expansion joint – Used for ducts that require large amount of axial compression or large axial movement. The design of expansion joint for the equipment with complicated design is mostly customized by the supplier of the equipment.

One such design for FCCU is shown in Fig. 13.18 with refractory lining inside.

13.4.3 Thermal Shock

Refractories in the furnaces or reactors experience temperature fluctuations in spite of its continuous type of operation. These variations may be due to fluctuations of feed to the furnace or reactor, due to variation in supply of primary or secondary air, variation in fuel input rate to burners, etc. In rotary kiln during rotation, the lining has an alternate exposure to higher temperature (when in top) and lower temperature (insulated by material cover when in bottom). This periodic temperature fluctuation of the lining can go as high as 400 °C and may take its toll on refractory performance, unless the refractory is resistant to thermal shock.

The origin of thermal shock varies widely from process to process. For an example, in an unplanned feed outage in a FCCU, when catalyst circulation is also stopped, restoration of some catalyst circulation, where possible, will prevent a large drop in reactor-side temperatures and avoid a complete thermal cycle from operating temperature to steam temperature.

The resistance to thermal shock becomes very important during heat up and shut down of the unit. Faster cooling is more detrimental to the life of refractories, especially the basic refractories, than faster heating. The reasons are as follows: refrac-

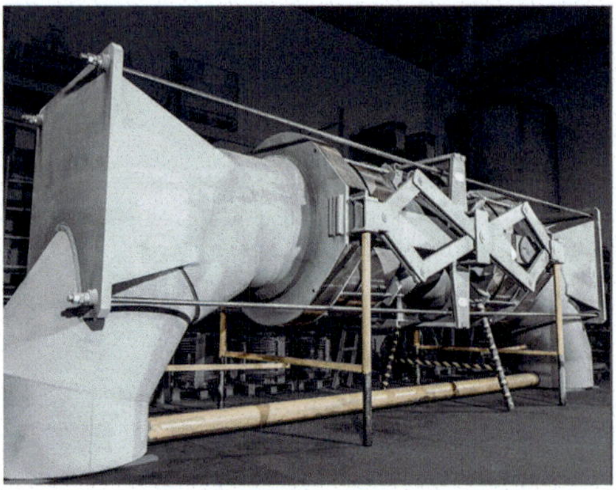

a) The expansion joint used in FCCU
(Courtesy- Belman Group)

b) The refractory lining inside the joint

Fig. 13.18 (**a**) The expansion joint used in FCCU. (**b**) The refractory lining inside the joint. (Courtesy – Belman Group)

tory lining experiences tensile stress on cooling and compressive stress on heating, and refractory materials are weak in tensile stress by nature.

13.4.4 Abrasion

There is a continuous countercurrent flow of gas and solid material in many of the furnaces and reactors. The flow of high-velocity dust-laden gases and the solid materials in the kiln system has a very abusive effect on the refractory life, because of the continuous abrasion. As an example, Table 13.11 shows the gas velocity in the different application areas in a cement plant.

In some reactors, the coating of the processed materials, over the refractory lining, for example, inside the cement rotary kiln, saves the refractory from direct exposure to the stream of dust-laden gas. But in reactors like FCCU, the refractories are to withstand continuous severe abrasion from the abrasive catalyst particles. The abrasion of TAD lining by the high-velocity dust-laden gases and abrasion of refractory lining by flow of solid material in calcining zone of the kiln are shown in Figs. 13.19 and 13.20, respectively.

13.4.5 Effect of Kiln Diameter

A very common failure in cement rotary kiln is falling down of bricks from the lining rings of rotary kiln, which leads to development of hot spots in the kiln (Fig. 13.21) [1].

Table 13.11 Velocity of gases in different equipment in a cement plant

Area in the cement plant	Velocity of the gas (m/s)
Hood	6.0
Bull nose	15.0
Burning zone	9.5
Feed and transition zone	13.0
Riser	24.0
Preheater gas duct	18.0
Tertiary air duct	25

Fig. 13.19 Abrasion of refractory in TAD

This kind of problem occurs mainly with aluminosilicate refractories and mainly in the calcination zone to kiln entrance end. The reason of this failure is insufficient tightness of the bricks in the ring.

The refractory bricks in the cement rotary kiln are laid without any mortar. The bricks are tightened by the key effect of the bricks. During the rotation of the kiln, the bricks in the ring, not tightened properly, may come down slowly and fall down. This happens in the part of the kiln where temperature is comparatively less and expansion of the refractory is also less to tighten the bricks in the ring. Aluminosilicate refractories, because of the lower thermal expansion coefficient, compared to basic refractories, are prone to this problem. This phenomenon happens after the upper transition zone to kiln entrance, because no clinker coating forms there, which can hold the bricks together from falling down.

After the installation of a brick ring, therefore the insertion of steel shims, in two to three points in the ring, is very essential. The problem is also more frequent in case of kilns with large diameter, because with larger diameter, the key effect of the side arch bricks get lessened. The frequent stoppage of the kiln aggravates the prob-

Fig. 13.20 Abrasion of refractory in calcination zone

Fig. 13.21 Cavity forms in the lining of 6 m diameter kiln, because the 60% alumina bricks fell down in the calcining zone after 30 days of running the kiln

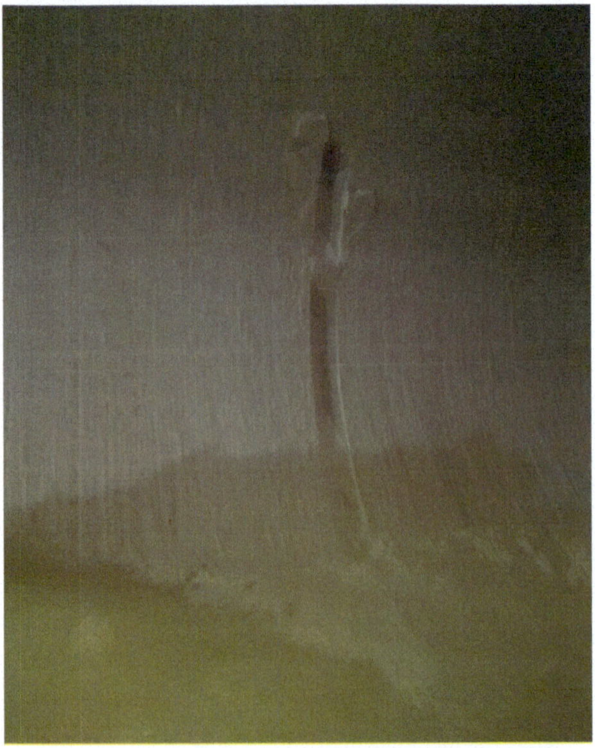

lem, because the jerk, in stop and start of the kiln (especially with bigger diameter), that further helps the falling of the brick from the ring.

Similar problem of falling of bricks happens in case of SRU reactor also.

13.5 Lining Tightness

If the joint of the bricks remains loose during installation, the lining will definitely underperform. The problems will be different depending upon the furnace and process. For example, in case of the glass or coal gasification, the lining deterioration due to liquid penetration and corrosion will be much faster. In case of lining not in contact with liquid, the gas will penetrate through joints and will condense behind and will corrode the steel shell anchors, etc. and may cause refractory failure.

In case the furnace is dynamic one, like rotary kiln, the brick lining and the kiln shell may not rotate at the same rpm but two different rpm, and as a result, the brick will be rubbed continuously at the cold face (Fig. 13.22) by the shell and will lose its thickness [1]. The lining gets distorted and may form skew.

Fig. 13.22 The rubbing of the refractory cold face due to relative movement of brick ring and kiln shell

13.6 Refractory Failure due to Anchor

Many a times the refractory lining fails because of anchor failure. Besides the proper design, the material of construction of the anchors also plays a very important role. Iron-chromium-based stainless steel anchors are commonly used for high-temperature service conditions. Unfortunately, these alloys are susceptible to embrittlement, when held within or cooled slowly through the temperature range of 400 °C to 980 °C. There are in fact two significant embrittlement phenomena that occur within this temperature range; the first one is known as 475 °C embrittlement (400–550 °C), and the second one is known as sigma (σ) phase embrittlement (565–980 °C).

The 475 °C embrittlement phenomenon only occurs in ferritic and duplex stainless steels and not with austenitic stainless steels and can be reversed by heating to 675 °C or above. Therefore 475 °C embrittlement phenomenon is not expected to cause problems for austenitic stainless steel refractory anchor systems.

13.6.1 Sigma-Phase Embrittlement

Sigma-phase embrittlement occurs because of long exposure to temperatures in the range 565–850 °C, although this temperature range varies with the composition and processing of the stainless steel. Sigma phase is an iron-chromium intermetallic phase that is extremely hard and brittle [31, 32].

Sigma phase forms more readily in ferritic than in austenitic stainless steels. Elements such as silicon, molybdenum, and to a lesser degree aluminum, tungsten, vanadium, titanium, and niobium all promote σ-phase formation. Large additions of Ni and Mn retard σ-phase formation. Nitrogen, which is now used extensively in the austenitic and duplex stainless steels to stabilize austenite, also helps to retard σ-phase formation. Carbon additions decrease σ-phase formation.

Sigma phase reduces steel's toughness, creep resistance, and fatigue strength. However it is not always solely the cause of degradation of properties, and therefore each situation must be evaluated to optimize the performance. In a typical refractory lining, there is always a thermal profile which exists between the hot and cold faces (Fig. 13.23). There is nearly always a portion of the lining and any support anchors exposed to temperatures within the range of sigma-phase formation.

The illustration below shows a typical lining configuration, and the resulting temperature profile, highlighting the zone in which σ-phase would be expected to form.

In many common austenitic steels, it is impossible to avoid the formation of this phase entirely. What is possible, however, is to choose a steel or alloy, and to

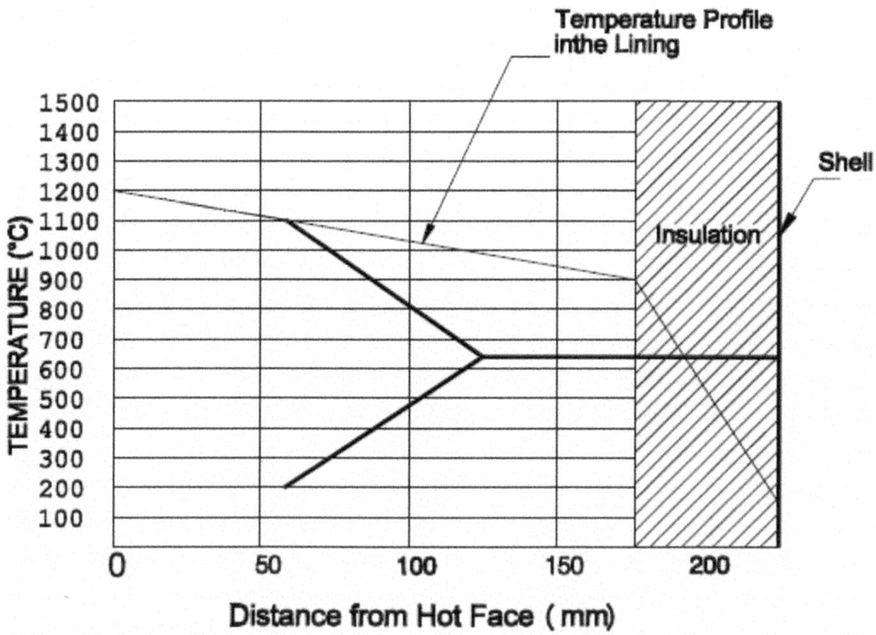

Fig. 13.23 Temperature profile across a steel anchor embedded in the castable lining

minimize the formation of the phase to a level, where the effect does not significantly compromise the mechanical properties required by the lining design.

13.6.2 Remedy

One of the tools available to the design engineer, for that purpose, is the Schaeffler-DeLong diagram, which scans the range of alloys and steel by their nickel and chromium equivalents [33, 34]. By plotting these parameters for the common steels/alloys, their susceptibility to σ-phase formation can be assessed to a preliminary degree.

In Fig. 13.24, several common steels have been plotted by these parameters. Resistance to σ-phase formation can be assessed by the "ferrite number" ascribed to the area of the chart in which the steel is plotted. In general, steels occupying the upper left zone of this diagram can be expected to show better resistance to sigma-phase formation. The best selection of the anchor quality is always a compromise between the suitability of different metal quality for different operating conditions as shown in Table 13.12.

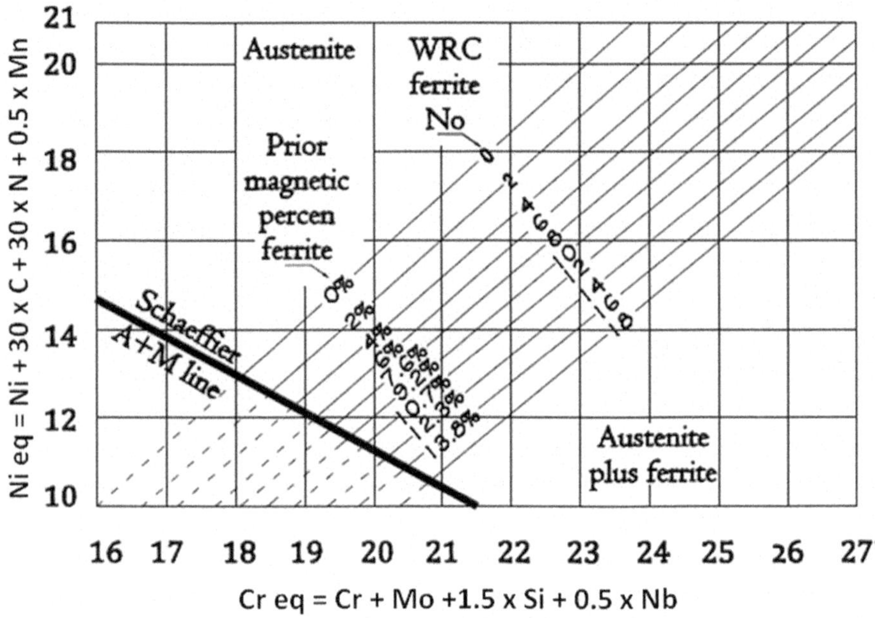

Fig. 13.24 Schaeffler-DeLong diagram of several steel qualities

Table 13.12 Recommendation of anchor quality for different operational conditions

Operational condition	Effect on anchor	Best choice	Constrictions
High temperature	Oxidation	Cr and Cr + Ni, i.e., SS 310 or I.4845	σ-Phase embrittlement at 700–900 °C
High-sulfur-containing environment	Sulfidization	Mn + C and no Ni, i.e., SS 310 and 309	σ-Phase embrittlement at 700–900 °C and embrittlement due to aging
High chloride content	Salt attack	Ni, I 4876, 4856	Reduced sulfate resistance and cost
Presence of C and CO	Carbonization	Ni + Si, I 4876, 4856	Reduced sulfate resistant and cost

13.6.3 *High-Temperature Attack of Hydrogen (HTHA)*

Similar to anchor failure, where refractory fails actually due to failure of steel, there are instances where refractory failed due to hydrogen attack (HTHA) on steel. High-temperature hydrogen attack (HTHA), also called hot hydrogen attack, is a problem which concerns steels operating at elevated temperatures (typically above 400 °C) in hydrogen environments, refinery, petrochemical and other chemical facilities, and, possibly, high-pressure steam boilers. HTHA is the result of hydrogen dissociating and dissolving in the steel and then reacting with the carbon in solution in the steel to form methane. This can result in either surface decarburization, when the reaction mostly occurs at the surface and draws carbon from the steel, or internal decarburization when atomic hydrogen penetrates the material and reacts with carbon to form methane, which accumulates at grain boundaries and/or precipitates at interfaces, and cannot diffuse out of the steel. This causes the fissures and cracking which are typical of HTHA.

Surface decarburization results in a decrease in hardness and increase in ductility of the material near the surface. This is usually only a minor concern for these types of application. However, internal decarburization, and in particular the formation of methane and consequent development of voids, can lead to substantial deterioration of mechanical properties due to loss of carbides and formation of voids and may lead to catastrophic failure. The list of steel qualities, most affected by HTHA, is given below, in order of increasing resistance to HTHA, which indicates that carbon steel is least and SS is most resistant to HTHA [34–36].

Carbon steel, C-0.5Mo, Mn-0.5Mo, 1Cr-0.5Mo, 1.25Cr-0.5Mo, 2.25Cr-0.5Mo, 2.25Cr-Mo-V, 3Cr-1Mo, 5Cr-0.5Mo, and 300 Series SS (austenitic grade)

Factors which influence the attack are:

1. Temperature
2. Partial pressure of hydrogen
3. Time of exposure
4. Stress

HTHA is preceded by a period of time till any noticeable change occurs. Nelson curve shown in Fig. 13.25 [37] can be used as a tool to predict the HTHA on any metallic body.

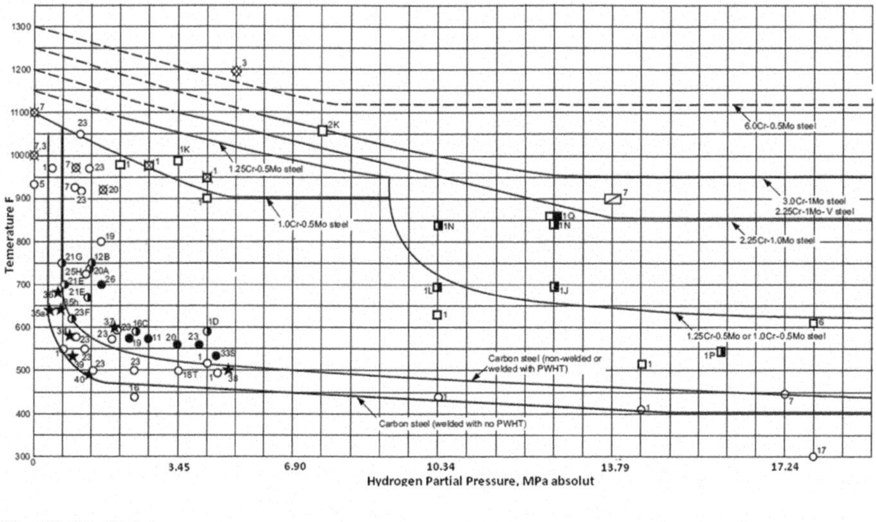

Fig. 13.25 Nelson curve

The broken-line curves at the top of Fig. 13.1 represent the tendencies for surface decarburization of steels while they are in contact with hydrogen. The solid-line curves represent the tendencies for steels to decarburize internally with resultant fissuring and cracking.

The HTHA is very important for considering material of construction for equipment in petroleum refineries, syngas producing units, catalytic crackers, and even boiler tubes. The failure of a secondary reformer of Haldor Topsoe design was reported because of crack of welding in the 80–92-mm-thick plate of the reformer vessel [38].

13.7 Refractory Performance

Refractory performance depends on design of the equipment also. A paper [39] reported, that, in a secondary reformer of Humphreys & Glassgow (Jacob) design in a fertilizer plant in India, the vertical cylindrical conical vessel having two layers of refractory lining inside (hot face 99% Al2O3 brick and backup insulating castable) had severe erosion of Refractory lining at air mixer ring bricks and top cone area. After review and analysis of both operational parameters and design, modifications were made in the design which solved the Refractory erosion without any change in Refractory quality.

13.8 Refractory Lining Failure Statistics

The studies on the contribution of different factors, e.g., refractory quality, operational parameters, etc., on the refractory failure in cement rotary kiln were analyzed [39], and the result is shown in Fig. 13.26.

It clearly shows that 77% or the 3/4 of all the refractory failures in the cement rotary kiln are due to improper operation or improper lining of the refractories. If these two factors are well taken care of, then 3/4 of the all refractory failures can be avoided. This observation is well corroborated to the concept given in Fig. 13.1.

Although this study is for cement industries, this holds good with all kinds of chemical process industries as well. A proper collaboration and coordination of the manufacturer, installer, and user only can generate the best performance out of any refractory lining.

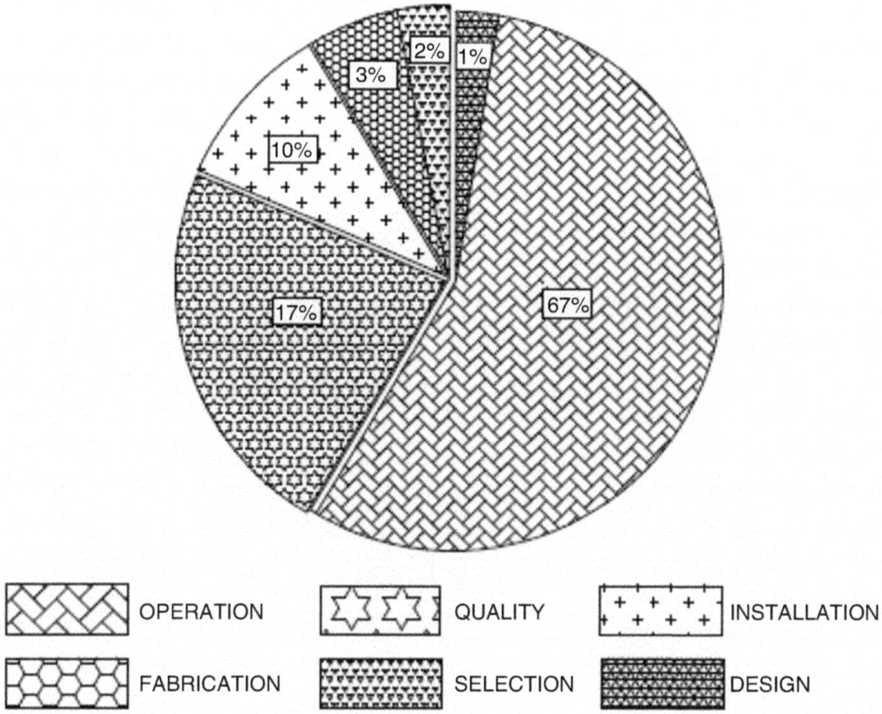

Fig. 13.26 Factors (with probabilities) responsible for refractory failures in cement rotary kiln. (Courtesy of Refratechnik)

13.9 Analysis of Refractory Lining Failure

Whenever any refractory lining undergoes through a premature campaign life, the normal tendency in the industry is to hold the refractory quality responsible for the failure. But this is not correct, and this hinders to the solution of the problem. Any problem arising in a refractory lining must be looked at without any bias. As shown in Fig. 13.1, refractory premature life may be because of so many reasons also, other than refractory quality. Most of the time, the failure is not due to a single reason and is a cumulative effect of number of factors. It is very common in the industry that two furnaces having the same design and same operational parameters to behave differently, with respect to refractory performance. Therefore a systematic study and investigation should be made to analyze the probable reasons of low life or a premature failure of a refractory lining. Schemes for the investigation of refractory failure for the lining with shaped and unshaped refractory products are shown in Figs. 13.27 and 13.28. The problems and failure pattern may vary widely depending upon the industry and the type of refractory lining, but the philosophy behind the investigation remains the same.

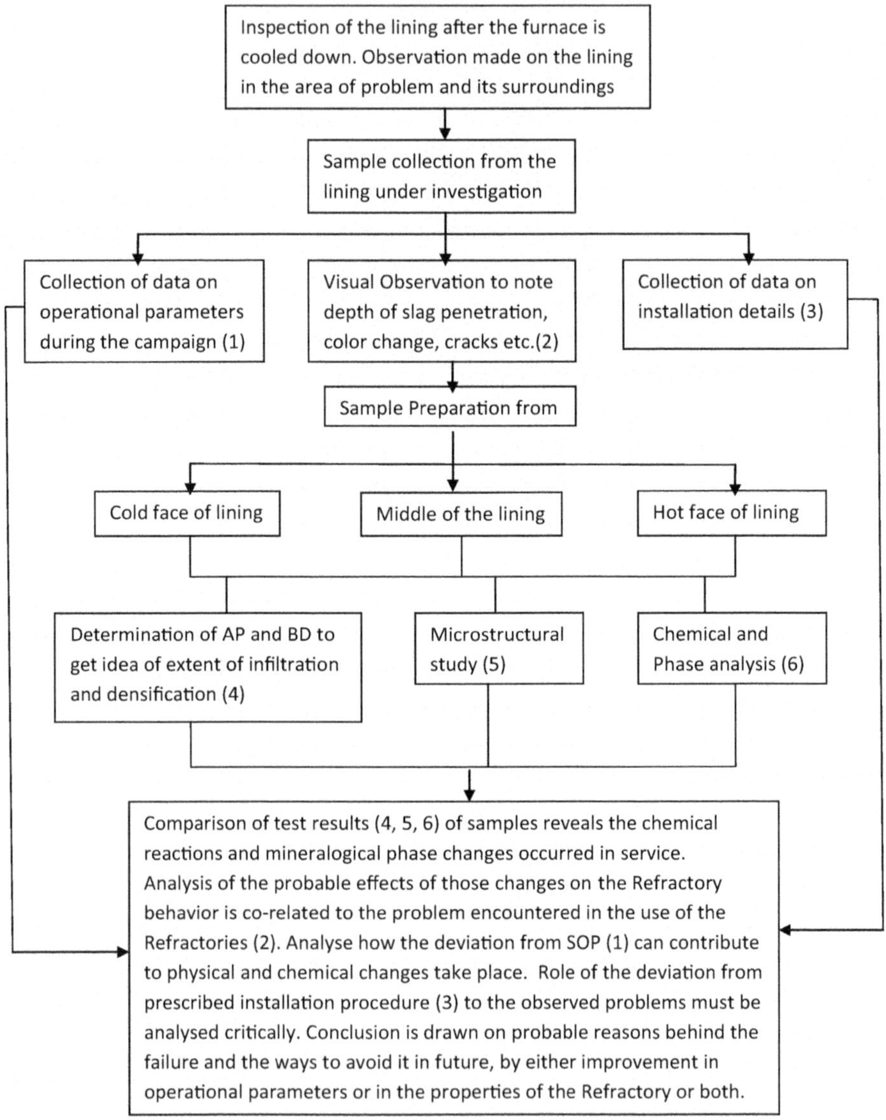

Fig. 13.27 Scheme for investigation of a shaped refractory lining failure

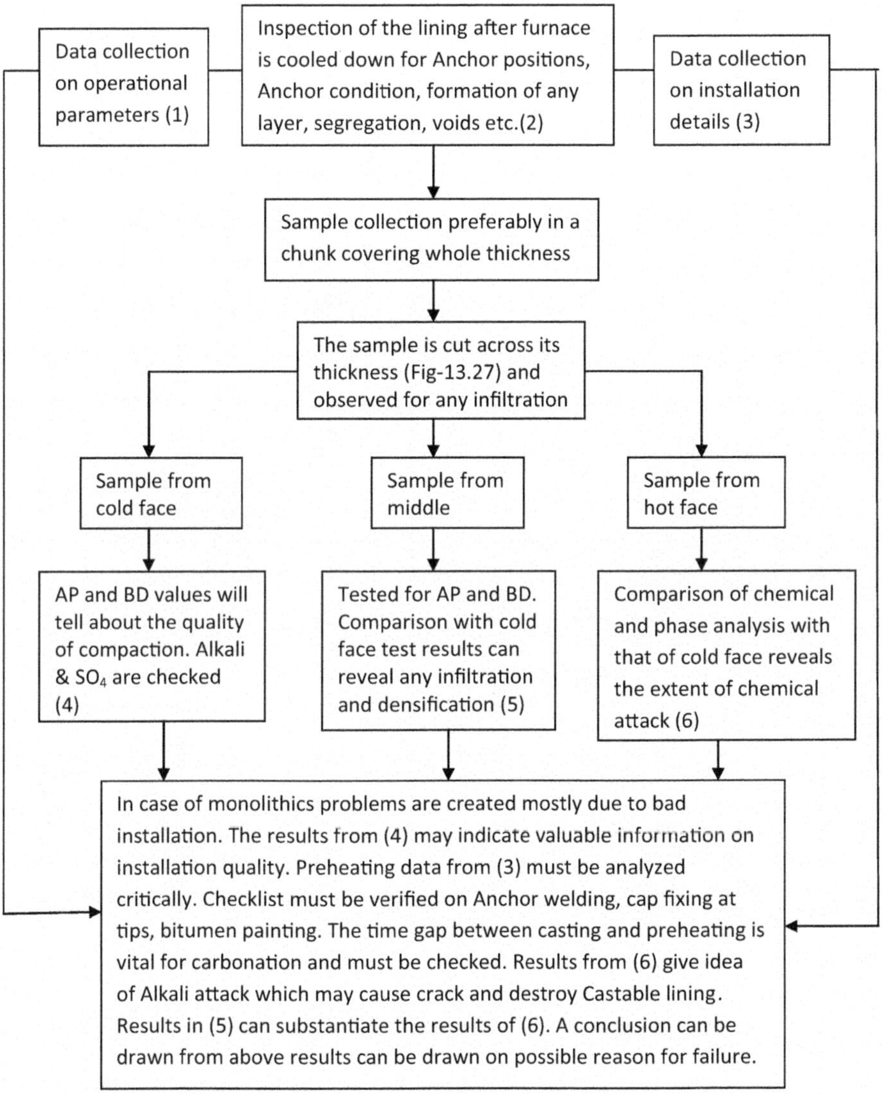

Fig. 13.28 Scheme for investigation of an unshaped refractory lining failure

References

1. Sengupta P. Refractories for the cement industry. Springer. ISBN:978-3-030-21339-8
2. Refractory materials and coating formation. Holderbank Cement Seminar 2000,
3. Bartha P. The cement rotary kiln and its refractory lining. Interceram. Refractories manual, 2004
4. Stresses on refractories – challenges and solution – ECRA News letter, 2/2014
5. Lee WE, Zhang S (2004) Direct and indirect slag corrosion of oxide and oxide-C refractories. In: VII international conference on molten slag, fluxes and salt. The South African Institute of Mining and Metallurgy
6. Brosnan DA Corrosion of refractories. In: Schacht CA (ed) A chapter in the refractories handbook. Mercel Dekker Inc, New York
7. Muhammed MA et al (2009) Corrosion mechanism of refractory bricks lining in the cement kiln. Iraq J Mech Mater Eng 9(2)
8. Weigand R et al (2011) Way to reduce the interaction between glass melt and refractory. Refractory Worldforum 3
9. Weigand R et al (2013) Bubbles as criteria for Refractory corrosion by glass. Refractory Worldforum 5(3)
10. Tschope K, Rutlin J, Grande T (2000) Chemical degradation of map for sodium attack on refractories. Dept of Material Science and Engineering, Norwegian University of Science and Technology, Trondheim, Light Metals
11. Alkali attack on cement plant refractories – an alternative perspective. Michael Walton, Principal, Refmet, Melbourne
12. Demirsan V, Buyukcayir I (2013) Study of Refractory bricks with improved alkali resistance for the cement industry. Refractory Worldforum 5(1)
13. Schlegel E, Aneziris CG, Fischer U. Alkali corrosion of refractories in cement kiln – PPT. Bergakademie, Technische Universitat, Freiberg
14. Bemdiek K, Klischat H-J (2011) Installation and performance of monolithic materials with graded alkali resistance for the cement industry. In: UNITECR proceedings, 2011. 2-B2-1
15. Sengupta P, Gupta N (2010) Development of alkali resistant refractories for cement industry. In: AUCBM conference proceedings
16. Beerkens RGC (1986) Deposits and condensation from the flue gases in glass furnace. Doctoral thesis, Eindhoven Technische Universitet
17. Spear KE, Allendorf MD (2002) Thermodynamic analysis of alumina refractory corrosion by sodium or potassium hydroxides in glass melting furnaces. J Electrochem Soc 149(12)
18. Lugisani P. Identification of refractory material failure in cement kiln. M.Sc thesis, University of Witwatersrand, Johannesburg, South Africa
19. del Mar Kotada Mut M (2014) Sulfur release during alternate fuel combustion in cement rotary kilns. Ph.D thesis, Technical University of Denmark
20. Fotoyi NZ, Eric RH. Interaction of MgO-MgR_2O_4 refractories with SO3 containing gases. Minteq, South Africa and University of Witwatersrand, Johannesburg, South Africa
21. Wei GC, Tennery VJ (1976) Impact of alternate fuels on industrial refractories and refractory insulation applications – ORNL report, Metals and Ceramics Division
22. Steencamp JD (2011) Magnesia refractory dry out- managing the risk of hydration. J South Afr Inst Mining Metall 111
23. Caldon RC, McNamara PD (2009) Case study involving the hydration of magnesite containing brick. In: UNITECR proceedings
24. Hydration mechanism of magnesite based refractory bricks. Thesis submitted by Shuxin Zhou, Metals and Materials Engineering, University of British Columbia, 2004
25. Test of the degree of hydration of magnesia chrome and magnesia spinel bricks. RHI AG Refractories
26. Pasaribu D et al (2002) Hot spot and Ruby formation in secondary reformer. Ammonia Technical Manual

27. Gronwall F. Optimization of burner kiln 7, Cementa site. SLU, Swedish University of Agricultural Science, Faculty of Natural Resources and Agricultural Sciences, Department of Energy and Technology
28. Ramaneka D (2015) Numerical equation of brick lining status in rotary kiln. Thesis, Lulea University of Technology
29. Switalski M (2010) The measurement of shells elastic ovality as essential element of diagnostic of rotary drum's technical state. Diagnostyka 1(53)
30. Ono M, Kozuka H (2004) Damage of refractories lined in cement rotary kiln. In: Tehran international conference on refractories, 4–6 May, 2004
31. Walton MC, Plater PA. When a refractory failure isn't – an anchor issue. Refmet, PRAHRAN, Victoria 3181, Australia
32. Anchor selection for monolithic lining. Technical bulletin, Harbison-Walkar Refractories Company
33. Hau J, Seijas A. Sigma phase embrittlement of stainless steel in FCC service. Corrosion Nacexpo, 61st conference and exposition
34. Poinescu A et al (2018) Microstructural and mechanical characterization of 301 stainless steel welded joints. J Arts Sci 1(42):275–282
35. High Temperature hydrogen attack (HTHA) – Anup Ray, A PPT presentation, Swiss Re
36. Benac DJ, Mc Andrew P (2012) Reducing the risk of High temperature Hydrogen attack (HTHA) failures. J Fail Anal Prevent 12:624–627
37. Ibrahim SE (2018) Effect of high temperature hydrogen attack on dissimilar materials. Inspection J 24(3)
38. Bali RK et al (2002) Failure and repair of a secondary reformer – an experience. Ammonia Technical Manual
39. Installation of refractory materials –PPT by Reiner Gocht. Refratechnik GmbH

Glossary

Amphoteric Molecule of an amphoteric compound can react both as an acid as well as base. Metal oxides which react with both acid and the base to produce salt are known as amphoteric oxide, e.g. Aluminium oxide.

Andalusite It is also naturally occurring Alumino-silicate mineral containing about 58% Al2O3. Kyanite, Sillimanite and Andalusite have almost same Alumina content but they differ in their crystal structure.

Bauxite it is a sedimentary rock containing hydrated Aluminium oxide. Refractory grade and metallurgical grade Bauxites have different characteristics. Bauxite is one of the major raw material for production of high alumina refractories.

Calcium- Aluminate Cement Calcium Aluminates like CaO.Al2O3 and CaO.2Al2O3 etc have hydraulic properties and they set and becomes hard when comes in contact with water at ambient temperature.

Chromite It is a naturally occurring mineral. It is Iron chromium oxide: $FeCr_2O_4$. It is an oxide mineral belonging to the spinel group. Magnesium can substitute for iron in variable amounts as it forms a solid solution to form Magnesio-Chromite ($MgCr_2O_4$) and substitution of Chromium by Aluminium occurs leading to formation of Herceynite ($FeAl_2O_4$). All these minerals present in Chromite.

Clinkerization The process of clinker formation when the raw meal in cement rotary kiln is rotated along with kiln shell at high temperature because some liquid formation at high temperature which binds the powder to form cement clinkers.

Cristobalite It is a high temperature polymorph of Silica. Chemical formula is SiO2. The difference from Quartz or other polymorphs are in its crystal structure.

Contact angle The **contact angle** is the **angle**, conventionally measured through the liquid, where a liquid–vapor interface meets a solid surface. It quantifies the wettability of a solid surface by a liquid.

Diffusion It is a process by which the atom or molecule are transported from one part of the body to other due to thermal agitation to homogenize the composition throughout the body.

P. Sengupta, *Refractories for the Chemical Industries*,
https://doi.org/10.1007/978-3-030-61240-5

Emissivity The **emissivity** of the surface of a material is its effectiveness in emitting energy as thermal radiation. Thermal radiation is electromagnetic radiation and it may include both visible radiation (light) and infrared radiation, which is not visible to human eyes. The thermal radiation from very hot objects (see photograph) is easily visible to the eye. Quantitatively, emissivity is the ratio of the thermal radiation from a surface to the radiation from an ideal black surface at the same temperature as given by the Stefan–Boltzmann law. The ratio varies from 0 to 1.

Eutectic point The point in a phase diagram indicating the chemical composition and temperature corresponding to the lowest melting point of a mixture of components.

Feedstock Bulk raw material (input) fed into a process to convert it to some other material (output). The term mostly used in Hydrocarbon and Petroleum industry.

Granulometry It is the measurement of the size distribution in a collection of grains.

Hexavalent Chrome Chromium can occur in different mineral and chemicals with different valencies, The Chromium with +6 valency is called hexavalent chrome.

Kiln An equipment or installation used for high temperature treatment a material to carry out certain Chemical transformation.

Kyanite It is a naturally occurring Alumino-Silicate mineral with about 58% A12O3. It

Liquidus It is the lowest temperature at which, a multi component system of metals or its compounds, remain in liquid state.

Mullite It is a mineral phase developed in Alumino-silicate refractories during its treatment at high temperature. Alumina and Silica reacts at high temperature to form Mullite.3A12O3 + 2SiO2 = 3A12O3.2SiO2.

Periclase It is the major component of dead burnt Magnesia or fused Magnesia. Chemically it is Magnesium Oxide (MgO) with cubic crystal habit.

Phase A component in a material system that is distinguished by chemical composition and or physical state (solid/liquid/gas) and /or crystal structure.

Plastic Clay The Plastic clay is used as binder for the manufacture of Refractory shapes because it has got the characteristic property that moist condition permits it to be deformed without cracking and to retain its deformed shape when the deforming stress is removed.

Pore size distribution Is the volume percent of pores with different diameter

Pyrolysis It is a process of high temperature decomposition of a compound due to heat, in absence of Oxygen.

Redox reaction Redox reactions are reactions in which one species is reduced and another is oxidized. Therefore the oxidation state of the species involved must change.

Sillimanite It is also naturally occurring Alumino-silicate mineral containing about 58% A12O3.

Sintering The general term for the Densification by heat of a powder compact to produce a polycrystalline body. Densification may be by the formation of liquid phases which fill pores; by solid state diffusion or other mechanisms which also reduce surface energy.

Silica Sol Colloidal Silica where SiO2 is suspended in colloidal form in an aqueous medium.

Sol-Gel Process It is the co-precipitating two or more oxides by mixing them in the form of sol and transforming it to Gel to ensure the homogeneous mixing in the molecular level.

Surface tension Surface tension is the elastic tendency of a fluid **surface** which makes it acquire the least surface area possible. ... The net effect is an inward force at its **surface** that causes the liquid to behave as if its surface were covered with a stretched elastic membrane.

Thixotropy It is a process of isothermal gel to sol transformation under the influence of a shear stress and the retransformation to gel on withdrawal of the shear stress.

Torque is rotational force. Torque is a measure of how much force acting on an object causes that object to rotate

Viscosity Viscosity is a measure of a fluid's resistance to flow. It describes the internal friction of a moving fluid.

Vitreous It means Glassy. In the Vitreous material the atoms are not arranged in regular pattern like in crystal but in same way as in glass.

Index

A
Abrasion, 11, 12, 348, 350
Accelerator, 105
Acoustic method, 131
Air-cooled wall, 190, 191, 193
Air quenching cooler (AQC), 314
Alkali metals, 77
Alkali resistance, 22, 48
Alkaline earth silicate (AES), 76
Alkalis, 322–326
Alloy 253MA, 112
Alumina, 165
Alumina cement-bonded castables, 65
Alumina refractories, 48, 49, 159
  manufacturing flow sheet, 48
Alumina-zirconia-silica (AZS), 243, 244, 246, 257, 258, 260, 261, 264, 265, 273
Aluminosilicate monolithic refractories, 63
Aluminosilicate refractory, 15, 321, 322
  alumina, 48
  category, 46
  fireclay, 46, 47
  high alumina, 47–48
Aluminum carbide, 3
Anchor
  austenitic stainless steels, 352
  ceramic, 114
  construction material, 111
  design engineer, 353, 354
  ferritic and duplex stainless steels, 352
  material of construction, 352
  metal, 113, 114
  selection criteria, diameter, 112
  shape, 111–113
  sigma-phase embrittlement, 352
  size design, 111–113
  steel grades, 111
Anchor spacing, 117
  factors, 114, 115
  lining thickness *vs.* support spacing, 116
  positions, 116
  support spacing, strength
    anchors, 115
    castable, 115, 116
Anchors orientation, 118
Andalusite, 48
Arch and roof design
  building construction, 93
  castable, 95
  parameters, 95
  skewback bricks, 93–94
  suspended roofs, 94, 96
  vertical wall, 93
  wooden former, 93
Arch/crown, 93
Ash cooler, 311, 312
Ash fusibility temperature (AFT), 179, 182
Ash hopper, 303
Autothermal reforming (ATR), 156
AZS fused cast blocks, 56

B
Barium oxide, 3
Barracuda's erosion model, 219
Basic refractories
  cement rotary kiln, 52, 55
  high alumina, 49
  MgO, 50
  spinels, 51, 52, 54
  usages, 49

P. Sengupta, *Refractories for the Chemical Industries*,
https://doi.org/10.1007/978-3-030-61240-5

Batch carryover, 246, 248, 263, 265
Bauxite-based refractories, 48
Boiler
chemical industries, 303
types, 303
Boudouard reaction, 23
Brick joints, 89
Brick lining, 117
Bubble alumina, 70
Bulk density, 5
Burner pipe, 299
Burning zone, 293, 294, 299

C
Calciner, 285–286
Calcium aluminate, 36, 63, 98, 124, 125, 253
Calcium carbide, 3
Calcium silicates, 70, 71
blocks, 71
Carbon black
global production, 197
industrial chemical, 197
manufacturing processes, 198
process principle, 197
reactor design, 198, 200
refractories selection, 201–203
temperature, 200
Carbon disintegration, 332
Carbonation, 124, 125
Carcass reactor, 198
Castable refractory, 108
Castables, 189
alumina cement, 58
ambient conditions, 58
applications, 58
conventional, 59
installation methods, 59
LCC, 59, 60
manufacturing flow sheet, 59
mixing quality, 59
NCC, 61
phosphate-bonded (*see* Phosphate-bonded castable)
properties, 58, 64
recommendation, 62
refractory aggregates, 58
refractory manufacturer, 58
shotcreting, 63
technology, 58
types, 60, 61
ULCC, 60
Catalytic partial oxidation (CPO), 155, 156
Cement industry, 313, 315
Cement moduli, 320
Cement rotary kiln, 55
Ceramic anchors, 113, 114
Ceramic blanket, 76, 77, 90
Ceramic ferrule, 230–234
Ceramic fiber board, 159
Ceramic fiber lining, 126–129, 210
Ceramic fiber modules, 208
Ceramic fibers
alumina and silica, 73
amorphous structure, 73
categories, 73
characteristics, 76
chemical stability, 77
environmental issue, 77
heat-treated, 75
insulation performance, 76
manufacturing, 73, 75
properties, 76
recrystallization, 74, 75
shots, 73
sol-gel process, 75, 76
thermal conductivity, 74
types, 75
vitreous quality, 73
Ceramic welding method, 274
Checker wall, 224, 227–231
Chemical corrosion, 127
Chemically bonded *vs.* cement-bonded castables, 62
Chloride, 322, 323, 326–329
Circulating fluidized bed (CFB) boiler
components, 303, 304
refractories
ash cooler, 311, 312
bottom lining, 308, 309
combustor upper wall and ceiling, 306–308
combustor wall, 305, 306
cyclone, 309, 310
flue gas duct, 312, 313
operational parameters, 305
properties, 305
seal pot, 310, 311
work function, 303, 304
Classification of refractories
basicity, 31
form, 32
insulating property, 40, 41
parameters, 31
Claus reactor, 224, 226, 227, 229, 234
Clinker, 278, 279, 288, 289, 293, 294, 299, 300
Clinker coating, 55

Coal gasification
advantage, 173
air-cooled wall, 190, 191, 193
capacity, 174
carbon, 173
chemical engineering processes, 173
chemicals, 173, 174
corrosion of refractory, 191, 193–195
design
acidic refractories, 182
air-cooled refractory, 179
average velocity, 183
basicity, 182, 183
characterization, 182
coal ash, 182
corrosion, 185
gasifier, 183
geological formation, 180
impurities, 180
mechanisms, 181
operational parameters, 183
oxygen potential, 183
properties, 183
refractory lining, 179
slag behavior, 183
slag quality, 180
slag residence time, 183
temperature profile, 179
viscosity, 183, 185
water-cooled refractory lining, 179, 180
freeze lining, 186–189
H_2 and CO, 173
liquid fuels, 173
membrane wall, 185, 186
nitrogen, 173
power, 174
process, 175, 177, 179
products, 173, 174
sulfur, 173
syngas, 173, 175
wall type, 185
water-cooled/air-cooled wall, 185
Coal slag of gasifier, 191
Cold crushing strength (CCS), 7
Combustion chamber, 303, 305
Computational fluid dynamics (CFD) modeling, 230
Conventional castables, 59, 97
Cooler, 288
Corrosion resistance, 19–22, 55
Corundum refractory, 49
Cracking, 197, 198, 200
Cristobalite, 45, 46
Crystalline fiber manufacturing, 75
Cyclones, 218–223
castables, 310
ceiling, 310
cylinder and cone wall, 309
dense castable, 310
dense refractory layer, 309
operating conditions in target area, 309, 310
refractories, 310
solids, 309
target area, 309
Cyclone system
alkali and acid vapors, 283
atmosphere, 283
brick and monolithic lining, 284
brick shapes, 296, 298
buildups, 283
burner pipe, 299
calciner, 285–286
chemical attack, 283
cooler, 288, 289
cost-effective method, 283
ducts, 283
factors, 282
flue gas, 282
higher operating temperatures, 283
inlet cone, 290, 291
insulation, 287
kiln, 287–290
kiln refractory performance, 299–301
meal pipes, 287
operating temperature, 283
outlet area, 294–296
preheating zone, 291
properties, 283
riser duct, 286
shaped and unshaped refractories, 285
shell temperature, 284
smoke chamber, 286–287
TAD, 288
temperature, 283
transition zone, 291, 293, 294
zircon, 284
Cylindrical vessel, 165

D

Dead burnt magnesite (DBM), 50
Deformation temperature (DT), 182
Dense cast, 57
Dense firebricks, 159
Dense refractory layer, 309

Depressurization, 123
Dolomite brick, 55
Dolomite refractory, 55
Dome, 93
Drying, 118

E
Earth elements, 111
Elastifiers, 54
Electron microscopy, 26, 27
Electron probe microanalyzer (EPMA), 26
Ellingham diagram, 23
Endoscopic method, 274
Endothermic reaction, 164
Entrained flow gasifier, 177, 179, 185
Expansion joint, 91–94, 108, 215–217

F
Feedstock, 199
Fiber index, 73
Finite element analysis (FEA), 129
Fireclay refractories, 46, 47
Fixed bed gasifier, 177
Flexibilization, 54
Float glass process
  hydrogen, 251
  manufacturing, 250
  nitrogen, 251
  refractory selection, 252–254
  temperature, 250
  tweels, 254, 255
Floating anchor, 113
Flow temperatures (FT), 182
Flue gas, 126, 222
Fluid catalytic cracking unit (FCCU)
  cyclones, 218–223
  flue gas line, 222
  process description, 212–213
  processes, 212
  reactor and regenerator wall, 217
  refining process, 212
  riser pipe, 215–217
  thermal conversion process, 212
  transfer lines, 219
  unit, 213
Fluidized bed gasifier, 177–179
Fluidized bed heat exchanger (FBHE), 311
Forsterite, 54
Fracture, 9, 10
Free quartz, 45
Freeze lining, 186–189
Fuel gas, 199
Furnace atmosphere, 86
Furnace black method, 198
Furnace/reactor, 86
Fused cast refractory, 246, 248, 260
  abrasion resistance, 56
  blocks, 57
  cavity, 56, 57
  chemical and physical properties, 56
  conventional brick-making processes, 56
  crystalline and glassy phases, 57
  dense cast, 57
  graphite electrode, 56
  manufacturing process, 56
  molten material, 56
  regular cast, 57
  solidification process, 56
  tilt cast, 57
  void free, 57
  zero porosity, 57

G
Galaxite, 52
Gaseous atmosphere, 23, 24
Gibbs free energy, 23
Glass manufacturing
  batch-type furnaces, 238
  bulk glass products, 238
  characteristics, 242
  chemical compositions, 237
  chemistry, 237
  compositions, 240
  continuous glass melting plant, 239
  cross-fired furnace, 240
  crown– refractory lining, 248, 249, 251
  defects
    composition, 260
    electrochemical reaction, 260
    fused cast block, 261
    gaseous inclusion, 259
    glass tank furnace, 259
    glassy phase, 260
    polyvalent impurities, 261
    refractory qualities, 261
    silica drip, 261
    temperature, 260
    types, 258, 259
    ZrO_2 crystal, 260
  distribution of production, 238
  downward drilling, 242
  end-fired furnace, 240
  flow sheet, 237
  furnace melting capacity, 239
  fused cast and zircon refractory, 243

glass melting furnace, 238–241
glass tank furnace, 239
glass tank refractory lining, 245
ingredients, 237
maintenance, 272–274
metal penetration, 242
molten glass
Arrhenius equation, 256
boundary layer, 257, 258
chemical and physical processes, 255
convection flows, 257
corrosion rate, 256
direct dissolution, 255
effect of temperature, 256–258
forced convection, 257
free laminar density convection, 257
indirect dissolution, 256
influence, 257, 258
mathematical model, 258
mechanism, 255
physical properties, 258
refractory corrosion, 259
surface tension, 255
temperature, 256
thermal radiation, 256
viscosity, 255
monitoring, 272–274
operating parameters, 242
refractories, 241
side wall lining, 244
superstructure, 246–248
tank furnace, 241
types, 237
versatile material, 237
zircon bricks, 243
Glass tank furnace, 238, 239, 241, 242, 244, 259
Gunning machine, 106
installation, 63
machine, 62
mass, 62
material, 62, 63

H

HA cement-bonded castables, 66, 67
Heat capacity, 15
Heaters
designs, 208
fuel, 206
furnace, 206
hot flue gas, 206
process, 207
refractory selection, 207–212
shapes, 207
shield section, 206
steel tubes, 206
temperature, 206
tubular furnaces, 206
Hemispherical temperature (HT), 182
Hercynite, 53
High alumina refractories
alkali resistance, 47, 48
bauxite-based, 48
composition, 47
manufacturing flow sheet, 47
mullite, 47
phosphate-bonded, 48
pyroplasticity, 48
raw materials, 47, 48
target properties, 47
usage, 48
High-temperature hydrogen attack (HTHA), 354–356
Hooke's law, 8
Hot modulus of rupture (HMOR), 16
Hot spot detection, 132
Hydration, 55, 333–335
Hydrocarbon gas, 198
Hydrogen gas, 165
Hydrogen pressure vessels, 165
Hydrogen transfer line, 159

I

Installation methods, shaped refractories
arch and roof design, 93–95
chemical industry, 88
expansion joints, 91–94
quality, 89
rules, refractory lining, 89
vertical wall, 90
Insulating castables, 71, 73
Insulating refractories, 40
Insulation bricks, 71
Insulation refractory, 125–127, 129, 144
ASTM classification, 69, 70
calcium silicates, 70, 71
ceramic fibers (*see* Ceramic fibers)
features, 69, 70
insulating castables, 71
microporous (*see* Microporous insulation)
service temperature, 70
shapes, 70
thermal insulating materials, 70
uses, 69
ISO, 296

K
Kiln hood, 287, 288
Kyanite, 48

L
LCC *vs.* ULCC, 61
Lime calcination
 industrial chemical, 135
 industrial sectors, 136
 metallurgical processes, 135
 process, 135
Liquid phases, 16
Loop seal, 303
Low bio-persistent (LBP), 77
Low cement castable (LCC), 59, 60
 alumina cement, 98
 calcium aluminate hydrates
  chemical reactions, 98
  properties, 98
  temperature, 98–100
  water, 98
 castable mix, 98
 consistency, 102
 drying process, 103
 mixing temperature, workability, 102
 precautions, 100, 102
 recommendation, water quality, 103
 soundness of installation, 104
 vibration, 102
 water addition, strength, 103
 water temperature, 101

M
Mag-alumina spinel, 53
Magnesia (MgO)
 DBM, 50
 formation, 50
 high-temperature calcinations, 50
 magnesite-chrome refractory, 50
 properties, 50–52
 seawater, 50
 shock resistance, 51
 spinel, 52
Magnesia chrome refractories, 51
Magnesia-chromite brick, 51
Magnesia-hercynite refractory, 53
Magnesia-spinel bricks, 55
Magnesia-spinel refractories, 53
Magnesia-zircon, 54, 55, 264, 270–272
Magnesite-chrome refractories
 cement rotary kiln, 50
 environmental pollution, 51
 manufacturing flow sheet, 51, 52
Maximum service temperature (MST), 13
Meal pipes, 287
Membrane wall, 179, 185, 186, 306
Merwinite, 329, 334
Metal anchors, 111–114
Microcracks, 9
Microporous effect, 79
Microporous insulation
 benefits, 79
 hydrogen, 78, 82
 microporous effect, 79
 nuclear reactor, 78
 opacifiers, 79
 pyrogenic silica, 78, 79
 radiation, 79
 SAS, 78
 selection, 80
 solid conduction, 79
 thermal conductivity, 79–81
Microstructure
 aluminosilicate category, 25
 electron microscopy, 26, 27
 optical microscopy, 25
 physical properties, 25
Modulus of elasticity, 8
Modulus of rupture, 7
Monitoring of lining
 acoustic method, 131
 airway pipe, 131
 convenient method, 131
 damaged refractories, 131
 FCCU refractory lining, 131
 information, 130
 mechanical/electrical fault, 130
 poly-diacetylene polymers, 133
 refractory thickness, 131
 regenerator cone area, 131
 thermographic survey, 131
 thermometric methods, 130, 132
Mono aluminum phosphate (MAP), 63, 64
Monoclinic/stabilized zirconia, 54
Monolithic refractories
 anchor (*see* Anchor)
 application, 97
 castables, 97
 coarse aggregates, 97
 conventional castables, 97
 fine powders, 97
 gunning, 106
 installation method
  budget, 110
  castable refractory, 108

ease and speed, 109
environmental condition and equipment, 109
factors, 108
installation team skill, 110
location and site condition, 109
storage condition, 109
time consumption, 110
volume, castable, 109
joints
expansion, 108
separation, 108
low cement castable (*see* Low cement castable)
mixer, 97
monitoring and safety, 107
products, 97
pumping, 107
semifinished refractories, 97
shotcreting, 104–106
sound installation, 97
Monticellite, 329, 331
Mortars, 69
Moving bed gasifier, 186
Mullite, 47, 248, 249, 264, 265

N

Natural gas and oil, 121
Nernst equation, 256
No cement castable (NCC), 61

O

Opacifiers, 79, 81
Optical microscopy, 25
Overheating, 337, 338
Oxides, 77

P

Parallel Flow Regenerative (PFR), 137
Partial oxidation (POX), 150, 155, 156
PCW fibers, 76
Peeling, 19
Periscopic methods, 274
Permanent linear change (PLC), 14, 15
Petrochemical industry, 127
Petroleum refinery
downstream sector, 205
fuels, 205
gasoline, 205
hydrocarbons, 205
operational parameters, 205
refining process, 206
units, 205
Phosphate-bonded castables
additives, 65
advantages, 65
binders, 65
CO attack resistance, 66
CO disintegration, 65
exothermic, 65
fluidized coking environment, 65
installation, 65
pack system, 65
refractory surfaces, 65
SFRC, 66, 67
thermal expansion, 66
Phosphate-bonded high alumina refractories, 48
Phosphate-bonded plastics, 221
Phosphoric acid, 63
Plastic mass
mortars, 69
property, 68, 69
Plastic refractory, 306
Plastics, 36
Pleonaste, 52
Poly-diacetylene polymers, 133
Polymorphic forms, 44
Polymorphism, 44
Porosity, 5, 6
Portland cement
acidic gases, 279
areas of application, 282
chemical reactions, 278
clinker, 278
electrical energy, 279
energy, 279
fuels, 279
gases, 280
heat exchanger, 279
heavy chemical industry, 277
hypothetical relationship, 281
lining tightness, 351
manufacturing process, 277
operating temperatures, 279
operational conditions, 279
raw material, 278
refractory characteristics, 281
solids, 280
suitability, 282
tailor-made products, 281
thermal energy, 279
wear factors, 281

Preheating
- castable lining
  - calcium aluminates, 121
  - decomposition temperatures, 121
  - heat up curve, 123
  - low cement castables, 121
  - low thermal conductivity, 122
  - organic fibers, 122
  - physical water, 121
  - reformer operating, 123
  - temperature, 122, 123
- drying, 117
- heat source, 118
- hot gases, 118
- moisture removal, 117
- refractories
  - bricks, 121
  - cement rotary kiln, 120, 121
  - gas/liquid fuel, 121
  - installation, 119
  - mechanical stress, 120
  - moisture, 119
  - reactor/furnace, 119
  - thermal expansion, 120, 121
- refractory hot face, 118
- temperature, refractory, 118, 119
- thermal equilibrium, 117
- thermal expansion, 119
- thermal stress, 119
- water removal, 117

Preheating zone, 291
Pressure-swing adsorption (PSA), 154
Pressurization, 123
Process Gas Cooler (PG Cooler), 165
Pyrogenic silica, 78, 79
Pyrometric cone equivalent (PCE), 13
Pyroplasticity, 48

Q

Quartzite, 44

R

Ramming mass, 35, 68
RCF-type fibers, 76
Reaction furnace, 224
Reactor design, 198, 200
Recrystallization, 74, 75
Redox reactions, 331–333
Refractories
- carbon black reactor, 199, 201, 202
- cement rotary kiln, 85
- design, 85
- hot face, 201
- installation, 85
- mineral phase, 201
- properties, 43, 201, 202
- reactors, 201
- selection, 85, 86

Refractoriness, 13
Refractoriness under load (RUL), 16, 17
Refractory aggregates, 58
Refractory castable, 107
Refractory ceramic fiber (RCF), 75
Refractory lining, 69
Refractory materials
- abrasion resistance, 11, 12
- alkali resistance test, 22
- apparent porosity, 5, 6
- Bayer's process, 1
- bulk densities, 5
- cement industry, 4
- chemicals, 1
- corrosion resistance, 19–22
- gaseous atmosphere, 23, 24
- heat capacity, 15
- in-depth studies, 2
- industrial scale, 2
- influences, 2
- inorganic solid materials, 2
- mechanical properties
  - CCS, 7
  - fracture, 9, 10
  - modulus of elasticity, 8, 9
  - modulus of rupture, 7, 8
- mechanical strength of steel, 2
- melting point, 3
- modern civilization, 1
- PCE, 13
- permeability, 6, 7
- PLC, 14, 15
- process flow sheet, 1
- properties, 4
- purposes, 1
- raw materials, 4
- RTE, 13, 14
- smelting of iron, 1
- specific gravity, 4
- thermal conductivity, 15
- thermal diffusivity, 16
- wedge-splitting test, 10, 11

Refractory microstructure, 53
Refractory performance
- chlorides, 326–329
- factors, 319
- hydration, 333–335
- industrial furnaces, 317

influence, 317
operational conditions, 319
process equipment, 317
quality, 317
redox reactions, 331–333
refractory lining failure, 357
responsibility distribution, 318
rotary kiln refractory lining, 318
ruby formation, 336, 337
storage quality, 317
sulfur, 326–329
vanadium oxides, 329–331
Refractory raw materials, 43
Regenerator
behavior, 262
checker
alkali attack, 267
checker bricks, 267
chimney blocks, 266
conditions, 268
efficiency, 266, 268
flow of air, 266
flue gas, 266
glass melting, 265
high-purity magnesite bricks, 269
liquid phase, 269
magnesia-zircon refractory, 270–272
magnesite refractory, 269
MgO-CaO-SiO2 phase system, 269
microstructure, 269
middle course, 270
mineral phases, 269, 270
parameters, 268
properties, 270
quality, 266
requirements, 266
static load, 267
types of design, 266
functions, 263
glass furnace, 263
properties, 264
recommendation, 262
side wall, 265
silica melting, 262
superstructure, 264
target wall, 265
thermal conductivity, 263
thermal efficiency, 262
thermo-mechanical and thermo-chemical stability, 263
tridymite crystal, 262
Regular cast, 57
Reversible thermal expansion (RTE), 13, 14
Rider bricks, 94
Riser duct, 286
Rotary kilns
calcium carbonate, 142
concentric burner, 141
dynamic unit, 142
energy consumption, 142
feed and combustion, 141
insulation, 144, 145
paper mill, 143
refractory, 142–144
static unit, 142
temperature, 141
thermal energy, 142
Ruby formation, 336, 337

S

Scanning electron microscopy (SEM), 26
Schaeffler-DeLong diagram, 354
Seal pot, 310, 311
Seawater, 50
Semifinished refractories, 97
Separation joints, 108
Shaped refractories
advantages, 88
aluminosilicate, 46–49
basic (*see* Basic refractories)
bricks, 87, 88
crushed materials, 43
disadvantages, 88
furnace, 88
fused cast, 34, 56–57
installation (*see* Installation methods of shaped refractories)
manufacturing process, 43
precast shape, 34
pressed shape, 32, 34
silica, 44–45
vacuum formed shape
advantages, 34
aluminosilicate aggregate, 36
basic and nonbasic composition, 35
calcium aluminate cement, 36, 37
calcium oxide, 38
castable products, 37
castables, 36, 37
cement-based and no-cement castables, 38
ceramic anchors, 36
ceramic fiber insulation, 34
classification, 33
conventional castables, 37
disadvantages, 34
features, unshaped refractories, 39

Shaped refractories (*cont.*)
gunning material, 39
liquid, 38
low cement, 38
mixing quality, 36
mortars, 35
organic/inorganic binders, 34
plastic refractories, 36
powder, 38
size fractions, 35
steel former, 39
traditional casting process, 39
ultralow cement, 38
water/liquid binder, 35
Shotcreting, 36, 39, 63, 104–106
Shots, 73
Silica refractories
Al_2O_3 and TiO_2, 44
CaO, 44
free quartz, 45, 46
impurities, 44
manufacturing flow sheet, 44, 45
polymorphic transformation, 44, 45
polymorphism, 44
quartzite, 44, 45
raw material, 44
temperature profile, 44
thermal shock resistance, 45
tridymite, 45
Silica refractory, 248–250
Silicon carbide (SiC), 47
Sillimanite, 48
Slabbing, 19
Smoke chamber, 286–287
Sodium carbonates, 124
Softening temperature (ST), 182
Sol-gel process, 75, 76
Solid conduction, 79
Solidification process, 56
Solid-state reactions, 14
Soluble oxides, 124
Spinels, 52, 53
Steam methane reforming (SMR), 150–152, 154
Steel fiber reinforced castable (SFRC), 66, 67
Stokes-Einstein equation, 255
Strain-controlled load, 345–348
Stress distribution, 162
Stress relaxation, 120, 121
Stress-strain relationship, 8
Structural spalling, 19
Sulfur, 326–329
Sulfur recovery units (SRU)
checker wall, 227–231
chemical reaction, 224
Claus reactor, 234
compounds, 224
feed composition, 226
feed gas, 224
flame temperature, 225
heat energy, 234
high-grade alumina refractories, 235
layers, 226–227
properties, 235
reaction furnace, 224
refractory lining, 225
refractory/ceramic ferrule, 230–235
tail gas, 234
thermal section, 224, 225
waste heat boiler, 235
Superstructure, 241, 246–248, 259, 261, 263
Suspended roofs, 94, 96
Syngas manufacturing
chemical process
ATR, 156
carbon dioxide reforming, 150
CPO, 155, 156
hydrocarbons, 150
non-catalytic oxidation reaction, 150
POX, 155, 156
reforming, 157, 159
SMR, 151, 152, 154
technologies, 150
feedstocks, 149
process, 149
refractories
depressurization, 167
pressurization, 167
primary reformers, 159, 160, 162, 163
reformer, 169
refractory, 166, 168
secondary reformer, 163–167, 170
synthetic gas, 149
Synthetic amorphous silica (SAS), 78

T

Target wall, 309
Temperature, 92, 199, 200
Temperature indicating paint (TIP), 131
Tertiary air duct (TAD), 288
Thermal conductivity, 15, 74
Thermal cracking, 197
Thermal diffusivity, 16
Thermal expansion, 92, 120
Thermal gradient, 90, 91
Thermal insulation, 69
Thermal load
flame, 338, 339
overheating, 337, 338

Thermal parameters, 317
Thermal shock, 55, 159, 347
Thermal stress, 119
Thermo-chemical effects
  alkalis, 322–326
  aluminosilicate refractories, 321, 322
  cement industry, 320
  chemical corrosion, 320
  depth of slag penetration, 321
  feed, 320
  fuel, 320
  liquid phases, 321
  optimum control, 320
  refractory lining, 320
  solid-liquid and solid-gas phase reactions, 320
Thermo-mechanical load
  abrasion, 348, 350
  cement manufacturing and lime calcinations, 340
  ceramics, 340
  compression, 342
  compressive and tensile stress, 343
  compressive stresses, 342
  distribution of forces, 344
  horizontal and vertical deviations, 341
  kiln diameter, 343, 348, 349, 351
  kiln shell, 342
  linear relationship, 344
  mechanism of destruction, 345
  metal, 340
  ovality, 341, 343
  properties, 340
  relaxation, 342
  steel loses, 340
  strain-controlled load, 341, 345–348
  stress development, 340
  thermal shock, 347
Thermo-mechanical property, 215
  creep in compression, 17
  HMOR, 16
  RUL, 16, 17
  structural spalling, 19
  thermal shock resistance/spalling, 18, 19
Thermometric methods, 130, 132
Thermometry, 273
Tilt cast, 57
Transfer line, 219
Transition zone, 290, 291, 293, 294
Transmission electron microscopy (TEM), 26
Tread reactor, 198
Tridymite, 45

U
Ultralow cement castable (ULCC), 60
Unshaped refractories, 32, 35, 36, 39
  castables (*see* Castables)
  industrial applications, 58
  plastic mass, 68, 69
  ramming mass, 68

V
Vanadium oxides, 329–331
VDZ, 296, 298
Velocity, 219
Vertical shaft kilns
  refractory, 140, 141
  single shaft kiln, 137
  twin shaft kiln, 137–139
Viscosity, 322
Void free, 57

W
Waste heat temperature, 313
Waste heat utilization, power generation
  cement industry, 313
  cement plant, 315
  industry, 312
  petroleum refinery, 313, 314
  recovery system, 314
  technical parameters, 313
  temperature, 313, 314
Water wall/water panels, 306, 307
Wedge-splitting test, 10, 11
Wooden former, 93

X
X-ray, 26, 27
X-ray diffraction pattern, 78
X-ray fluorescence method (XRF), 27

Y
Young's modulus, 54, 215

Z
Zirconia, 55
Zirconia-enriched dolomite refractories, 55

The manufacturer's authorised representative in the EU is Springer Nature Customer Service Centre GmbH, Europaplatz 3, 69115 Heidelberg, Germany. If you have any concerns regarding our products, please contact ProductSafety@springernature.com

Printed and bound by CPI Group (UK) Ltd, Croydon, CR0 4YY
24/07/2026
02175923-0010